A NOTE FROM THE AUTHORS

Congratulations on your decision to take the AP Chemistry exam! Whether or not you're completing a year-long AP Chemistry course, this book can help you prepare for the exam. In it you'll find information about the exam as well as Kaplan's test-taking strategies, a targeted review that highlights important concepts on the exam, and Practice Tests. Take the Diagnostic Test to see which subjects you should review most, and use the two full-length exams to get comfortable with the testing experience. Don't miss the strategies for answering the free-response questions: You'll learn how to cover the key points AP graders will want to see.

By studying college-level chemistry in high school, you've placed yourself a step ahead of other students. You've developed your critical-thinking and time-management skills, as well as your understanding of the practice of chemical research. Now it's time for you to show off what you've learned on the exam.

Best of luck,

David Wilson
Anaxos Inc.

AP® CHEMISTRY

2013-2014

Kaplan offers resources and options to help you prepare for the PSAT, SAT, ACT, AP exams, and other high-stakes exams. Go to www.kaptest.com or scan this code below with your phone (you will need to download a QR code reader) for free events and promotions.

snap.vu/m87n

RELATED TITLES

AP Biology

AP Calculus AB & BC

AP English Language and Composition

AP English Literature and Composition

AP Environmental Science

AP European History

AP Human Geography

AP Macroeconomics/Microeconomics

AP Physics B & C

AP Psychology

AP Statistics

AP U.S. Government and Politics

AP U.S. History

AP World History

SAT Premier with CD-Rom

SAT: Strategies, Practice, and Review

SAT Subject Test: Biology E/M

SAT Subject Test: Chemistry

SAT Subject Test: Literature

SAT Subject Test: Mathematics Level 1

SAT Subject Test: Mathematics Level 2

SAT Subject Test: Physics

SAT Subject Test: Spanish

SAT Subject Test: U.S. History

SAT Subject Test: World History

AP® CHEMISTRY

2013-2014

David Wilson

PUBLISHING

New York

© 2012 by Kaplan, Inc.

Published by Kaplan Publishing, a division of Kaplan, Inc.
395 Hudson Street
New York, NY 10014

Printed in the United States of America

10 9 8 7 6 5 4 3 2

ISBN-13: 978-1-60978-715-8

Kaplan Publishing books are available at special quantity discounts to use for sales promotions, employee premiums, or educational purposes. For more information or to purchase books, please call the Simon & Schuster special sales department at 866-506-1949.

TABLE OF CONTENTS

PART THREE: AP CHEMISTRY REVIEW

ABOUT THE AUTHORS

Anaxos Inc. Founded in 1999 by Drew and Cynthia Johnson, Anaxos is a leading provider of educational content for print and electronic media.

David Wilson earned a PhD in organic chemistry from Columbia University and works as a medical doctor in Hawaii. He taught chemistry at the university level for five years.

KAPLAN PANEL OF AP EXPERTS

Congratulations! You have chosen Kaplan to help you get a top score on your AP exam.

Kaplan understands your goals, and what you're up against—achieving college credit and conquering a tough test—while participating in everything else that high school has to offer.

You expect realistic practice, authoritative advice, and accurate, up-to-the-minute information on the test. And that's exactly what you'll find in this book, as well as every other in the AP series. To help you (and us!) reach these goals, we have sought out leaders in the AP community. Allow us to introduce our experts:

AP CHEMISTRY EXPERTS

Lenore Hoyt teaches chemistry at Idaho State University. She has done postdoctoral studies at Yale University and holds a PhD from the University of Tennessee. She has been a reader for the AP Chemistry exam for more than 5 years.

Jason Just has been teaching AP, IB, and general chemistry for more than 15 years at suburban high schools around St. Paul, Minnesota. He has contributed to science curriculum and teacher training through St. Mary's University, the Science Museum of Minnesota, as well as Lakeville Area, South St. Paul, and North St. Paul public schools.

Lisa Zuraw has been a professor in the Chemistry Department at The Citadel in Charleston, South Carolina, for more than 13 years. She has served as a faculty reader for the AP Chemistry exam and as a member of the AP Chemistry Test Development Committee.

| Part One |

THE BASICS

CHAPTER 1: INSIDE THE AP CHEMISTRY EXAM

If you're holding this book, chances are you are gearing up for the AP Chemistry exam. Your teacher has spent the year cramming your head full of the chemistry know-how you will need to have at your disposal. But there is more to the AP Chemistry exam than chemistry know-how. You have to be able to work around the challenges and pitfalls of the test—and there are many—if you want your score to reflect your abilities. You see, studying chemistry and preparing for the AP Chemistry exam are not the same thing. Rereading your textbook is helpful, but it's not enough.

That's where this book comes in. We'll show you how to marshal your knowledge of chemistry and put it to brilliant use on test day. We'll explain the ins and outs of the test structure and question format so you won't experience any nasty surprises. We'll even give you answering strategies designed specifically for the AP Chemistry exam.

Preparing yourself effectively for the AP Chemistry exam means doing some extra work. You need to review your text *and* master the material in this book. Is the extra push worth it? If you have any doubts, think of all the interesting things you could be doing in college instead of taking an intro course filled with facts you already know.

Advanced Placement exams have been around for half a century. While the format and content have changed over the years, the basic goal of the AP program remains the same: to give high school students a chance to earn college credit or advanced placement. To do this, a student needs to do two things:

- Find a college that accepts AP scores
- Do well enough on the exam to earn credit

The first part is easy, since a majority of colleges accept AP scores in some form or another. The second part requires a little more effort. If you have worked diligently all year in your course work, you've laid the groundwork. The next step is familiarizing yourself with the test itself.

OVERVIEW OF THE TEST STRUCTURE

Let's talk about the test itself. The AP Chemistry exam consists of two parts. In Section I, you have 90 minutes to answer 75 multiple-choice questions with five answer choices each. This section is worth 50 percent of your total score.

Section II of the exam consists of six free response questions (three quantitative problems, one set of reactions, and two essay questions) that are worth 50 percent of your score. The free-response questions are long, multi-step, and involved; you will spend the 95 minutes of Section II answering all six problems. Although these free-response problems are long and often broken down into multiple parts, they usually don't cover an obscure topic. Instead, they take a fairly basic chemistry concept and ask you several questions about it. Expect to use some formulas (many of which are provided) and crunch some numbers. It's a lot of chemistry work, but it's fundamental chemistry work.

The College Board—the company that creates the AP exams—releases a list of the topics covered on the exam. The College Board even provides the percentage amount that each topic appears on the exam. Since this information is useful to anyone considering taking the test, check out the breakdown below.

Topics Covered on the AP Chemistry Exam

I. **Structure of Matter (20 percent)**
Topics include atomic theory, atomic structure, chemical bonding, and nuclear chemistry

II. **States of Matter (20 percent)**
Topics include gases, liquids, solids, and solutions

III. **Reactions (35–40 percent)**
Topics include reaction types, stoichiometry, equilibrium, kinetics, and thermodynamics

IV. **Descriptive Chemistry (10–15 percent)**
Topics include relationships in the periodic table, chemical reactivity, and an introduction to organic chemistry

V. **Laboratory (5–10 percent)**
Questions in this category seek to determine the skill level of students in observing reactions, recording data, interpreting results, and communicating these results effectively.

All of these topics are covered, in detail, in the review section of this book. For a more detailed outline of each topic, go online to http://apcentral.collegeboard.com. This is the official website for the AP tests, and it contains a great deal of general information that you may find useful.

CHEMICAL CALCULATIONS

The AP Chemistry exam includes several different types of calculations, including percentage composition, empirical and molecular formulas, molar masses, gas laws, stoichiometry, mole fractions, Faraday's Law, equilibrium constants, standard electrode potentials, thermodynamic and thermochemical calculations, and kinetics. When completing these calculations, pay careful attention to significant figures, precision of measured values, and logarithmic and exponential relationships. Finally, be sure to check that your results seem reasonable. For example, if you start with a few grams of reactants, you should not end up with several thousand kilograms of products. Checking the order of magnitude of your answers is a quick way to confirm that your results are reasonable.

HOW THE EXAM IS SCORED

Beginning with the May 2011 administration of AP exams, the method for scoring the multiple-choice section has changed. Scores are based on the number of questions answered correctly. **No points are deducted for wrong answers.** No points are awarded for unanswered questions. Therefore, you should answer every question, even if you have to guess.

When your three hours of testing are up, your exam is sent away for grading. The multiple-choice part is handled by a machine, while qualified graders—current and former chemistry teachers—grade your responses to Section II. After an interminable wait, your composite score will arrive. Your results will be placed into one of the following categories, reported on a five-point scale:

5 = Extremely well qualified (to receive college credit or advanced placement)

4 = Well qualified

3 = Qualified

2 = Possibly qualified

1 = No recommendation

Depending on the college, a score of 4 or 5 on the AP Chemistry exam will allow you to leap over the freshman introductory chemistry course and jump right into more advanced classes. These advanced classes are usually smaller in size, better focused, more intellectually stimulating and, simply put, just more interesting than a basic course. If you are solely concerned about fulfilling your science requirement so you can get on with your study of pre-Columbian art or Elizabethan music or some other non-chemistry-related area, the AP exam can help you there, too. Ace the AP Chemistry exam and, depending on the requirements of the college you choose, you may never have to take a science class again.

Some colleges will give you college credit for a score of 3 or higher, but it's much safer to get a 4 or a 5. If you have an idea of which colleges you would like to attend, check out their websites or call the admissions office to find out what their particular rules regarding AP scores are.

REGISTRATION AND FEES

To register for the exam, contact your school guidance counselor or AP coordinator. If your school does not administer the AP exam, contact the College Board for a listing of schools that do.

As of the printing of this book, the fee for the exam is $87. For those qualified with acute financial need, the College Board offers a $26 credit. In addition, many states offer subsidies to cover all or part of the cost of the exam.

ADDITIONAL RESOURCES

For more information on the AP Program and the AP Chemistry exam, contact:

AP Services
P.O. Box 6671
Princeton, NJ 08541–6671
Phone: (609) 771–7300 or 888-225–5427 (toll-free in the U.S. and Canada)
Email: apexgrams@info.collegeboard.org
Website: www.collegeboard.com/student/testing/ap/about.html

CHAPTER 2: STRATEGIES FOR SUCCESS: IT'S NOT ALWAYS HOW MUCH YOU KNOW

In education today, standardized tests have become a common occurrence. Following closely on the heels of standardized tests are *strategies* to help you succeed on standardized tests. You are probably familiar with some of the general strategies that help students increase their scores on a standardized exam. Even so, a quick review will help to jog your memory.

GENERAL TEST-TAKING STRATEGIES

1. **Pacing.** Since many tests are timed, proper pacing allows students to attempt every question in the time allotted. Poor pacing causes students to spend too much time on some questions to the point that they run out of time before getting a chance at every problem.

2. **Two-Pass System.** Using the two-pass system is one way to help your pacing on a test. The key idea is that you don't simply start with question one and trudge onward from there. Instead, you start at the beginning, but take a first pass through the test, answering all the questions that are easy for you. If you encounter a tough problem, you spend only a small amount of time on it and then move on in search of easier questions that might exist after that problem. This way, you don't get bogged down on a tough problem when you could be earning points answering later problems that you do know. On your second pass, you go back through the section and attempt all the tougher problems that you passed over the first time. You should be able to spend a little more time on them, and this extra time might help you answer the problem. Even if you don't reach an answer, you might be able to employ techniques like the process of elimination to cross out some answer choices or just take a guess, since wrong answers are no longer penalized.

3. **Process of Elimination.** On every multiple-choice test you ever take, the answer is given to you. The only difficulty resides in the fact that the correct answer is hidden among incorrect

choices. Even so, the multiple-choice format means you don't have to pluck the answer out of the air. Instead, if you can eliminate answer choices you know are incorrect and if only one choice remains, then that must be the correct answer.

4. **Patterns and Trends.** The key word here is the *standardized* in "standardized testing." Being standardized means that tests don't change greatly from year to year. Sure, each question won't be the same, and different topics will be covered from one administration to the next, but there will also be a lot of overlap from one year to the next. That's the nature of standardized testing: If the test changed wildly each time it came out, it would be useless as a tool for comparison. Because of this, certain patterns can be uncovered about any standardized test. Learning about these trends and patterns can help students taking the test for the first time.

5. **The Right Approach.** Having the right mind-set plays a large part in how well people do on a test. Those who are nervous about the exam and hesitant to make guesses often fare much worse than students with an aggressive, confident attitude. Students who start with question one and plod on from there don't score as well as students who pick and choose the easy questions first before tackling the harder ones. People who take a test cold have more problems than those who take the time to learn about the test beforehand. In the end, factors like these determine if people are good test takers or if they struggle even when they know the material.

These points are all valid for every standardized test, but they are quite broad in scope. The rest of this chapter will discuss how these general ideas can be modified to apply specifically to the AP Chemistry exam. These test-specific strategies—combined with the factual information covered in your course and in this book's review—are the one-two punch to help you succeed on this specific test.

HOW THIS BOOK CAN HELP

Kaplan's *AP Chemistry* contains precisely the information you will need to ace the test. There's nothing extra in here to waste your time—no pointless review of material you won't be tested on, no rah-rah speeches. Just the most potent test preparation tools available:

1. **Test Strategies Geared Specifically to the AP Chemistry Exam.** Many books give the same talk about Process of Elimination that's been used for every standardized test given in the past 20 years. We're going to talk about Process of Elimination as it applies on the AP Chemistry exam, and only on the AP Chemistry exam. There are several skills and general strategies that work for this particular test, and these will be covered in the next two chapters.

2. **A Well-Crafted Review of All the Relevant Subjects.** The best test-taking strategies alone won't get you a good score if you don't know the difference between oxygen and helium. At its core, this AP exam covers a wide range of chemistry topics, and learning these topics

is necessary. However, chances are good you're already familiar with these subjects, so an exhaustive review is not needed. In fact, it would be a waste of your time. No one wants that, so we've tailored our review section to focus on how the relevant topics typically appear on the exam and what you need to know in order to answer the questions correctly. If a topic doesn't come up on the AP Chemistry exam, we don't cover it. If it appears on the test, we'll provide you with the facts you need to navigate the problem safely.

3. **Two Full-Length Practice Tests to Cut Your Teeth On.** Few things are better than experience when it comes to standardized testing. Taking a practice AP exam gives you an idea of what it's like to answer chemistry questions for three hours. Granted, that's not a fun experience, but it is a helpful one. Practice exams give you the opportunity to find out what areas are your strongest and what topics you should spend some additional time studying. And the best part is that it doesn't count! Mistakes you make on our practice exams are mistakes you won't make on the real test.

Those three points describe the general outline of this book: strategies, review, and then practice. This chapter will help you learn some specific skills you can use on the AP Chemistry exam.

HOW TO APPROACH THE MULTIPLE-CHOICE QUESTIONS

Although you might not like multiple-choice questions, there's no denying the fact that guessing is easier on a multiple-choice question than it is on an essay question or a problem set. On a multiple-choice problem, the answer is always there in front of you; the trick is to find it amongst the forest—okay, thicket—of incorrect answers. Contrast this with Section II of the AP Chemistry exam. On that section, if you don't know how to work a problem, you have to write down what you know and hope that the mythical King of Partial Credit is feeling kindly toward you that day.

We'll talk more about Section II later. For now, let's talk about what you'll encounter on the AP Chemistry exam before you even see the first question.

HELPFUL INFORMATION IN AN UNHELPFUL FORMAT

Imagine that your parents are forcing you to invite someone to go to a party that you are hosting. You don't actually like this person, but your parents are insistent, so you have no choice but to comply. You give the acquaintance directions to the party, but you intentionally make the directions vague and difficult to use, filled with bland phrases like, "Turn left at the light and then turn right a couple of miles down from there." Your hope is that you've provided the required information, but done so in a manner that will not help much.

This isn't very nice of you, but it does help provide a useful example. Before each section of the AP Chemistry exam, you will be given some information that you can use throughout the section. This information is presented just like those directions in the story above. It's useful, but it doesn't go out of its way to be easy to use. The periodic table presented before Section I is a stripped-down version that consists primarily of letters, numbers, and blocks. Any additional information that might be found on a regular periodic table is NOT there.

Now, if you're familiar with the periodic table, you probably don't need any of that extra information anyway. In other words, if you know enough about the periodic table to understand the periodic table at the front of the test, then you probably won't need that periodic table very often.

So, in the end, the facts given at the front of the section are a bit of a wash. Still, keep in mind that they are there. There might be a question or two that requires you to grab some specific information—such as the atomic number—from the periodic table, in which case having the table there is very handy. Most likely, though, there won't be any overt indication that you need to use the table. In other words, the question won't state, "Use the periodic table to help you on this problem," or anything remotely like that.

Now that the introductory material has been covered, let's talk about the 75 multiple-choice questions in Section I that you have 90 minutes to tackle. These questions come in three types.

1. **Cluster Questions.** These are the first problems you'll see. With Cluster Questions, you get some initial piece of information (often visual), and this information has the letters A through E within it. Two to three problems without answer choices follow this information, and each question describes one of the choices (A–E) in the initial information. You pick the right letter choice for each problem. Here's an example:

   Questions 1 and 2 refer to the following pure substances:

   (A) SiO_2
   (B) CH_4
   (C) H_2O
   (D) I_2
   (E) SiH_4

   1. Has the highest melting point
   2. Has the largest dipole movement

   This shows a typical group of Cluster Questions. The initial information lists five substances as answer choices, and then the two problems ask you to pick the substance that matches what is being requested in each question.

   Cluster Questions are great because they don't take very much time. One set of answers covers two to three simple, brisk problems. Sadly, there will only be about 10–15 Cluster Questions on the exam. Even so, knock out these problems and give yourself as much time as you can for the Stand Alones.

2. **The Stand Alone Questions.** These questions make up the bulk of the AP Chemistry exam. Each Stand Alone Question covers a specific topic, and then the next Stand Alone hits a different topic. Each stem provides you with the information you need to answer the problem. Here's a typical Stand Alone:

46. If 100 mL of a 1 M HCl solution and 100 mL of a 5 M NaCl solution are mixed, what is the final molarity of the chloride (Cl^- ion)?

   (A) 1 M
   (B) 2 M
   (C) 3 M
   (D) 4 M
   (E) 5 M

You get some information to start with, and then you're expected to answer the question. The number of the question, 46, makes no difference since there's no order of difficulty on the AP Chemistry exam. Tough questions are scattered between easy and medium questions with no pattern or reason.

3. **Double Data Questions.** If you were to take two related Stand Alone Questions and merge them together, you would get a pair of Double Data Questions. Typically, a graph is presented, and then two questions are asked about that graph. Depending on how well you understand the topic presented in the graph, Double Data Questions can be fairly quick and simple: You get to answer two questions, but you only have to deal with one graph. It's a nifty time-saving trick, but don't expect to get to use it very often since Double Data Questions are the rarest of all the question types. You'll probably see only four to six Double Data Questions in Section I of the test.

How you do answering the Stand Alones is going to make the greatest impact on your Section I score. This question type deserves some attention, yet it's easier to talk about what isn't in the Stand Alone Questions than what is there. Consider the following points:

- There's no order of difficulty. In other words, the problems don't start out easy and gradually become tougher.

- There's no system to explain the order in which chemistry concepts appear in the section.

The Stand Alones look like a bunch of disconnected chemistry questions that appear one after the other. That's just what they are. Since randomness rules the day, there's no sense in answering these questions in consecutive order. A two-pass system should be used, but you can tweak the general idea of the two-pass system and apply it specifically to the AP Chemistry exam.

THE TWO-PASS SYSTEM ON THE AP CHEMISTRY EXAM

If you wanted, you could take all the AP Chemistry questions and place them in a spectrum ranging from "fastest to answer" to "hardest or longest to answer." Questions involving Roman numerals often take the most time to solve. (Typically, you have to solve each Roman numeral, meaning that you answer three questions but only get credit for one.) Cluster Questions can be solved the fastest. Double Data Questions can also be solved quickly, since you often gain an advantage by using the same graph for more than one problem.

Those are the obvious ways to pace oneself, and many students do no more than that. But the more advanced your pacing system is, the more time you might have at the end of Section I to answer questions. To further refine your two-pass abilities, consider the following three points when scanning through Section I:

1. **Small Question Stems.** In test-speak, the portion of a problem that comes before the answer choices is typically called the question stem. Stem length varies on the AP Chemistry exam from eight-word simple sentences to elaborate 50-word descriptions of hypothetical situations. Either way, while you're reading the question stem, time is ticking away. The longer the stem, the more time it takes you to read it. If you're a fast reader, this might not be very much time. If you read at a more methodical rate, you might try first passing over the wordier problems.

2. **Topics You Like.** Draw up two lists using the topics covered on page 4 of chapter 1. Label one list "Concepts I Enjoy and Know About in Chemistry" and label the other list "Concepts That Are Not My Strong Points." When you get ready to tackle the Stand Alone section, keep these two lists in mind. On your first pass through the section, answer all the questions that deal with concepts you like and about which you are knowledgeable. When you come to a question that is on a subject that's not one of your strong points, skip it and come back to it later. The overarching goal is to use your available time to answer as many questions correctly as possible.

3. **Balanced Equations.** Balanced equations show up on the AP Chemistry exam in a number of places and in a variety of forms. Because these questions are easily recognizable, you might want to take a moment to zip through Section I and work all the equation questions in one go. (Of course, if balancing equations is one of those "Concepts That Are Not My Strong Points," then this course of action is not recommended.) Once these questions are finished, you can take off your Balancing Equation Hat and move on.

You don't have much time to ponder every tough question, so trusting your instincts keep can you from getting bogged down and wasting time on a problem. You might not get every educated guess correct, but again, the goal isn't to get a perfect score. The goal is to get a good score and to survive hard questions by going with your gut feelings.

On other problems, though, you might have no inkling of what the correct answer should be. In that case, turn to the following key idea.

THINK "GOOD SCIENCE"!

The AP Chemistry exam rewards good chemists. The test wants to foster future chemists by covering fundamental topics and sound laboratory procedure. What the test doesn't want is bad science. It doesn't want answers that are factually incorrect, too extreme to be true, or irrelevant to the topic at hand.

Yet these bad science answers invariably appear because it's a multiple-choice test and you have to have four incorrect answer choices around the one right answer. So if you don't know how to answer a problem, look at the answer choices and think "Good Science." This may lead you to find some poor answer choices that can be eliminated.

46. If 100 mL of a 1 M HCl solution and 100 mL of a 5 M NaCl solution are mixed, what is the final molarity of the chloride (Cl^- ion)?

   (A) 0.5 M
   (B) 2 M
   (C) 3 M
   (D) 4 M
   (E) 5 M

This is a Stand Alone problem, but even if you don't know how to answer this problem, you can use Good Science to give yourself a chance at guessing the right answer. Look at choice (E), 5 M. This value is used within the question stem. Because it appears in the stem, it's unlikely to be the right answer. Why? There are two reasons:

1. Even if you were from another planet, you could see the "5" symbol in the question stem and then in the answer choice. Picking it for that reason doesn't take any mental skills at all, and this AP exam is about testing your knowledge.

2. Choice (E), 5, could be the answer if the problem was a trick question of some kind. But the AP Chemistry exam doesn't generally use trick questions, so (E) is not the correct answer.

So eliminate (E). You could also use some Good Science and say, "If I'm adding 1 M to 5 M, the answer will probably be somewhere between these two values." That gets rid of choice (A), 0.5 M.

You now have a one-in-three shot of answering the problem. If you don't know how to tackle this problem, these odds are pretty good.

HOW TO APPROACH THE FREE-RESPONSE QUESTIONS

Section II is 95 minutes long, consists of six questions, and is divided into two parts, Part A and Part B. The diagram below is a breakdown of what you will encounter:

| Part A:
(55 minutes)
(Calculators permitted) | Q1: Equilibrium problem
(20% of section score) | ___ | Q2: Other problem*
(20% of section score) | ___ | Q3: Other problem*
(20% of section score) |

| Part B:
(40 minutes)
(No calculators) | Q4: Reactions question
(10% of section score) | ___ | Q5: Essay question*
(15% of section score) | ___ | Q6: Essay question*
(15% of section score) |

*One of the other problems or essays will be based on laboratory work.

Understand that diagram ahead of time, because you don't want to waste brainpower on test day just figuring out the *directions*. Here are some points to review:

- Section II is 95 minutes total.

- You can use a calculator for Part A but not for Part B.

- Part A accounts for 60 percent of the overall score, and the individual questions in Part A are worth more than the ones in Part B.

- Question 1 is a problem involving chemical equilibrium.

- For question 4, you will write balanced chemical equations for three different sets of reactants and will answer a short question about each of them.

- Questions 2, 3, 5, and 6 may involve the quantitative analysis of data in a laboratory-based question.

A SWAMP OF INFORMATION

We used the "vague party directions" to describe the information doled out before Section I. Section II goes for more of an "information overload" approach. There are four full pages of information before Section II begins. There's a periodic table, a reduction potential table, and numerous equations and constants covering:

- Atomic structure

- Equilibrium

- Thermochemistry

- Gases, liquids, and solutions
- Oxidation-reduction; electrochemistry

There are almost 50 equations and a slightly smaller number of constants displayed for you, not including all the information given on the two tables. See pp. 281–284 for examples of these tables.

Faced with all this, you might think, "I can't possibly remember to use all this stuff." Don't let the sheer amount of information overwhelm you and cause this reaction. Instead, understand that some of this information will be needed for several of the answers, but that the rest is there to act as a smoke screen. Consider things from the test-writer's standpoint. They can't place *only* the information needed in Section II; that would make things too simple, as everyone would know to use the facts provided. Therefore, they have to devise a system in which the chem-savvy people find the facts they need, while the less-fortunate students get lost in the swamp of additional information.

It's not exactly as simple as that, but for our purposes this description works fine. Look over the tables at the beginning of Section II and tell yourself, "I will need to use some, but certainly not all, of this information on several of the questions in this section." That will give you the right approach to handling the large amount of data that precedes Section II.

The topics covered in the free-response questions are usually fairly common chemistry topics. They aren't trick questions asking about an obscure subject. However, a well-known topic doesn't mean the problem will be simple. Most questions have many parts. You usually won't get one broad question like, "What is the real meaning and significance of chemistry?" Instead, you'll get an initial setup followed by questions labeled (a), (b), (c), and so on. Expect to write at least one paragraph (or provide a multi-step equation) for each letter.

For each sub-question on a free-response question, points are given for saying the right thing. The more points you score, the better off you are on that question. Going into the details about how points are scored would make your head spin, but in general, the AP Chemistry people have a rubric, which acts as a blueprint for what a good answer should look like. Every subsection of a question has one to five key ideas attached to it. If you write about one of those ideas, you earn yourself a point. There's a limit to how many points you can earn on a single subquestion. There are also other strange regulations, but it boils down to this: Writing smart things about each question will earn you points toward that question.

So don't be terse and don't rush. You have 20–30 minutes for each free-response problem. Use it to be as precise as you can be for each sub-question. Sometimes doing well on one sub-question earns you enough points to cover up for another sub-question for which your answer isn't as strong.

When all the points are tallied for that free-response problem, you come out strong on total points, despite the fact that you didn't ace every single sub-question.

Finally, since you are given a calculator to use on Part A of the Section II, expect to use it. There's no point in dividing a section into calculator/no calculator if they aren't going to make you press some digits in Part A. You can also guess that Part A will require you to use some of the formulas at the beginning of the section in your calculations. This isn't 100 percent certain, of course, because nothing on an unknown test ever is. But it is a very good guess, kind of like picking the most favored horse to win a race.

If you get a question on a subject in which you're weak, things might look grim for that problem. Still, take heart. Quite often, you'll earn some points on every question since there will be some sub-questions or segments with which you are familiar. Remember, the goal is not perfection. If you can ace four of the questions and get partial credit on the other two, you will put yourself in position to get a good score on the entire test. Don't lose sight of the big picture just because you don't know the answer to one sub-question on Part B.

Be sure to use all the strategies discussed in this chapter when taking the practice exams. Trying out the strategies there will get you comfortable with them, and you should be able to put them to good use on the real exam.

Of course, all the strategies in the world can't save you if you don't know anything about chemistry. The next part of the book will help you review the primary concepts and facts that you can expect to encounter on the AP Chemistry exam.

STRESS MANAGEMENT

1. PACING

Since the test is timed, proper pacing will ensure that you will get to every question. Don't get yourself into a situation where you run out of time before seeing each question at least once. Use your time wisely and don't dawdle. Make sure you take the time now to familiarize yourself with the style of questions that will be asked, so you can get going right away on test day.

2. PROCESS OF ELIMINATION

For the multiple-choice questions, the answer is listed. The difficulty resides in the fact that it is hidden among incorrect choices. If you can eliminate answer choices you know are incorrect, you'll be in better shape to guess if necessary.

AP EXPERT TIP

When you answer free-response questions in practice or homework, have a timer running for fifteen minutes. This is about the amount of time you have per question, and you'll begin to feel more comfortable as you get used to working under time restrictions.

3. PATTERNS AND TRENDS

Standardized tests don't change much from year to year. Yes, questions themselves change, but the format and the general content remains constant. That's the nature of standardized testing—if the test changed wildly each time it came out, it would be useless as a tool for comparison.

Learn what you can about the previous test administrations of AP Chemistry to understand the test you are about to take. Use the trends and patterns to your advantage.

You can familiarize yourself with these patterns and trends by taking one or both of the practice exams included in this book.

4. THE RIGHT APPROACH

Having the right mind-set can play a bigger role than you might think in a test. Those who are nervous about an exam and hesitant to guess often fare much worse than those bearing a proactive, confident attitude.

And students who start with question one and plod on from there tend not to score as well as those who carefully pick and choose their questions. Tackle the easy questions first before you take on the harder ones.

AP EXPERT TIP

When the pressures come knocking before test day, remember that if you can get 50 percent of the questions right, you'll get at least a 3 on the exam.

Part Two

DIAGNOSTIC TEST

AP CHEMISTRY DIAGNOSTIC TEST

Following is a 20-question Diagnostic Test, which is designed to cover most of the topics you will encounter on the AP Chemistry exam. (Be sure to take it under testlike conditions.) After you take it, you can use the results to give yourself a broad idea of what subjects you are strong in and what topics you need to review more. You can use this information to tailor your approach to the following review chapters. Hopefully you'll still have time to read all the chapters, but if pressed, you can start with the chapters and subjects you know you need to work on.

Give yourself 30 minutes for the 20 multiple-choice questions and 20 minutes for the free-response question. Time yourself, and take the entire test without interruption.

Be sure to read the explanations for all questions, even those you answered correctly. (This is something you should do on the two practice exams as well.) Even if you got the problem right, reading the breakdown of why it's right can give you insights that will prove helpful on the real exam.

Good luck on the Diagnostic!

DIAGNOSTIC TEST ANSWER GRID

To compute your score for the Diagnostic Test, calculate the number of questions you got right, then divide by 20 and multiply by 100 to get the percentage of questions that you answered correctly.

The approximate score range is as follows:

5 = 80–100% (extremely well qualified)

4 = 60–79% (well qualified)

3 = 50–59% (qualified)

2 = 40–49% (possibly qualified)

1 = 0–39% (no recommendation)

A score of 49% is approximately a 2, so in this case you can definitely do better. If your score is low, keep on studying to improve your chances of getting credit for the AP Chemistry exam.

1. Ⓐ Ⓑ Ⓒ Ⓓ Ⓔ   11. Ⓐ Ⓑ Ⓒ Ⓓ Ⓔ

2. Ⓐ Ⓑ Ⓒ Ⓓ Ⓔ   12. Ⓐ Ⓑ Ⓒ Ⓓ Ⓔ

3. Ⓐ Ⓑ Ⓒ Ⓓ Ⓔ   13. Ⓐ Ⓑ Ⓒ Ⓓ Ⓔ

4. Ⓐ Ⓑ Ⓒ Ⓓ Ⓔ   14. Ⓐ Ⓑ Ⓒ Ⓓ Ⓔ

5. Ⓐ Ⓑ Ⓒ Ⓓ Ⓔ   15. Ⓐ Ⓑ Ⓒ Ⓓ Ⓔ

6. Ⓐ Ⓑ Ⓒ Ⓓ Ⓔ   16. Ⓐ Ⓑ Ⓒ Ⓓ Ⓔ

7. Ⓐ Ⓑ Ⓒ Ⓓ Ⓔ   17. Ⓐ Ⓑ Ⓒ Ⓓ Ⓔ

8. Ⓐ Ⓑ Ⓒ Ⓓ Ⓔ   18. Ⓐ Ⓑ Ⓒ Ⓓ Ⓔ

9. Ⓐ Ⓑ Ⓒ Ⓓ Ⓔ   19. Ⓐ Ⓑ Ⓒ Ⓓ Ⓔ

10. Ⓐ Ⓑ Ⓒ Ⓓ Ⓔ   20. Ⓐ Ⓑ Ⓒ Ⓓ Ⓔ

DIAGNOSTIC TEST

Directions: Each of the questions or incomplete statements below is followed by five suggested answers or completions. Select the best choice in each case.

1. A solid sample of aluminum carbonate ($Al_2(CO_3)_3$) is strongly heated to drive off CO_2, which is then bubbled through NaOH, forming Na_2CO_3 and water. How many moles of NaOH are required to completely react with 0.60 moles of aluminum carbonate?

   (A) 0.1
   (B) 0.3
   (C) 0.6
   (D) 1.8
   (E) 3.6

2. A sample of NCl_3 (0.100 moles) explodes according to the equation below. What is the volume of the gaseous products, assuming ideal behavior, at 1 atm and 273 K?

   $$2NCl_3(l) \rightarrow N_2(g) + 3Cl_2(g)$$

   (A) 1.12 L
   (B) 2.24 L
   (C) 4.48 L
   (D) 6.72 L
   (E) 8.96 L

3. Under which set of conditions will a sample of gas deviate most from ideality?

Size of molecule	Temperature	Pressure
(A) large	low	high
(B) small	high	low
(C) large	low	low
(D) small	high	high
(E) large	high	low

4. Which of the following best explains why diamonds are hard and graphite is soft?

   (A) The carbon-carbon bonds in diamonds are stronger than those in graphite.
   (B) The melting point of diamonds is higher than that of graphite.
   (C) Diamonds have interlinked crystalline layers, while graphite has flat planes of crystals.
   (D) The carbon-carbon bonds in diamonds are shorter than those in graphite.
   (E) Diamonds are more stable than graphite.

GO ON TO THE NEXT PAGE

5. Rubidium is less electronegative than lithium. The electron that is lost in order to form a stable rubidium ion

(A) has a higher ionization energy than lithium's valence electron.

(B) is held tighter than lithium's valence electron.

(C) has a smaller charge than lithium's valence electron.

(D) has a lower ionization energy than lithium's valence electron.

(E) is part of radioactive decay.

6. Which of the following will dissolve in concentrated hydrochloric acid?

I. CuS

II. CaC_2

III. NaCl

(A) I only

(B) I and II only

(C) II and III only

(D) I and III only

(E) I, II, and III

7. Ozone reacts with ammonia according to the following overall reaction equation:

$$5O_3 + 6NH_3 \rightarrow 6NO + 9H_2O$$

If the decomposition of ozone, shown below, is the slowest step in this reaction, how will the reaction rate change if the pressure of ammonia is doubled?

$$O_3 \rightarrow O_2 + O \quad \text{(slow)}$$

(A) Increase by 2^6

(B) Increase by $2^{\frac{6}{5}}$

(C) Remain the same

(D) Decrease by $2^{\frac{6}{5}}$

(E) Decrease by 2^6

8. What is the pH of a solution made by combining 5 mL of 0.03 M NaOH and 5 mL of 0.01 M H_3PO_4?

(A) Strongly basic

(B) Weakly basic

(C) Neutral

(D) Weakly acidic

(E) Strongly acidic

9. For the following reaction, identify the element being reduced and the reductant.

$$CO + NO_2 \rightarrow CO_2 + NO$$

	Reduced	Reductant
(A)	Carbon	Nitrogen dioxide
(B)	Nitrogen	Carbon monoxide
(C)	Nitrogen	Nitrogen dioxide
(D)	Carbon	Carbon dioxide
(E)	Oxygen	Carbon monoxide

GO ON TO THE NEXT PAGE

10. If ΔH is negative for a spontaneous reaction, what is true about the value of ΔS?

    (A) ΔS must be positive.

    (B) ΔS must be negative.

    (C) ΔS must be zero.

    (D) ΔS can be positive or negative depending on the temperature.

    (E) ΔS can have no value that results in a spontaneous reaction regardless of temperature.

11. Lead nitrate is strongly heated, generating a red-brown gas. This gas is then bubbled into water, giving an acidic solution. Which pair of unbalanced equations best describes this process?

    (A) $PbNO_3 \rightarrow Pb + NO_3$
    $NO_3 + H_2O \rightarrow H^+ + NO_3^-$

    (B) $Pb(NO_3)_2 \rightarrow Pb + NO_2 + O_2$
    $NO_2 + H_2O \rightarrow H^+ + NO_2^-$

    (C) $Pb(NO_2)_2 \rightarrow PbO + NO_2 + O_2$
    $NO_2 + H_2O \rightarrow H^+ + NO_3^- + NO$

    (D) $Pb(NO_3)_2 \rightarrow PbO + NO + O_2$
    $NO + H_2O \rightarrow H^+ + NO_2^-$

    (E) $Pb(NO_3)_2 \rightarrow PbO + NO_2 + O_2$
    $NO_2 + H_2O \rightarrow H^+ + NO_3^- + HNO_3$

12. How many physically distinct isomers of dichloropropane ($C_3H_6Cl_2$) are possible?

    (A) 1

    (B) 2

    (C) 3

    (D) 4

    (E) 5

13. A 0.1 M aqueous solution of an unknown white crystalline substance shows a precipitate when treated with a silver nitrate solution and turns a phenolphthalein solution pink. The unknown substance could be

    (A) NH_4Cl.

    (B) K_3PO_4.

    (C) KNO_3.

    (D) $NaCl$.

    (E) NH_4ClO_4.

14. Radionuclides such as ^{241}Am, which emit nonpenetrating radiation, are less dangerous with respect to external exposure than radionuclides such as ^{60}Co, which emit penetrating radiation. What is the non-penetrating type of radioactive particle called?

    (A) Thermal particle

    (B) Beta particle

    (C) Gamma particle

    (D) Alpha particle

    (E) Kappa particle

15. Which of the following aqueous solutions would have a boiling point most similar to that of a 0.2 M solution of sucrose ($C_{12}H_{22}O_{11}$)?

    (A) 0.2 M NH_4Cl

    (B) 0.1 M K_3PO_4

    (C) 0.1 M acetic acid

    (D) 0.4 M $NaCl$

    (E) 0.1 M $LiClO_4$

GO ON TO THE NEXT PAGE

16. Which reagent is most likely to effect the following transformation?

$$3I^- + ? \rightarrow I_3^-$$

(A) $KMnO_4$

(B) H_2S

(C) KNO_3

(D) $Fe(NO_3)_2$

(E) $K_2S_2O_3$

17. Identify which solution would function as a weakly acidic buffer.

(A) 0.10 M NaCl/0.10 M HCl

(B) 0.10 M CH_3CO_2Na/0.10 M CH_3CO_2H

(C) 0.10 M NH_4Cl/0.10 M HCl

(D) 0.10 M H_2S

(E) 0.11 M HCl/0.10 M NaOH

18. Which of the following does not represent a ground state electronic configuration for the given element or ion?

(A) Fe : $1s^2 2s^2 2p^6 3s^2 3p^6 4s^2 3d^6$

(B) N : $1s^2 2s^2 2p^3$

(C) Al^{3+} : $1s^2 2s^2 2p^4 3s^2$

(D) O^{2-} : $1s^2 2s^2 2p^6$

(E) Cl^+ : $1s^2 2s^2 2p^6 3s^2 3p^4$

19. A compound is most likely to have a 1:1 ratio of elements if it is formed from which two groups of elements?

(A) Alkali metals and chalcogens

(B) Alkaline earth metals and halogens

(C) Alkaline earth metals and chalcogens

(D) Transition metals and halogens

(E) Transition metals and noble gases

20. The Hall process produces aluminum metal from Al_2O_3 and carbon in an electrochemical cell.

In this process, carbon is consumed at one of the terminals of the electrochemical cell (either the anode or the cathode). Which of the following describes what happens to the aluminum?

(A) It is reduced.

(B) It becomes the elecytrolyte.

(C) It is oxidized.

(D) It is inert in this reaction.

(E) It fuses with carbon to form an alloy.

SECTION II

Time: 20 Minutes

Directions: Write out answers to the following questions. Clearly show the method used and steps involved in arriving at your answers. Partial credit can only be given if your work is clear and demonstrates an understanding of the problem.

Calculators are permitted, except for those with typewriter (QWERTY) keyboards.

1. In an automobile, power is derived from the explosive combustion of gasoline and oxygen from the air in an engine's cylinder. Use isooctane (C_8H_{18}) to represent gasoline for this problem.

    (a) Write a balanced equation describing the complete combustion of isooctane.

    (b) Using the information below, calculate the amount of heat released from the combustion of 0.10 g of isooctane (assume standard conditions).

    Isooctane: $\Delta H_{Combustion} = -5,491$ kJ/mol

    (c) If oxygen makes up 20% of the atmosphere by volume, what volume of air is required to completely burn 0.10 g of isooctane under standard conditions?

    (d) What is the final pressure in a 80 mL cylinder after 0.10 g of isooctane is combusted with the minimum amount of air? Assume a final temperature of 1,500°C and ideal gas behavior.

    (e) What is the volume of the same gas as it leaves the tailpipe? Assume a pressure of 1 atm and a temperature of 200°C.

IF YOU FINISH BEFORE TIME IS CALLED, YOU MAY CHECK YOUR WORK ON THIS SECTION ONLY. DO NOT TURN TO ANY OTHER SECTION IN THE TEST.

STOP

Answer Key

1.	E	8.	B	15.	E
2.	C	9.	B	16.	A
3.	A	10.	D	17.	B
4.	C	11.	E	18.	C
5.	D	12.	D	19.	C
6.	E	13.	B	20.	A
7.	C	14.	D		

ANSWERS AND EXPLANATIONS

SECTION I

1. E

First write out full, balanced equations for the reactions, keeping the number of CO_2 molecules the same for each equation:

$$Al_2(CO_3)_3 \rightarrow Al_2O_3 + 3CO_2$$
$$3CO_2 + 6NaOH \rightarrow 3Na_2CO_3 + 3H_2O$$

So for each mole of aluminum carbonate reacted, 6 moles of NaOH are required. We have 0.6 moles given in the question, so 6×0.6 mol = 3.6 mol. [Stoichiometry]

2. C

For each mole of reactant, two moles of gas are produced. From 0.100 moles of NCl_3, we would expect 0.200 moles of gas. Each mole of ideal gas occupies 22.4 L at STP, so the answer is $22.4 \times 0.200 = 4.48$ L. [Gas Laws]

3. A

The ideal gas law assumes that gases do not have intermolecular attractive forces and have negligible volume. As the size of a molecule increases, so does its interaction with other gas molecules, thus deviating from ideality. As temperature decreases, the kinetic energy of the gas drops and intermolecular attractive forces become important. Finally, as the pressure increases, the gases are more likely to interact with each other and also deviate from ideal behavior. [Gas Laws and Stoichiometry]

4. C

Diamonds are hard due to their crystal structure, which makes it difficult to scratch off any of the interlinked crystalline layers. Graphite has flat planes of crystals in which the forces holding the planes of carbon atoms together are much weaker than those in diamonds, therefore it is much easier to remove layers of graphite. [Bonding/Intermolecular Forces]

5. D

The least electronegative, or most electropositive, elements are in the lower left of the periodic table. The valence ionization energy decreases as we go down the periodic table because the valence electrons are farther away from the attractive, positively charged nucleus. Therefore, the electron that is lost to form a stable rubidium ion will have a lower ionization energy than the electron lost to form a stable lithium ion. [Atomic Theory]

6. E

Both I and II dissolve with reaction in concentrated hydrochloric acid according to the following equations:

$$CuS + 2HCl \rightarrow H_2S(g) + Cu^{2+} + 2Cl^-$$
$$CaC_2 + 2HCl \rightarrow Ca^{2+} + C_2H_2(g) \text{ (acetylene)} + 2Cl^-$$

Sodium chloride will still dissolve to an appreciable extent in concentrated HCl even though the common ion effect will decrease its solubility relative to that of pure water. [Solutions]

7. C

Because ammonia is not involved in the rate-determining step, a change in its concentration would not affect the overall rate of the reaction. [Rates and Equilibrium]

8. B

Begin by writing the balanced chemical equation:

$$3NaOH + H_3PO_4 \rightarrow Na_3PO_4 + 3H_2O$$

There are an equal number of moles of H_3PO_4 and Na_3PO_4 in the balanced chemical equation, so the product of the reaction is going to be a 0.005 M solution of Na_3PO_4. Because the salt is formed from a strong base (NaOH) and a weak acid (H_3PO_4), the solution will be weakly basic. [Acid/Base]

9. B

Nitrogen is reduced because it is gaining electrons (+4 to +2). Carbon monoxide is the reductant because it is losing electrons (+2 to +4). [Redox]

10. D

Spontaneous reactions have a negative ΔG and $\Delta G = \Delta H - T\Delta S$. If ΔH is negative, then ΔS can be either positive or negative provided that the temperature doesn't make the term $T\Delta S$ larger than ΔH. [Thermodynamics]

11. E

Nitrogen dioxide (NO_2) is a red-brown gas that is responsible for creating nitric acid (HNO_3) in acid rain. It can be formed in the laboratory by the decomposition of certain metal nitrates according to the equations in choice (E). [Descriptive]

12. D

The four compounds are: $CHCl_2CH_2CH_3$, $CH_2ClCHClCH_3$, $CH_2ClCH_2CH_2Cl$, and $CH_3C Cl_2CH_3$.

13. B

Solutions of salts NH_4Cl, K_3PO_4, and NaCl will all show precipitates when treated with a soluble silver salt, but only K_3PO_4 is basic (see question 8) and therefore able to turn a phenolphthalein solution pink. Ammonium chloride is acidic and sodium chloride is neutral. [Laboratory/Spot Tests and Acid/Base]

14. D

Alpha particles are helium nuclei and interact strongly with matter. Only thin shielding is required to stop these particles and thus they present a smaller hazard with external exposure. Gamma rays are high-energy electromagnetic radiation and can penetrate thick layers of shielding. Thus, they present a significant danger from external exposure. [Nuclear]

15. E

Dissolving a solute in a solvent both increases the solvent's boiling point and decreases its melting point. The colligative effects are dependent on the molal concentration of solute particles (ions or molecules) that are present. Sucrose does not ionize in water and thus represents 0.2 M of solute particles. Choice (C) is a weak acid and does not fully ionize, so the best answer is choice (E). [Solutions/Colligative Properties]

16. A

The iodide anion is being oxidized in this reaction from a -1 oxidation state to a $-\frac{1}{3}$ oxidation state, so we are looking for an oxidizing reagent. Nitrate is not a strong oxidizing reagent in solution and choices (B) and (E) are reducing reagents. Permanganate is a strong oxidizing reagent in solution. [Redox]

17. B

Acidic buffers are most effective when composed of an equal concentration of a weak acid and its conjugate base. The pH of the solution in choice (B) is equal to the pK_a of the weak acid. [Buffers]

18. C

The correct electronic configuration of the ion is $Al^{3+}: 1s^2 2s^2 2p^6$. [Atomic Theory]

19. C

Alkaline earth metals normally have an oxidation state of $+2$ whereas chalcogens (oxygen, sulfur) are -2. A compound formed from these two groups will likely have a 1:1 ratio of elements. [General]

20. A

In this reaction, aluminum is being reduced (at the cathode) while carbon is being oxidized at the anode. [Electrochemistry]

SECTION II

1. (a) $2C_8H_{18} + 25O_2 \rightarrow 16CO_2 + 18H_2O$

    (b) -4.82 kJ

    First, calculate the number of moles of isooctane in 0.10 g. The formula weight is required first:

$FW = (12 \times 8) + (1 \times 18) = 114$ g/mol.

The number of moles is

$$\frac{0.1}{114} = 0.000877 \text{ mol.}$$

Multiplying this number by the $\Delta H_{\text{Combustion}}$ gives us the answer: $0.000877 \text{ mol} \times -5{,}491 \text{ kJ/mol} = -4.82 \text{ kJ.}$

(c) 1.25 L

From (a), we know that for each mole of isooctane, 12.5 moles of oxygen are required. So 0.1 g of isooctane requires $0.000877 \times 12.5 = 0.011$ moles of oxygen. Assuming ideality, this represents 22.4 L/mol $\times$ 0.011 mol $= 0.25$ L of oxygen. Because oxygen is only 20 percent of the atmosphere by volume, we must divide this number by 0.2. This gives us $\frac{0.25}{0.2} = 1.25$ L of air for each 0.10 g of isooctane.

(d) 73 atm

First, calculate the total amount of gas in the end of the reaction. The end gas comes from two sources: the inert gases (mostly nitrogen) that came along with the oxygen in the air and the products of the combustion. The inert gas is 80 percent of the initial air intake, or 0.8×0.82 L $= 0.66$ L at STP. The products of the combustion are formed from isooctane. We know from (a) that, for each mole of isooctane burned, 17 moles of gases are produced. So $0.000877 \times 17 = 0.015$ moles of gas. This at STP is equivalent to 22.4 L/mol $\times 0.015$ mol $= 0.33$ L. Adding these two sources together gives 0.33 L $+$ 0.66 L $= 0.99$ L at $0°C$ and 1 atm. This must now be converted to $1{,}500°C$ and 80 mL, and we must solve for pressure. Because the number of moles of the gas is not being changed, the following equation can be used:

$$\frac{P_1 \times V_1}{T_1} = \frac{P_2 \times V_2}{T_2}.$$

The temperature must be in Kelvin. Plugging in our known values gives:

$$\frac{1 \text{ atm} \times 0.99 \text{ L}}{273 \text{ K}} = \frac{P_2 \text{ atm} \times 0.08 \text{ L}}{1,773 \text{ K}}.$$

Solving for the final pressure P_2 gives 80 atm.

(e) 1.55 L

The same equation applies with $P_2 = 1$ atm and $T_2 = 473$ K (200ºC); solving for V_2:

$$\frac{1 \text{ atm} \times 0.99 \text{ L}}{273 \text{ K}} = \frac{1 \text{ atm} \times V_2 \text{ L}}{473 \text{ K}}$$

$V_2 = 1.7$ L.

DIAGNOSTIC TEST CORRELATION CHART

Use the following table to determine which AP Chemistry topics you need to review most. After scoring your test, check to find out the areas of study covered by the questions you answered incorrectly.

Area of Study	Question Number
Stoichiometry	1, 3
Gas Laws	2, 3
Bonding/Intermolecular Forces	4
Atomic Theory	5, 18
Solutions	6
Rates and Equilibrium	7
Acid/Base	8, 13
Redox	9, 16
Thermodynamics	10
Descriptive	11
Organic	12
Laboratory/Spot Tests	13
Nuclear	14
Solutions/Colligative Properties	15
Buffers	17
General	19
Electrochemistry	20

| Part Three |

AP CHEMISTRY REVIEW

CHAPTER 3: ARITHMETIC AND DIMENSIONS

IF YOU LEARN ONLY FIVE THINGS IN THIS CHAPTER . . .

1. Use MKS units (meters, kilograms, and seconds) and corresponding derived units whenever possible on the AP exam. One important exception is when you are dealing with pressure, for which atm is typically used.

2. Dimensional analysis will help you verify that you have performed the correct calculation any time you are using a formula. Make sure that you match the correct constants with the correct formulas. For example, the equations $PV = nRT$ and $\Delta G° = -RT\ln K$ use different values of R with different units.

3. Always express your answers using the correct number of significant figures. When you are performing a long calculation, carry as many decimal places as possible until you present the final result.

4. Scientific or exponential notation is a convenient way of representing numbers. Review the mathematical operations you may need to perform when dealing with numbers in scientific notation.

5. You will need to use common (base 10) logarithms in acid-base problems. Make sure you are comfortable calculating logs on the calculator you will be using on test day. Also, make sure you know how to calculate logs without a calculator for the multiple-choice portion of the exam.

Every last point counts on the AP Chemistry exam. So be sure to present your answers in the correct format and to use units correctly in your answers; on test day, these things are just as important as understanding the concepts being tested. In this section, we will present the skills necessary to collect these "easy points." Chemistry is like a language such as French, Spanish, or Latin. In order to communicate with each other clearly, chemists must use a common language, especially with respect to numbers and physical quantities.

In this chapter, we will start by addressing the units attached to any quantity you will encounter on the AP Chemistry exam. We will discuss some useful tricks, such as the use of units to arrive at the correct answer. We will also discuss significant figures, which allow chemists to communicate the precision with which physical quantities are measured. Finally, we will present some common mathematical operations you will be asked to perform, namely scientific notation and logarithms.

THE MKS SYSTEM

The MKS system refers to a system of measurement based on meters (m), kilograms (kg), and seconds (s). Focusing on meters, kilograms, and seconds versus other similar units of measure—such as kilometers, grams, or hours—will help you to approach calculations from the most likely starting place. These basic units are SI units, which are part of the metric system. Table 1 shows the basic SI units you will most likely encounter on the AP Chemistry exam.

Physical Quantity	Name of Unit	Symbol
Length	Meter	m
Mass	Kilogram	kg
Time	Second	s
Temperature	Kelvin	K
Amount of Substance	Mole	mol
Electric Current	Ampere	A

There is also a set of derived units, all of which can be represented by combinations of these basic units, as shown in Table 2:

Physical Quantity	Name of Unit	Symbol	SI Units
Density		ρ	kg/m^3
Force	Newton	N	$kg\ m/s^2$
Pressure	Pascal	Pa	N/m^2
Energy	Joule	J	$kg \cdot m^2/s^2$
Electric Charge	Coulomb	C	$A \cdot s$
Electric Potential Difference	Volt	V	J/C

Always use SI units (or their derived counterpart) when in doubt on the AP exam. If you express all given quantities in terms of SI units and present your answer in these units as well, you will never be penalized by the test graders. However, keep in mind that some important constants are not presented in terms of SI units, the most important of which is **pressure**. Units of pressure are usually encountered when performing calculations involving ideal gases that use the ideal gas constant $R = 0.0821$ L·atm/mol·K. Notice that in this case, you must use pressure in terms of atmospheres (atm), not pascals (Pa), in order to perform the appropriate calculations.

In Section II of the exam, many of these constants and units will be provided to you. Nevertheless, you should try to memorize as many as possible to increase your test-taking efficiency.

DIMENSIONAL ANALYSIS

The basic principle of dimensional analysis is that numbers and units must follow the rules of algebra. Therefore, units may be used in calculations to check whether or not you are using the correct formula. For example, let's suppose you wish to calculate the mass of a gallon (3,784 mL) of lead (density = 11.34 g/mL). Density is calculated by the following formula:

$$\text{Density} = \frac{\text{Mass}}{\text{Volume}}$$

Manipulating this equation,

$$\text{Mass} = \text{Volume} \times \text{Density}$$

$$\text{Mass} = 3,784 \text{ mL} \times \left(\frac{11.34 \text{ g}}{1 \text{ mL}} \right) = 42,910 \text{ g}$$

Notice that in this case the "mL" units appeared in the numerator of the expression for volume and in the denominator of the expression for density. The result was that this unit was cancelled, leaving only g, or mass units. This result implies that our original expression for density was correct, as well as our manipulation of this equation. Again, analysis of the units assisted us in arriving at the correct answer. Remember the following general expression, which you will use again and again on the AP exam:

$$\text{Known units} \times \left(\frac{\text{Desired units}}{\text{Known units}} \right) = \text{Desired units}$$

REVIEW QUESTION

Later in this book, you will see that the average kinetic energy of a sample of gas $\left(\text{in } \frac{\text{J}}{\text{mol}} \right)$ can be represented by the following:

$$(KE)_{AVG} = \left(\frac{3}{2} \right) RT$$

In this equation, the value of R is 8.3145. What are the units of this constant?

SOLUTION

Since we know the units of kinetic energy $\left(\frac{J}{mol}\right)$, as well as temperature (K), we can determine the units of R by using algebra. If we look at the previous equation and consider only the units:

$$\left(\frac{J}{mol}\right) = (R) \times (K)$$

$$R = \frac{J}{mol \cdot k}.$$

Therefore, the units of R are $\frac{J}{mol \cdot k}$.

AP EXPERT TIP

Nearly every free response question on the exam uses three significant figures, and credit is given for being within one. So stick with three, and stop worrying about significant figures!

AP EXPERT TIP

Pay attention to multiple-choice questions: significant figure questions do appear there. You can usually identify these because the math is fairly simple, and there is more than one answer that is mathematically correct, but they vary by significant figures.

SIGNIFICANT FIGURES

There are some very simple rules for correct presentation of significant figures. Remember the following rules:

To determine the number of significant figures in a measurement, read the number from left to right. Begin counting digits at the first number that is not zero. For example, 2.0045 has five significant figures while 0.00032 has two significant figures.

When adding or subtracting, the number of significant figures is limited by the number with the fewest decimal places. For example, when adding the three quantities:

```
   12.3
 + 0.045
 + 2.23
  14.575
```

However, because 12.3 only includes decimal places to the tenths, this sum can be reported as 14.6 (rounding up).

When multiplying or dividing, the number of significant figures will be the same as the quantity with the fewest significant figures. Therefore:

$$(1.20) \times (45.876) = 55.0512$$

This number should be reported as 55.1 (three significant figures).

When a number is rounded, the retained digit is increased by one only if the following digit is greater than five. If the following digit is equal to five, use the even-odd rule, which states that odd numbers round up while even numbers don't. Therefore, 12.35 and 12.45 both round to 12.4.

Assess significant figures only at the end of the problem. Carry as many decimal places as possible up to this point, and then look back to decide how many you need to keep (usually just by looking at the quantities given at the beginning of the problem).

EXPONENTS AND SCIENTIFIC NOTATION

In exponential or scientific notation, a number is expressed using the following form:

$$N \times 10^m$$

The first number N must be between 1 and 10, and 10^m represents the exponential term. Remember that m may be a positive or negative number and reflects the number of decimal places you must count either to the left or to the right to place the decimal point. For example, for 1.34×10^5, the 5 in the exponential term implies that you must move the decimal place five places to the right, adding zeros as needed. Therefore:

$$1.34 \times 10^5 = 134,000$$

In contrast, for 4.6×10^{-3}, you must move the decimal place three positions to the left, again adding zeros as needed. Therefore:

$$4.6 \times 10^{-3} = 0.0046$$

There are some basic rules you will need to know for calculations involving exponents:

- For adding and subtracting numbers in scientific notation, remember that each number must be represented in the same powers of 10. Therefore, when adding 5.2×10^{-4} and 1.23×10^{-3}, we must present both in terms of the same powers of 10. We can express 5.2×10^{-4} in terms of 10^{-3} by moving the decimal place one spot to the left (you can prove this to yourself by writing the whole number out). Therefore:

  $$0.52 \times 10^{-3} + 1.23 \times 10^{-3} = 1.75 \times 10^{-3}$$

  In this case, the final answer is in the correct format (with N between 1 and 10), so we are done.

- For multiplying exponents, we need to multiply the N values and also add the m values to determine the new exponential term. In other words:

  $$(N \times 10^m) \times (P \times 10^q) = (N \times P) \times 10^{m+q}$$

For example, let's try multiplying 6.3×10^4 and 3.4×10^5. In this case, we get:

$$(6.3 \times 3.4) \times 10^{4+5} = 21.42 \times 10^9$$

Notice that in this case, we did not get a number with correct scientific notation, so we need to move the decimal place to the left and add one to the exponent. With two significant figures, the correct result is 2.1×10^{10}.

- For dividing exponents, we need to divide the N values and subtract the m values to determine the new exponential term. Therefore:

$$\frac{(N \times 10^m)}{(P \times 10^q)} = \left(\frac{N}{P}\right) \times 10^{m-q}$$

- When raising exponents to a power, we need to raise the N value to that power and multiply the m value by that power. In other words:

$$(N \times 10^m)^q = N^q \times 10^{mq}$$

For example, let's raise 2.0×10^3 to the third power:

$$(2.0 \times 10^3)^3 = (2.0)^3 \times 10^9 = 8.0 \times 10^9$$

LOGARITHMS

On the AP exam, you will have to deal with common logarithms (base 10) and occasionally natural logarithms (base e). You will see common logarithms when you are working with pH values in acid-base chemistry. For starters:

$$\log x = n \qquad x = 10^n$$
$$\ln x = m \qquad x = e^m$$

We will focus on common logs since it's highly unlikely you will be asked to use natural logs. A common logarithm is the power to which you must raise 10 to obtain the number.

You should familiarize yourself with the way your calculator works when using logs. On some calculators, you calculate the log of a number by typing in the number and hitting the "log" button on your calculator; on others, you press "log" first and then enter the number. If you hit "inverse log" (or the analogous button labeled 10^x), you are actually calculating 10^n for the number you put in. For example, look at the expression for pH:

$$pH = -\log [H^+]$$

If we wished to calculate the pH of an acidic solution for which the concentration of H^+ ions was equal to 1.0×10^{-4}, we would just type this value into our calculator and hit "log." The result is -4, so the pH of this solution is 4. Notice that you could have also arrived at this value by remembering the definition of logs. Remember the main consequence of the above logarithmic relationship: A change in pH by a factor of 1 corresponds to a tenfold change in the concentration of H^+.

What if we were given a pH of 10 and asked to calculate the concentration of H^+? In this case, we would type in -10 and hit "inverse log" on the calculator. Remember that performing this operation means taking 10 to the -10th power, so the corresponding concentration of H^+ is 1.0×10^{-10}.

REVIEW QUESTION

The Richter scale for determining the power of earthquakes is a logarithmic scale. How much more powerful is a 7.0 earthquake than a 5.0 earthquake?

SOLUTION

Each unit on the Richter scale corresponds to a tenfold change in the power of the quake. Therefore, the 7.0 earthquake is 100 times as strong as the 5.0 earthquake.

CHAPTER 4: ATOMIC STRUCTURE

IF YOU LEARN ONLY SIX THINGS IN THIS CHAPTER . . .

1. Atoms are composed of neutrons, protons, and electrons. The majority of atomic mass is contained in the nucleus, whereas the electrons are the most important determinants of chemical behavior. The Rutherford gold-foil experiment was essential in determining the structure of the atom.

2. Isotopes represent variations in the number of neutrons in an atom. Unstable isotopes may undergo radioactive decay, including the emission of α and β particles.

3. Remember that radioactive decay obeys first-order kinetics and that you can predict how fast a radioactive sample will decay using half-life.

4. Electrons in an atom occupy atomic orbitals, which are defined by quantum numbers. There are four types of atomic orbital shapes: s, p, d, and f. You must be able to represent the ground-state electron configuration for an atom. Electrons fill orbitals in the exact same pattern every time. The presence of unpaired electrons corresponds to a paramagnetic state, whereas lack of unpaired electrons corresponds to a diamagnetic state.

5. The electromagnetic spectrum defines the various types of electromagnetic energy and the wavelengths associated with it. The relationship between frequency and wavelength is defined by the following equation:

   $$v\lambda = c$$

   where v is frequency in $\sec^{-1}$, λ is wavelength in meters, and c is the speed of light in meters/sec.

6. The Bohr model allows us to predict the energies of electron transitions for hydrogen and hydrogen-like atoms. When energy is absorbed, it is possible for an electron to jump to a higher energy level. In contrast, when an electron falls from a higher energy level to a lower one, energy is released in the form of electromagnetic radiation. The energy of the principal energy levels in the Bohr model are given by:

$$E = -2.178 \times 10^{-18} \text{ J} \left(\frac{Z^2}{n^2} \right)$$

where Z is the nuclear charge and n is an integer corresponding to the principal energy level.

Atoms are the basic building blocks of matter. They are the smallest units that can combine with other elements to form chemical compounds. A thorough understanding of atomic structure may seem a daunting task, but fortunately there is a limited range of questions about this topic that you will encounter on the AP Chemistry exam. We will leave the muons and gluons to the physicists! In this section, we will focus on the most essential aspects of atomic structure, with emphasis on those topics that will be on the exam.

Atomic structure is at the core of chemistry. This chapter will present events at the atomic level in a way that makes sense. We will start by discussing the composition of atoms. We will also discuss the ways in which chemists represent elements in terms of atomic mass, atomic number, and **isotopes**, which are variations of each element; they have the same atomic number but different atomic masses. At this point, we will also talk about some important reactions that are nuclear rather than chemical in nature, such as the various forms of radioactive decay.

We will also address quantum theory in a way that allows us to predict the behavior of atoms. Our main concern here is not structure but energy and the way electrons are organized in order to minimize the energy of a system. We will address atomic orbitals and how they are filled in stable atoms, and we will discuss the Bohr model, a simple model for predicting the energy of electronic transitions. Understanding energy will also allow us to predict physical properties, for example magnetism, color, and the photoelectric effect.

THE BASICS OF ATOMIC STRUCTURE

The basic structure of an atom is shown following. **Protons** have a positive charge, whereas **electrons** have a negative charge that is equal in magnitude. While the electron cloud represents the vast majority of atomic volume, the protons and neutrons in the atom's nucleus account for almost the entire atomic mass. What might be tricky on the AP exam is a question involving the famous Rutherford experiment that showed atomic structure.

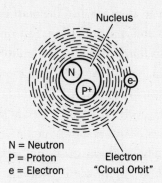

Nucleus

N = Neutron
P = Proton
e = Electron

Electron
"Cloud Orbit"

In the Rutherford experiment, a thin layer of gold foil was bombarded with **alpha particles** (an alpha particle consists of two protons and two neutrons, equivalent to a helium nucleus). It was noted that the vast majority of alpha particles passed through the foil without deflection. However, some particles were deflected after passing close to or colliding with the heavier, positively charged gold nuclei (with a +79 charge). This showed that the atoms contained a very small but massive nucleus. The experiment is described schematically here:

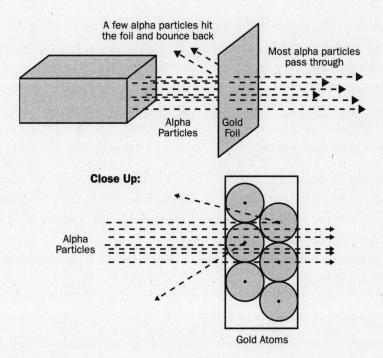

A few alpha particles hit
the foil and bounce back

Most alpha particles
pass through

Alpha
Particles

Gold
Foil

Close Up:

Alpha
Particles

Gold Atoms

ELEMENTS AND ISOTOPES

As we have just learned, the nucleus of an atom is composed of neutrons and protons. The total number of protons and neutrons in an atom is called the **mass number** (A). Since an electron represents only 1/1,837th of a mass unit, the mass of an electron is not included in the mass number. An element is defined by its number of protons, or **atomic number** (Z), which in a neutral atom will be equal to the number of electrons. Each element has one, and only one, atomic number associated with it. Elements may be represented in one of the three ways displayed below. On the Periodic Table, you often see an element represented as its atomic symbol with its mass number above and its atomic number below, as shown on the left. Alternatively, in text, an element may be represented as its atomic symbol with the mass number superscripted and the atomic number subscripted to the left, as shown in the center. Since the atomic number is always the same, the element may also be represented as its atomic symbol with only the mass number superscripted to the left, as shown on the right.

$$\text{Mass Number} \longrightarrow 195$$
$$\text{Element Symbol} \longrightarrow \mathbf{Pt} \quad \leftrightarrow \quad {}^{195}_{78}\mathbf{Pt} \quad \leftrightarrow \quad {}^{195}\mathbf{Pt}$$
$$\text{Atomic Number} \longrightarrow 78$$

In contrast, the mass number may vary among atoms of a given element. **Isotopes** represent atoms of the same element that have a different number of neutrons and therefore have different mass numbers. For example, uranium 238 and uranium 235 are two naturally occurring isotopes of the radioactive element uranium. We know that both atoms, by definition, have the same number of protons; by looking at the periodic table, we see that the atomic number of uranium is 92. Thus, uranium 238 must have ($238 - 92$), or 146 neutrons, while uranium 235 has ($235 - 92$), or 143 neutrons. While isotopes of an element may have different nuclear properties, these isotopes have almost identical chemical behavior.

Isotopes will also vary in abundance in nature. This is why the periodic table has some funny-looking entries for atomic mass. For example, the atomic mass of chlorine, Cl, is listed as 35.45. This is because there are two naturally occurring isotopes of chlorine in nature: ${}^{35}Cl$ and ${}^{37}Cl$. Of all the chlorine that exists on planet Earth, it turns out that 76 percent of it is ${}^{35}Cl$ and 24 percent of it is ${}^{37}Cl$. In order to calculate the atomic mass of chlorine based on this data, we need to perform a "weighted" average:

Atomic mass = (Mass of isotope #1) × (Fractional abundance of isotope #1) +
(Mass of isotope #2) × (Fractional abundance of isotope #2)

Thus for Cl, atomic mass = (35) × (0.76) + (37) × (0.24) = 35.45 amu

NUCLEAR STABILITY AND RADIOACTIVE DECAY

Certain radioactive isotopes, for example plutonium (Pu) and uranium (U), can be used in atomic weapons and nuclear reactors; they can also emit damaging radioactive particles. For the purpose of the AP Chemistry exam, it will be important to 1) know the factors that contribute to nuclear stability, 2) describe the different types of radioactive decay, 3) understand the basics of fusion and fission, and 4) solve problems involving the kinetics of nuclear decay.

Nuclear stability can be described as the probability that a nucleus will undergo **radioactive decay** and in the process form a new nucleus. The AP exam is unlikely to ask you why some nuclei are stable and others aren't, since even chemists can't agree on the answer. However, there are a few things you should remember. First, all atoms with an atomic number of **84 or greater** are unstable (think uranium, with an atomic number of 92). Second, for light nuclei, atoms are stable in which the number of protons equals the number of neutrons (thus, ^{4}He is stable but ^{6}He isn't). Third, there is a **zone of stability** for atoms with higher molecular weights, such that the higher the atomic weight, the more the neutron to proton ratio will deviate from 1:1 in stable atoms. For example, ^{202}Hg (with a neutron:proton ratio of 1.53:1) is stable. This zone of stability is depicted in the figure below.

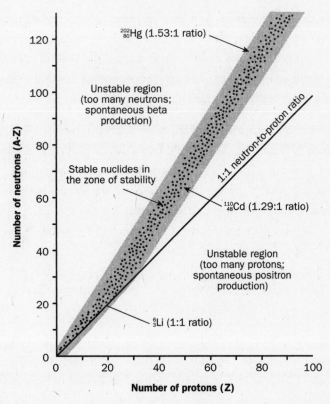

It's far more likely that you will be tested on your ability to represent the different types of radioactive decay using chemical equations. Unstable isotopes emit α, β, and γ radiation. Understand that while α and β radiation represents the release of **radioactive particles**, γ radiation is the emission of electromagnetic energy and thus cannot be represented chemically.

One way a radioactive isotope can decay is by ejecting an α **particle** (represented as a helium atom) from the nucleus. For example, consider the decay of ^{242}Pu (plutonium):

$$^{242}_{94}\text{Pu} \rightarrow\ ^{4}_{2}\text{He} +\ ^{238}_{92}\text{U}$$

Notice that this reaction is "balanced" with respect to neutrons and protons. This will always be the case for a nuclear reaction. For α particle emission, the atomic number will decrease by two units and the mass number will decrease by four mass units. Most elements with an atomic number greater than 83 will decay by ejecting an alpha particle.

A radioactive isotope can also decay by emitting a β **particle**, represented by an electron. The decay of ^{32}P, an important isotope used in the biochemistry laboratory, is represented by the following expression:

$$^{32}_{15}\text{P} \rightarrow\ ^{0}_{-1}\text{e} +\ ^{32}_{16}\text{S}$$

As this reaction proceeds, notice that the number of protons increases while the mass number does not change. This result indicates that a neutron is being converted to a proton as the nuclear reaction occurs. As a result, the atom is moving closer toward the "belt of stability." This makes sense since we started with an isotope that had a disproportionate number of neutrons, and therefore emission of a beta particle generated a more stable species.

In addition to alpha, beta, and gamma radiation, other decay processes are known. These are less likely to appear on your AP exam, but you should know two important types: **positron emission** and **electron capture**. Fluorine 18 (or ^{18}F) is an important isotope used in nuclear medicine. This isotope can be used to image various types of cancer in the human body. Fluorine 18 emits a positron as represented by the following nuclear reaction:

$$^{18}_{9}\text{F} \rightarrow\ ^{0}_{+1}\text{e} +\ ^{18}_{8}\text{O}$$

Electron capture corresponds to the capture of an electron by a nucleus:

$$^{7}_{4}\text{Be} +\ ^{0}_{-1}\text{e} \rightarrow\ ^{7}_{3}\text{Li}$$

Now that we have discussed the types of radioactive decay that you are most likely to see on the AP Chemistry exam, let's turn to some reactions that are extremely important for energy production and warfare. The most important type is **nuclear fission**, which involves the "bombarding" of an unstable nucleus by a slow neutron, resulting in a splitting of that nucleus and the release of a tremendous amount of energy. Most importantly, the fission of an unstable nucleus typically results in the release of multiple neutrons, which can result in a **chain reaction**. During

World War II, scientists in both the United States and Germany were working feverishly to obtain pure uranium 235 for use in the following reaction:

$$^{235}_{92}\text{U} + ^{1}_{0}\text{n} \rightarrow ^{141}_{56}\text{Ba} + ^{92}_{36}\text{Kr} + 3^{1}_{0}\text{n}$$

If this reaction were allowed to take place uncontrolled, the result would be a nuclear weapon. However, the above reaction can take place under controlled conditions in a **nuclear reactor**. In order to keep the reaction under control, the excess neutrons must be absorbed in order to avoid a runaway chain reaction. This may be accomplished using **control rods**, usually cadmium-based. The rate of the fission reaction can be reduced by inserting the rods or increased by withdrawing them.

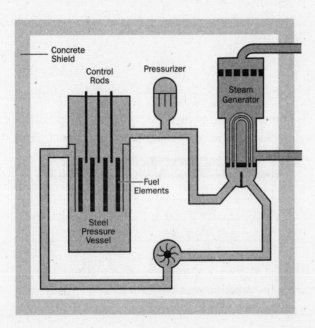

Believe it or not, fission is not the most powerful nuclear reaction available to us. In the sun, hydrogen nuclei combine to form helium, with a large amount of energy released:

$$4^{1}_{1}\text{H} \rightarrow ^{4}_{2}\text{He} + 2^{\,0}_{-1}\text{e}$$

This is called **controlled fusion**. Controlled fusion has been extraordinarily difficult to achieve, because fusion will only take place under extremely high temperatures.

On the AP exam, you will also be expected to understand the kinetics of radioactive decay. Radioactive decay generally obeys **first-order kinetics**, meaning that the rate of decay is proportional to the amount of radioactive material remaining. The most important concept here is the **half-life**, which is the time required for half of the radioactive particles in a given sample to decay. Most questions on the AP exam regarding nuclear chemistry can be solved using the concept of half-life, without needing to use a lot of complicated equations. First-order kinetics will be discussed in further detail in chapter 11.

For now, let's have a look at a typical question:

REVIEW QUESTION

An 8.0 gram sample of ^{32}P, a β-emitter, takes six weeks (42 days) to arrive from a laboratory in Europe. If the half-life of ^{32}P is approximately 14 days, what amount of ^{32}P remains when it arrives?

(A) 8.0 g

(B) 4.0 g

(C) 3.0 g

(D) 2.0 g

(E) 1.0 g

SOLUTION

The correct answer is choice (E). The easiest way to solve this problem is to note that the 42 days it takes the sample to arrive represents three half-lives for this isotope. Thus:

Days	Half-Lives	Sample Remaining
14	1	4 g
28	2	2 g
42	3	1 g

Therefore, there will be 1 g of the ^{32}P sample remaining after 42 days. The rest would have decayed to ^{32}S, via the reaction we saw earlier: $^{32}_{15}P \rightarrow {}^{0}_{-1}e + {}^{32}_{16}S$.

ATOMIC ORBITALS AND ELECTRONIC CONFIGURATION

This topic is tested frequently on the AP exam, so read this section carefully. The key is to remember that orbitals for a given atom are always filled by *exactly the same pattern* and that the periodic table represents a shortcut in answering most of the questions you will be asked.

Fortunately, you won't need to know much about the quantum mechanical model of the atom. Atomic orbitals (*s, p, d, f*) have different shapes that represent the probability that an electron in those orbitals will be found in a given place. These orbitals are characterized by a series of four **quantum numbers**. The **principal quantum number**, represented as *n*, has integral values 1, 2, 3 … etc. and is related to the size and energy of the orbital. Remember, it's all about energy—electrons will fill the lowest energy atomic orbitals available to them in the **ground-state** configuration. In **excited states**, you can trick electrons into jumping to higher-energy, more unstable atomic orbitals.

The second quantum number is the **angular momentum quantum number**, represented as l. This number indicates the shape or type of orbital that is occupied by an electron in a given sublevel. When $l = 0$, the sublevel is called an s orbital. An angular momentum quantum number of $l = 1$ corresponds to a p orbital; $l = 2$ corresponds to a d orbital; and $l = 3$ corresponds to an f orbital.

The **magnetic quantum number**, represented as m_l, indicates the numbers and orientations of the orbitals surrounding the nucleus. There are one s orbital, three p orbitals, five d orbitals, and seven f orbitals.

Finally we have the **spin quantum number**, represented as m_s, which can only assume the value $+\frac{1}{2}$ or $-\frac{1}{2}$, sometimes represented by an up or down arrow respectively. The spin quantum number indicates the orientation of an electron's magnetic field relative to an external magnetic field. Each orbital can only hold two electrons with opposite spins.

You will often be asked to represent the ground-state electronic configuration of an atom and to draw conclusions about unpaired electrons, stability, etc. To accomplish this, use the following strategy:

1. Calculate **the total number of electrons** in the atom you are trying to represent. For neutral atoms, this is simply the atomic number. For example, chlorine (Cl) has 17 electrons. However, for charged species you need to add an electron for every negative charge and subtract an electron for every positive charge. Thus, Cl^- has 18 electrons while Cl^+ has 16 electrons.

2. Start filling atomic orbitals using the following diagram, which summarizes the order of relative energies (and thus which orbitals are filled first):

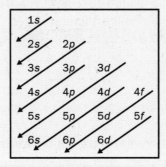

3. Every orbital can hold a maximum of two electrons. An s sublevel has one orbital, and so each s sublevel can hold up to 2 electrons. Each p sublevel has three p orbitals and can hold up to 6 electrons; and each d sublevel has five d orbitals and can hold up to 10 electrons. It's unlikely that you will be asked about f orbitals, which can hold up to 14 electrons.

4. For the electrons in the highest energy level, remember **Hund's rule**, which states that electrons avoid "pairing up" in degenerate orbitals (orbitals that are in the same sublevel, and therefore have the same energy) until each orbital has an electron.

5. The **valence electron configuration** refers to the electrons in the outermost principal energy level of an atom. Remember that the concept of valence relates to the organization of the periodic table—elements in the same vertical column of the periodic table (or group) have the same valence electron configuration.

REVIEW QUESTION

Consider the species O^- and Fe.

a) Write the ground-state electronic configuration for each atom.

b) Determine the number of valence electrons present for each atom.

c) Determine the number of unpaired electrons present for each atom.

SOLUTION

Since oxygen has an atomic number of 8, there should be 8 electrons in the uncharged atom and 9 electrons for O^-. For uncharged Fe, there should be 26 electrons.

a) Filling orbitals according to the diagram, the electronic configuration for O^- is $1s^2 2s^2 2p^5$ while the configuration for Fe is $1s^2 2s^2 2p^6 3s^2 3p^6 4s^2 3d^6$.

b) To find out the valence number of an electron, just look at the outermost principal energy level (second for oxygen and third/fourth for Fe). The valence electron configuration for O^- is $2s^2 2p^5$, so there are seven valence electrons. The valence electron configuration for Fe is $4s^2 3d^6$, so iron has two valence electrons.

c) To determine the number of unpaired electrons in each case, look at the outermost electrons and remember how atomic orbitals are filled.

$$\underset{2p}{\uparrow\downarrow} \quad \underset{2p}{\uparrow\downarrow} \quad \underset{2p}{\uparrow} \qquad\qquad \underset{3d}{\uparrow\downarrow} \quad \underset{3d}{\uparrow} \quad \underset{3d}{\uparrow} \quad \underset{3d}{\uparrow} \quad \underset{3d}{\uparrow}$$

$$\mathbf{O^-} \qquad\qquad\qquad\qquad \mathbf{Fe}$$

For O^-, the highest energy electrons are in a p-orbital, which can accommodate six electrons in three degenerate orbitals. So that leaves one unpaired electron. For Fe, the highest energy electrons are in a d-orbital, which can accommodate ten electrons in five degenerate orbitals. There are four unpaired electrons. Notice that this follows Hund's rule, which states that the most stable electronic configuration is the one with the largest number of unpaired electrons.

DIAMAGNETISM AND PARAMAGNETISM

These sound complicated, but magnetism is a simple molecular property that relates to the electronic configurations we have just reviewed. Basically, magnetism refers to how a material behaves when placed in a magnetic field. **Paramagnetism** is associated with **unpaired electrons** and causes a substance to be attracted to the applied magnetic field. **Diamagnetism** causes a substance to be repelled by a magnetic field. Many transition metals and their compounds are paramagnetic since they tend to contain unpaired electrons.

ELECTROMAGNETIC RADIATION AND THE QUANTUM THEORY

Visible light, X-rays, microwaves, and radio waves are all examples of **electromagnetic radiation**. These forms of radiant energy all have **wavelike characteristics** and travel at the **speed of light** (3.00×10^8 m/s in a vacuum). What distinguishes these forms of energy from each other is their **wavelength**, which is inversely related to **frequency**. The electromagnetic spectrum demonstrates different types of radiation and their corresponding wavelengths. On the AP exam, you must be able to calculate wavelength when given frequency and vice versa. The equation that relates these two is $c = \lambda v$, where v is frequency in seconds^{-1}, λ is wavelength in meters, and c is the speed of light in meters/seconds. This equation is provided on Section II of the AP exam, under "Atomic Structure."

Wavelength and frequency are inversely related, so electromagnetic radiation with higher frequencies (for example, X-rays and gamma-rays) will have a relatively short wavelength.

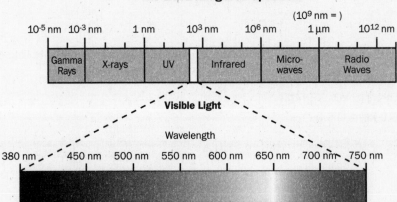

The Electromagnetic Spectrum

AP EXPERT TIP

Diamagnetism is not really covered on the AP exam, but you should know that paramagnetism is important in two areas covered on the test:

1. Magnetic properties of elements
2. Color of transition elements

Another important equation you will need to know relates frequency to energy. Energy cannot be gained or lost by a given atom in just any quantity. Rather, energy is **quantized**—it can be gained or lost only in whole number multiples of the quantity hv, where h is Planck's constant (6.626×10^{-34} J $\cdot$ s). Each discrete unit of energy is called a **quantum of energy**. The equation that relates frequency to energy is $\Delta E = nhv$, where n is an integer (1, 2, 3,...), h is Planck's constant, and v is the frequency of the electromagnetic radiation absorbed or emitted. This equation is provided on Section II of the AP exam, under "Atomic Structure," in the form $\Delta E = hv$. Frequency and energy are directly related—radiation of higher frequency carries more energy. Another way of thinking about a **quantum of energy** is as a tiny energy packet or **photon**.

Although it's unlikely that you will be asked a tough question on the AP exam regarding Einstein's special theory of relativity or wave-particle duality, you should know two basic things: 1) energy and mass are different forms of the same thing, and 2) electromagnetic radiation can be thought of as a particle or as a wave phenomenon. The net result is the following two equations:

$$E = mc^2 \quad \text{and} \quad \lambda = \frac{h}{mv},$$

where m is mass and v is velocity. These equations are also provided in the information that precedes Section II of the AP Chemistry exam. Energy and mass are equivalent and anything can be considered to have a wavelength as seen in the following example.

REVIEW QUESTION

What is the wavelength of a 70.0 kg skier traveling down a mountain at 15.0 m/s?

(A) 2.24×10^{-23} m

(B) 4.31×10^{-26} m

(C) 6.31×10^{-37} m

(D) 8.22×10^{-45} m

(E) 6.55×10^{-52} m

SOLUTION

The correct answer is choice (C). To solve this question we just need to substitute these values into the equation above, also known as de Broglie's equation:

$$\lambda = \frac{h}{mv} = \frac{(6.63 \times 10^{-34} \text{ kg} \cdot \text{m}^2/\text{s})}{(70.0 \text{ kg}) \times (15.0 \text{ m/s})} = 6.31 \times 10^{-37} \text{ m}$$

THE PHOTOELECTRIC EFFECT

The **photoelectric effect** refers to the phenomenon that occurs when electrons are knocked out of a metal by electromagnetic energy. When photons strike a metal, their energy is transferred to the valence electrons of the metal. There are two possible outcomes. If the photon has enough energy, it can allow the metal electrons to escape the attractive forces that hold it within the metal. If the photon lacks the minimum energy to free an electron, the metal electrons cannot escape the surface, but will be excited to higher energy levels. The key thing here is to understand *qualitatively* what's going on:

1. Photons of high frequency radiation have high energies and are therefore able to "knock" electrons out of the metal.

2. For electrons in the metal, energy must be absorbed in order to allow electrons to jump to higher energy levels.

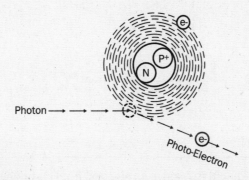

THE BOHR MODEL

On the AP exam, questions you will be asked regarding excited atomic states will relate to the **Bohr model** of the atom, which relates qualitatively to all atoms but quantitatively only to hydrogen or hydrogen-like atoms (i.e., atoms with only one electron). This model was developed based on the fact that photons emitted from hydrogen during electronic transitions had only a few **discrete wavelengths**. Thus, a model was proposed with the following features:

1. Electrons in the hydrogen atom move around the nucleus only in certain allowed circular orbits.

2. The allowed orbits correspond to principal energy levels ($n = 1, 2, 3\ldots$) with $n = 1$ corresponding to the ground state. $n = \infty$ refers to an electron that has escaped these orbits.

3. The energy in a given orbit is given by $E = -2.178 \times 10^{-18}$ J $\left(\dfrac{Z^2}{n^2}\right)$ where Z is the nuclear charge and n is the integer corresponding to the principal energy level.

This sounds easy enough, but you must be able to do multiple calculations on the AP exam. You must also possess a qualitative understanding of what is going on in order to collect points. It is essential to remember the relationship between **changes in energy** and **electron transitions**. Specifically, if an electron moves from a lower energy level to a higher one, for example from *n* = 1 to *n* = 6, energy must be absorbed, and we would expect the sign of the energy change to be ***positive***. If an electron moves from a higher energy level to a lower one, energy is released, and the sign of the energy change should be ***negative***. The **line spectrum** of hydrogen shows the wavelengths of photons emitted when an electron moves from an excited energy level to a lower one.

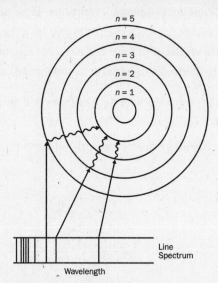

REVIEW QUESTION

What is the energy required to remove an electron from hydrogen in its ground state?

(A) 2.178×10^{-18} J

(B) 3.342×10^{-21} J

(C) 2.875×10^{-25} J

(D) 8.761×10^{-27} J

(E) 5.639×10^{-33} J

SOLUTION

The correct answer is choice (A). $Z = 1$ for hydrogen; in this case, we are moving an electron from $n = 1$ to $n = \infty$ and thus the energy required is:

$$\Delta E = E_{final} - E_{initial} = -2.178 \times 10^{-18} \text{ J} \left(\frac{1^2}{\infty^2} \right) - (-2.178 \times 10^{-18} \text{ J}) \left(\frac{1^2}{1^2} \right)$$

$$= 0 - (-2.178 \times 10^{-18} \text{ J}) \left(\frac{1^2}{1^2} \right)$$

$$= 2.178 \times 10^{-18} \text{ J}$$

Does this answer make sense? Yes, because the sign is positive, indicating that energy must be absorbed to remove the electron. The energy we just calculated is also known as the **ionization energy** of hydrogen.

REVIEW QUESTIONS

1. Which of the following regarding the Rutherford experiment is *NOT* correct?

   (A) Most alpha particles passed through the gold foil without being deflected since the nuclei of the gold atoms represent such a small portion of the total atomic volume.

   (B) Coulomb's law, which states that like charges repel each other, accounts for the deflection of alpha particles passing close to gold atom nuclei.

   (C) Most alpha particles passed through the gold foil without being deflected since the nuclei of the gold atoms represent such a small portion of the total atomic mass.

   (D) Because of the electrons' small masses, they did not deflect the alpha particles.

   (E) The high charge of the gold nuclei helps to account for their ability to deflect smaller helium nuclei passing close by.

2. Hydrogen can exist as ^1H, ^2H, and ^3H. Which of the following is true?

   (A) ^1H and ^2H both have identical numbers of neutrons.

   (B) Water (H_2O) in which all the hydrogen atoms are ^1H is unlikely to have a different boiling point from water in which all of the hydrogen atoms are ^2H.

   (C) ^1H and ^3H are likely to have similar radioactive properties.

   (D) ^3H and ^3He possess an identical number of neutrons.

   (E) ^1H does not possess a neutron.

3. With respect to electronic configuration, which of the following is *NOT* correct?

   (A) Se and O have an identical number of valence electrons.

   (B) B and Cl both have the same number of unpaired electrons in their ground-state electron configuration.

   (C) O^{2-} and F^- have an identical electron configuration.

   (D) Ne and Ar both have a filled p-orbital in the valence shell.

   (E) Se and Br are expected to have similar chemical properties since they are in the same row of the periodic table.

4. Which of the following statements is/are true?

   I. Fe is expected to be paramagnetic.

   II. Zn is expected to be diamagnetic.

   III. Zn^{2+}, which has a ground-state electron configuration of $4s^0 3d^{10}$, would be expected to be paramagnetic.

   (A) I only

   (B) III only

   (C) I and II only

   (D) II and III only

   (E) I, II, and III

5. With respect to the electromagnetic spectrum, which of the following is *NOT* correct?

    (A) Electromagnetic radiation of wavelength 700 nm falls in the red region of the visible spectrum.

    (B) A photon of wavelength 700 nm will have higher energy than a photon of infrared radiation.

    (C) Electromagnetic radiation of wavelength 700 nm has an identical speed to that of radiation of wavelength 400 nm.

    (D) Electromagnetic radiation of wavelength 700 nm has a higher frequency than that of X-rays.

    (E) Electromagnetic radiation of wavelength 700 nm has a lower frequency than that of X-rays.

6. Which of the following is true?

    (A) The Bohr model applies to ^{4}He.

    (B) The spectrum of hydrogen in the visible region of the electromagnetic spectrum is continuous.

    (C) Because the electron has a more negative potential energy for $n = 1$ than it does for $n = $ any integer > 1, the electron is most stable in its smallest allowed orbit.

    (D) The electron in hydrogen may assume either a circular or elliptical orbit.

    (E) To remove an electron from the $n = 2$ level requires the absorbtion of more energy than removal of a ground-state electron.

7. Fill in the missing reactant.

    $$^1_1H + ? \rightarrow 3\ ^4_2He$$

    (A) $^{11}_5Na$

    (B) ^3_1H

    (C) $^{13}_7Al$

    (D) $^{11}_5B$

    (E) ^3_1Li

8. Copper (Cu) can exist as several isotopes, but only two occur naturally, ^{63}Cu and ^{65}Cu. If ^{63}Cu has an abundance of 69.2 percent, and ^{65}Cu has an abundance of 30.8 percent, what is the atomic mass of copper?

    (A) 63.3 amu

    (B) 63.4 amu

    (C) 63.5 amu

    (D) 63.6 amu

    (E) 63.7 amu

9. Barium salts, when burned, emit a greenish light (often utilized in fireworks). One of the most common, $BaCl_2$, produces light with the wavelength of 511 nm. What would the corresponding frequency of this wavelength be?

    (A) 5.82×10^{10} s^{-1}

    (B) 5.87×10^{14} s^{-1}

    (C) 8.48×10^{16} s^{-1}

    (D) 2.32×10^{18} s^{-1}

    (E) 1.73×10^{20} s^{-1}

10. $^{235}_{92}U + ^1_0n \rightarrow X + ^{92}_{38}Sr + 2^1_0n$

    The reaction above represents the fission of a U-235 nucleus by a slow neutron. The identity of product X is

    (A) ^{142}Xe.

    (B) ^{144}Xe.

    (C) ^{142}Ba.

    (D) ^{144}Ba.

    (E) ^{142}Cs.

FREE-RESPONSE QUESTION

1. Consider the Bohr model of the hydrogen atom.

   (a) Calculate the change in energy when an electron in level $n = 6$ returns to ground state.

   (b) Calculate the wavelength of photons released.

   (c) What is the energy required to excite the hydrogen electron from level $n = 1$ to $n = 6$?

   (d) Calculate the energy required to remove an electron from a hydrogen atom in its ground state.

   (e) With respect to the helium atom:

      i. Can the Bohr model be used to calculate the ionization energy of He? Explain.

      ii. Would you expect the ionization energy for He to be higher or lower than the value calculated in (d)?

      iii. Calculate the ionization energy for He^+. Would you expect this value to be higher or lower than the ionization energy of He?

ANSWERS AND EXPLANATIONS

1. C

The statement is incorrect since the nucleus of an atom, consisting of protons and neutrons, represents the *majority* of atomic mass. The other statements regarding the Rutherford experiment are true.

2. E

Since the mass number of ^1H is 1, and by definition H has one proton, that means that since number of neutrons = mass number − atomic number = $1 - 1 = 0$, ^1H does not possess a neutron. Choice (A) is incorrect since ^1H and ^2H have an identical number of protons, but ^2H has one neutron while ^1H does not. Choice (B) is incorrect since the chemical properties of different isotopes are similar, but their physical properties vary. In fact, the boiling point of water composed of ^2H, or deuterated water, is 101.4°C. Choice (C) is incorrect since isotopes often have significantly different radioactive properties. In this case, ^3H emits β radiation, making it a useful tracer in the laboratory, while the ^1H nucleus is stable. Choice (D) is incorrect since ^3H has two neutrons (number of neutrons = mass number − atomic number = $3 - 1 = 2$), while ^3He has only one neutron ($3 - 2 = 1$).

3. E

Choice (E) is false since elements in the same row of the periodic table have a different number of valence electrons and thus are not expected to have similar chemical properties. In contrast, elements in the same column (or group) of the periodic table have an identical number of valence electrons and will behave similarly. Choice (A) is true since Se and O are in the same group of the periodic table. Choice (B) is true since B has a ground-state electron configuration of $1s^2 2s^2 2p^1$ and thus is expected to have one unpaired electron, while

Cl has a ground-state electron configuration of $1s^2 2s^2 2p^6 3s^2 3p^5$, with one unpaired electron in the outermost *p*-orbital. Choice (C) is true since both O^{2-} and F^- have 10 electrons. Choice (D) is true since Ne has a valence electron configuration of $2s^2 2p^6$ while Ar has a valence electron configuration of $3s^2 3p^6$. Both are noble gases and possess special stability due to their completely filled valence shell.

4. C

Since Fe has 26 electrons and is expected to have a valence configuration of $4s^2 3d^6$, it will have four unpaired electrons and be paramagnetic. In contrast, Zn has 30 electrons and a valence configuration of $4s^2 3d^{10}$ and will have no unpaired electrons. We would expect it to be diamagnetic. Zn^{2+} is also diamagnetic since Zn loses its two electrons from the 4*s* orbital to become Zn^{2+}. Since there are no electrons in the 4*s* orbital and ten electrons (a full house) in the 3*d* orbital, Zn^{2+} does not possess any unpaired electrons.

5. D

The electromagnetic spectrum is organized according to increasing wavelength, decreasing frequency, and decreasing energy per photon. Thus, X-rays, which lie to the left of visible light on the electromagnetic spectrum, have a shorter wavelength, higher frequency, and greater energy per photon than visible light. Therefore, choice (D) is incorrect and choice (E) is correct. Choice (A) is correct since red light is in the visible region of the electromagnetic spectrum and has a wavelength in the region of 700 nm. Choice (B) is correct since infrared radiation occurs above (i.e., to the right of) the visible spectrum and therefore has a longer wavelength, a lower frequency, and less energy per photon than visible light. Choice (C) is correct since all wavelengths of electromagnetic radiation travel at the speed of light, or 3.00×10^8 m/s.

6. C

An electron at $n = 1$ has more negative energy than an electron at any other value of n, implying that the electron is more stable and most tightly bound in its smallest orbit. Choice (A) is false since helium has two electrons and the Bohr model therefore cannot be solved exactly. Choice (B) is false since unlike a rainbow, the hydrogen line spectrum contains only a few discrete wavelengths, even for those energy transitions corresponding to the visible region of the electromagnetic spectrum. Choice (D) is incorrect since the Bohr model assumes that all orbits are circular. Choice (E) is incorrect since the electron is most tightly bound to the nucleus in the ground state, and therefore the energy required to remove the electron from the ground state will be the highest.

7. D

The answer to this question can be found by simple arithmetic. The top number to the left of the atomic symbol indicates the total number of protons and neutrons in the isotope, and the bottom represents the number of protons. These have to be equal on both sides of the equation, so the top number must be $(4 \times 3) - 1 = 11$ and the bottom $(2 \times 3) - 1 = 5$. The elemental symbol can be found by the bottom number, 5, which is also the atomic number, and the answer is boron, choice (D).

8. D

We must use the following equation:

Atomic mass = (Mass of isotope #1) × (Fractional abundance of isotope #1) + (Mass of isotope #2) × (Fractional abundance of isotope #2)

Thus, for copper, atomic mass = $(63) \times (0.692) + (65) \times (0.308) = 63.6$ amu.

9. B

The question asks us to relate wavelength and frequency. We should start by rearranging the equation

$$v\lambda = c \quad \text{to} \quad v = \frac{c}{\lambda}$$

We then need to put everything in the correct units:

$$5.11 \times 10^2 \text{ nm} \times \left(\frac{1\text{m}}{10^9\text{nm}}\right) = 5.11 \times 10^{-7} \text{ m}$$

So $v = \dfrac{c}{\lambda} = \dfrac{(3.00 \times 10^8 \text{ m/s})}{(5.11 \times 10^{-7} \text{ m})} = 5.87 \times 10^{14} \text{ s}^{-1}$.

10. A

In this case, we need to make sure that atomic mass and atomic number are "balanced" on both sides of the equation. The mass number in X must be $(235 + 1) - (92 + 2) = 142$. The atomic number of X must be $(92 + 0) - (38 + 0) = 54$. Atomic number 54 corresponds to Xe, so the identity of product X is ^{142}Xe.

FREE-RESPONSE QUESTION

1. **(a)** The change in energy is the final energy (in this case, it is the ground state energy corresponding to $n = 1$) minus the initial energy ($n = 6$). Since we are dealing with the hydrogen nucleus, which has one proton, $Z = 1$:

    $$\Delta E = E_{final} - E_{initial} = -2.178 \times 10^{-18} \, J \left(\frac{1^2}{1^2} \right) - (-2.178 \times 10^{-18} \, J) \left(\frac{1^2}{6^2} \right)$$
    $$= -2.118 \times 10^{-18} \, J$$

    Does our answer make sense? Yes, because when an electron moves from an excited state to a lower energy level, we expect energy to be released and the ΔE to be negative.

    (b) To calculate the wavelength of photon released, we must remember the equations that relate wavelength, frequency, and energy.

    Since $v\lambda = c$ and $\Delta E = hv$, we need to relate wavelength and energy. Rearranging both equations, we get:

    $$\lambda = \frac{c}{v} \text{ and } v = \frac{\Delta E}{h}$$

    By substitution, we get the expression:

    $$\lambda = \frac{hc}{\Delta E} = \frac{(6.626 \times 10^{-34} \, J \cdot s)(3.00 \times 10^8 m/s)}{2.118 \times 10^{-18} \, J} = 9.40 \times 10^{-8} \, m = 94 \, nm$$

    (c) The energy required to excite an electron from $n = 1$ to $n = 6$ should have equal magnitude, but opposite sign, to the energy calculated above for the reverse transition. Thus, the change in energy:

    $$\Delta E = E_{final} - E_{initial} = 2.118 \times 10^{-18} \, J$$

    (d) The energy required to remove an electron from H in its ground state is also known as the ionization energy. In this case, we are moving an electron from $n = 1$ to $n = \infty$. The change in energy is given by:

    $$\Delta E = E_{final} - E_{initial} = -2.178 \times 10^{-18} \, J \left(\frac{1^2}{\infty^2} \right) - (-2.178 \times 10^{-18} \, J) \left(\frac{1^2}{1^2} \right)$$
    $$\Delta E = 2.178 \times 10^{-18} \, J$$

    (e) i. The Bohr model of the hydrogen atom cannot be used to describe the energy levels of multiple-electron systems since it does not take into account electron shielding. Since He has two electrons, the Bohr model cannot be used to calculate the first ionization energy of He (though it could be used to calculate the second ionization energy).

ii.   We would expect the ionization energy of He to be significantly higher than that of H. The valence electrons in He are not completely shielded from the higher (+2) nuclear charge, making them more strongly bound than the electron of H. Therefore, they are more difficult to remove, and the ionization energy is higher.

iii.   Because the first electron has already been removed, leaving just one electron, the Bohr model does apply. However, we must incorporate the value of Z. The energy is given by:

$$\Delta E = E_{final} - E_{initial} = -2.178 \times 10^{-18} \text{ J} \left(\frac{2^2}{\infty^2} \right) - (-2.178 \times 10^{-18} \text{ J}) \left(\frac{2^2}{1^2} \right)$$

$$\Delta E = 8.712 \times 10^{-18} \text{ J}$$

We would expect this value to be higher than the first ionization value for He since there is now no electron-electron repulsion competing with the strong attraction to the nucleus. Subsequent ionization energies are always higher.

CHAPTER 5: THE ELEMENTS AND THE PERIODIC TABLE

IF YOU LEARN ONLY FIVE THINGS IN THIS CHAPTER . . .

1. The position of an atom in the periodic table is determined by its atomic number and valence shell electron configuration. Key properties of an atom, for example its size, ionization energy, electron affinity, and electronegativity, can be predicted by periodic trends.

2. The valence shell electron configuration may be determined by the position of an atom within the s, p, or d sections of the periodic table. For the s and p sections of the periodic table, the principal energy level n may be determined by the row, whereas the number of valence electrons can be determined by counting columns. For the d section of the periodic table, you must subtract one from the row to obtain the principal energy level.

3. Ionization energy and electronegativity increase as you go across a period and decrease as you go down a group. Electron affinity follows this same general trend but is less predictable.

4. Atomic radius decreases as you go across a period and increases as you go down a group.

5. In addition to recognizing the above trends, it's useful to understand how they relate to electronic structure. Atoms that gain or lose electrons wish to be isoelectronic with the noble gases, i.e., have an electron configuration with a completely filled p orbital in the outer shell.

General Chemistry is all about the periodic table. Topics related to periodic trends and the organization of the table come up again and again on AP exams. Use this to your advantage by knowing the periodic table to its full extent. In this chapter, we will discuss the more general principles relating to the periodic table; we'll leave in-depth discussion of group properties to the descriptive chemistry section.

Why does an atom behave the way it does in a chemical reaction? The valence electron configuration of an atom essentially defines its personality. This personality includes the atom's size, its appetite for electrons, and its willingness to associate with other atoms in chemical bonding. The trick to predicting the behavior of an atom is to figure out this personality, a task that's completed most easily by studying the periodic table.

We will start this section by describing how the periodic table is organized and how it can be used as a shortcut to determine the valence shell electron configuration of an atom. We will then discuss periodic trends, including ionization energy, atomic radius, electron affinity, and electronegativity. These quantities will be important in predicting the behavior of an atom in chemical bonding.

As we discussed in the last section, the behavior of atoms is defined by both structure and energy. Both are described by the atom's electron configuration. Atoms that have an identical valence electron configuration are organized according to groups (or columns) in the periodic table and will have similar physical and chemical properties. This section will serve as a foundation for understanding why atoms residing at different spots in the periodic table behave the way they do.

ELECTRON CONFIGURATIONS

As we've already discussed, the ground-state electron configuration essentially defines the chemical reactivity of an atom. When solving problems on the AP exam, remember that the electron configuration of an atom can generally be determined by looking at the periodic table. Most elements (certainly the ones we care about), are lumped into the s, p, and d sections of the periodic table. It's unlikely that you'll be asked anything about the f section of the periodic table. For elements in the s and p sections of the table, look at the **period** or row to determine the principal energy level n. For example, K (potassium) is in the fourth row of the periodic table and S (sulfur) is in the third row, so the highest principal energy level n will be 4 and 3, respectively.

To determine the number of electrons in a valence shell, look at which **group** or column the element is in within the s or p section. For example, K is in the first column of the s section while S is in the fourth column of the p section, so K will have one valence electron in its s orbital while S will have four valence electrons in its p orbital. Putting this all together, the valence electron configuration of K is $4s^1$ while the valence configuration of S is $3s^2 3p^4$. For S, notice that all electrons in the highest principal energy level ($n = 3$) are included in the valence configuration.

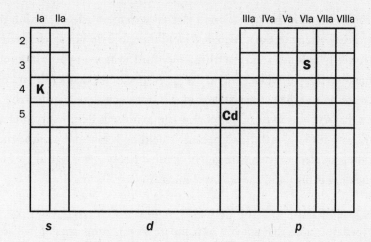

For the elements in the *d* section of the periodic table, or transition metals, the story is nearly the same except you must subtract 1 from the period or row number to determine the highest principal energy level *n*. Thus, although Cd (cadmium) is in the fifth row on the periodic table, most of its valence electrons are actually in the $4d$ orbital. However, since the $5s$ orbital fills before the $4d$ orbital, we include these electrons in the valence configuration. Therefore, the valence electron configuration of Cd is $5s^2 4d^{10}$. Remember that there are some exceptions to these rules for the transition metals. The AP exam is unlikely to ask you about these exceptions to the rules, but two that are worth remembering are Cr (chromium), which has a valence electron configuration of $4s^1 3d^5$, and Cu (copper) which has a valence electron configuration of $4s^1 3d^{10}$. Also, for cations of the transition metals, electrons are generally lost from the valence *s* orbital prior to being lost from the *d* orbital. Therefore, Zn^{2+} has a valence shell configuration of $4s^0 3d^{10}$ and not $4s^2 3d^8$.

The valence electron configuration of an atom or ion relates directly to its stability. More specifically, atoms with a completely filled *p* orbital possess special stability. Looking at the periodic table, we can see that neutral atoms with this configuration can be found in the sixth column of the *p* section of the periodic table (Group VIIIA). These elements, the **noble gases**, have a full set of six *p* electrons in their valence shells and possess special stability. For this reason, these elements are aloof and generally unreactive.

In general, other atoms try their best to be like noble gases. Therefore, they will pick up or lose electrons (usually to form chemical compounds) in order to have an electron configuration identical to that of a noble gas. Atoms or ions are said to be **isoelectronic** with each other when they possess an identical ground state electron configuration. In general, atoms as they exist in chemical compounds seek to be isoelectronic with the noble gases.

PERIODIC TRENDS

The AP Chemistry exam puts a priority on understanding the concept of effective nuclear charge—the net attractive force on an electron in a multi-electron atom. To receive full credit on a free-response question, a student must typically make three things explicit: first, that added

AP EXPERT TIP

There are four important
periodic trends:

1. Atomic radius

2. Electron affinity

3. Electronegativity

4. Ionization energy

All of these properties
generally increase moving
up and to the right on the
periodic table except for
atomic radius.

protons increase the attraction of electrons to the nucleus; second, that electrons are being added to the same n level and therefore do not shield each other as effectively as inner-shell electrons; and third, whatever point links effective nuclear charge to the particular property in question. In particular, the stability of noble gas atoms and their isoelectronic ions is not due to the fact that the p level is filled but that the effective nuclear charge is maximized. Removing an electron from a filled p sublevel is difficult because shielding is low and nuclear charge is high; adding an electron to an atom with a filled p sublevel is less likely because the shielding is high and the effective nuclear charge is low.

Remember that on the AP exam, the trend is your friend—there will probably be several questions that ask you to compare atomic properties of two atoms. You are particularly likely to be asked about ionization energy, atomic or ionic radius, electron affinity, and electronegativity. It's also likely that at some point you'll also be expected to understand the *rationale* behind these trends and how they relate to **electronic configuration**. All of the trends stem from the same phenomenon: as you go across a row, each added proton attracts the valence electrons more strongly, and the added electron doesn't repel strongly enough to overcome the extra attraction. The attractive force always increases more than the repulsive force, until an electron is forced to go into a higher energy level and start the next row.

IONIZATION ENERGY

Ionization energy is the energy required to remove an electron from an atom in its gaseous state. Since removing an electron always requires an input of energy, ionization energy is always a positive number. There are several key things to remember about ionization energy:

- The highest-energy electron (the one with the least negative potential energy) is always removed first.

- As electrons are removed sequentially from an atom, the ionization energy increases. Thus, the **first ionization energy** for Ca will be smaller than the **second ionization energy**, and so on. This is because it will be more difficult to remove a second electron from Ca^+ because the second electron is experiencing less electron-electron repulsion than the first and is therefore held more tightly.

- As you go across a period from *left to right*, the first ionization energy generally *increases*, though there are local deviations from the expected trend (most noticeably between ns^2 and np^1 and between np^3 and np^4). This is because as the number of protons increases, the positive charge that the electron experiences will be greater despite the added electrons. Since there is less repulsion between electrons, there is poor **shielding**. All electrons remain in some shell; as the charge of the nucleus increases, electrons will be held closer to the nucleus.

- As you go *down a group*, the first ionization energy *decreases*. Since the electrons in the valence shell are farther from the nucleus, they are less tightly bound and easier to remove.

These trends in ionization energy can be seen in the following figure:

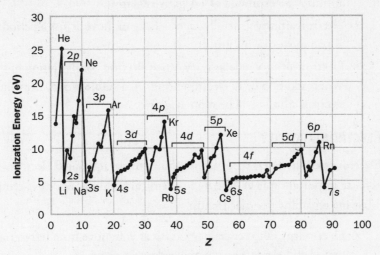

Ionization energies of neutral atoms of the elements

ATOMIC RADIUS

Atomic radius refers to the size of an atom, which is usually based on experimental data. For example, the atomic radius of Cl is 100 picometers since the distance between the nuclei in Cl_2 has been measured to be 200 picometers. You should know the following:

- Atomic radius *decreases* as you move from left to right across a period and *increases* as you move down a group.

- This trend can again be explained by nuclear charge. As you go across a period, the effective nuclear charge is increasing because you are adding more protons. However, the added electrons are in the same shell and do not repel each other very effectively, so the electrons are held more tightly. As you go down a group, the atomic radius increases since you are adding larger orbitals with successive principal energy levels.

Ion size is a key concept that relates to atomic radius. Adding electrons to a neutral atom will result in an anion larger than its parent atom. Removing electrons from a neutral atom will result in a cation smaller than its parent atom. For ions with an equal number of electrons, the higher the nuclear charge, the smaller the ion.

ELECTRON AFFINITY

Electron affinity refers to the energy change associated with the addition of an electron to an atom in the gaseous state. The periodic trends in this case are *less predictable* than those for ionization energy, so it's unlikely that you will be asked very many questions about relative electron affinities of atoms. However, you should know the following:

- Energy is released when an electron is added to the atom, and electron affinities are expressed as a negative energy.

- Electron affinity generally *increases* as you move across a period and *decreases* as you move down a group.

- Electron affinity reflects the valence electron configuration of the atom. When addition of an electron results in a **stable** anion, the value for the electron affinity will be more negative.

ELECTRONEGATIVITY

Electronegativity refers to an atom's ability to attract shared electrons to itself. It plays an important role in describing chemical bonding. For now, remember the following regarding periodic trends:

- Electronegativity generally *increases* as you move from left to right across a period and *decreases* as you move down a group.

- Almost all the elements have been assigned a **Pauling electronegativity value**, which varies between 0.7 (Cs) and 4.0 (F). Fluorine is the element with the highest electronegativity. Forget about the noble gases; they don't like to participate in bonding.

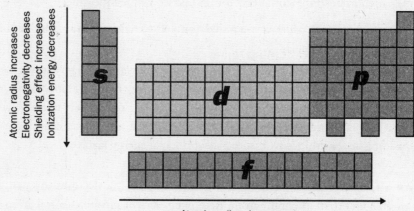

Atomic radius decreases
Electronegativity increases
Shielding effect is constant
Ionization energy increases

AP EXPERT TIP

Electron affinity and electronegativity are nearly the same and follow the same trend. Things to remember:

- Electronegativity only applies to elements that can make bonds.

- Electronegativity values are always positive. They are unitless.

- Electron affinity values can be negative and will be for those elements that tend to form anions. They have units of energy.

REVIEW QUESTIONS

1. Which of the following is/are true regarding Ca and Mg?

   I. Ca has two valence electrons in its outer shell.

   II. The valence electrons of Ca are in a principal energy level $n = 4$.

   III. Ca and Mg both have all electrons paired in the ground state.

   (A) I only

   (B) II only

   (C) I and II only

   (D) I and III only

   (E) I, II, and III

2. Which of the following is NOT true regarding the following atoms?

   (A) Cl^- and Ar are isoelectronic.

   (B) Cs is likely to exist in ionic compounds as Cs^{2+}.

   (C) O is as likely to exist in ionic compounds as O^{2-}.

   (D) He and Ne are unlikely to participate in covalent bonding with other atoms.

   (E) I^- and Br^- are stable ions in solution.

3. Which of the following regarding ionization energy is true?

   (A) The second ionization energy of Mg should be smaller than the first.

   (B) The first ionization energy of Cs is higher than that of Na.

   (C) The atom with the highest first ionization energy is H.

   (D) The noble gases as a group have the highest first ionization energies.

   (E) The atom with the lowest ionization energy is He.

4. Which of the following is expected to have the most negative value for its electron affinity?

   (A) Cl

   (B) S

   (C) O

   (D) Li

   (E) H

5. Which of the following shows Ca^{2+}, Ar, S^{2-}, Cl^-, and K^+ in order of increasing size?

   | Smallest | | | | Largest |

   (A) Ar, S^{2-}, Cl^-, K^+, Ca^{2+}

   (B) S^{2-}, Cl^-, Ar, K^+, Ca^{2+}

   (C) S^{2-}, Cl^-, Ar, Ca^{2+}, K^+

   (D) Ca^{2+}, K^+, Ar, Cl^-, S^{2-}

   (E) Ca^{2+}, K^+, Ar, S^{2-}, Cl^-

6. Which of the following is NOT true regarding Al?

   (A) Al has a higher electronegativity than Mg.

   (B) The second ionization energy for Al will have a larger value than the first ionization energy.

   (C) Al^{3+} is expected to be smaller than Al^{2+}.

   (D) Al is expected to have a more negative electron affinity than sulfur.

   (E) The third ionization of Al can be represented by: $Al^{2+}(g) \rightarrow Al^{3+}(g) + e^-$.

7. Which of the following statements regarding electronic configurations is true?

   (A) An atom with an electron configuration of $1s^2 2s^2 2p^6 3s^1$ has a higher second ionization energy than an atom with a configuration of $1s^2 2s^2 2p^6 3s^2$.

   (B) The neutral atom with the highest electronegativity has two unpaired electrons in its valence shell.

   (C) An atom with valence electron configuration $3s^2 3p^5$ has a small electron affinity.

   (D) Two ions of different elements with identical ground-state electron configurations are expected to be the same size.

   (E) An atom with electron configuration of $1s^2 2s^2 2p^6$ has a lower first ionization energy than an atom with electron configuration of $1s^2 2s^2 2p^5$.

8. An atom has the following ionization energies:

   $I_1 = 589.8$ kJ/mol

   $I_2 = 1,145.4$ kJ/mol

   $I_3 = 4,912.4$ kJ/mol

   $I_4 = 6,491$ kJ/mol

   These values most likely correspond to which of the following:

   (A) Ne

   (B) Li

   (C) I

   (D) Ca

   (E) Al

Questions 9–10 refer to the following atoms:

   (A) S

   (B) Ar

   (C) K

   (D) Mg

   (E) Rb

9. Which is the atom with the largest atomic radius?

10. Which is the atom with the highest first ionization energy?

FREE-RESPONSE QUESTION

1. $3 Br_2 (aq) + 2 Al (s) \rightarrow 2 Al^{3+} (aq) + 6 Br^- (aq)$

   The reaction between aluminum metal and bromine in aqueous solution occurs according to the reaction above.

   (a) Give the ground-state electron configurations of Al and Al^{3+}.

   (b) Give the ground-state electron configurations of Br and Br^-.

   (c) Are Al^{3+} and Br^- isoelectronic? Explain.

   (d) Which set of ionization energies most likely corresponds to Br? Which most likely corresponds to Al? Justify your answer.

Atom X	Atom Y
$I_1 = 12$ eV	$I_1 = 6$ eV
$I_2 = 22$ eV	$I_2 = 19$ eV
$I_3 = 36$ eV	$I_3 = 28$ eV
$I_4 = 47$ eV	$I_4 = 120$ eV

   (e) 2.0 grams of aluminum metal is submerged in aqueous solution, and 3.7 g of Br_2 is added to form 100 mL of solution. Assuming that the volume of the solution remains constant, what is the final concentration of Br^- in M?

ANSWERS AND EXPLANATIONS

1. E

By looking at the periodic table, we can see that Ca is in the fourth row and the second column of the s section of the table. The valence electron configuration is therefore $4s^2$. Since Mg is in the same group of the periodic table as Ca, we expect the two elements to have similar properties and an identical number of valence electrons (two). Since both Ca and Mg have a completely filled s orbital, both have all electrons paired in the ground state.

2. B

Although Cs$^+$ has an identical valence electron configuration to Xe ($5s^25p^6$), Cs^{2+} would have a valence configuration of $5s^25p^5$ and thus not have a filled p orbital. It would therefore possess no special stability. Choice (A) is true since Cl$^-$ and Ar have an identical number of electrons and share the ground-state electron configuration of $3s^23p^6$. Choice (C) is true since O^{2-} is isoelectronic with Ne and therefore would be a relatively stable ion. Choice (D) is true because noble gases generally do not participate in chemical bonding with other elements, due to their high stability. Choice (E) is true since these two anions both have a filled p orbital and are therefore quite stable.

3. D

Since the noble gases have a completely filled p orbital in their valence shell, they have exceptionally high first ionization energies. They have the maximum effective nuclear charge in each row of the periodic table. Choice (A) is incorrect since the second ionization energy for an atom is always greater than the first, even in this case when the second ionization results in a completely filled p orbital in the outer shell. Choice (B) is incorrect since Cs is below Na in the periodic table and

therefore will have a lower first ionization energy. Choice (C) is incorrect since ionization energy increases as you go across a period from left to right, so He is the atom with the highest first ionization energy. Choice (E) is incorrect for the same reason.

4. A

Since Cl has a ground-state electron configuration of $3s^23p^5$, the addition of an electron would complete the p orbital in the valence shell. The result would be the stable Cl$^-$ ion, so we expect the electron affinity of Cl to have a large negative value.

5. D

Notice that all five atoms/ions are isoelectronic; they all have 18 electrons and a valence shell electron configuration of $3s^23p^6$. We know that electrons will be more tightly held with higher (positive) nuclear charge. Therefore, Ca^{2+} will be the smallest ion in the series, and S^{2-} will be the largest.

6. D

Choice (D) is not true since electron affinity generally increases as you move from left to right across a period. Choice (A) is true since electronegativity increases as you move from left to right across a period. Choice (B) is true since the energy required to remove electrons in succession always increases. Choice (C) is true since Al^{3+} has one fewer electron than Al^{2+} and would therefore be expected to be smaller. Choice (E) shows a correct expression for the third ionization of aluminum.

7. A

Choice (A) is correct since the second ionization for the first atom will require removing an electron from a lower n level. Choice (B) is incorrect since fluorine has five p electrons and therefore a single unpaired electron in its valence shell. Choice (C) is incorrect since addition of an electron will result in a filled

p orbital and therefore the electron affinity is large (quite negative). Choice (D) is incorrect since two ions, for example Br^- and Rb^+, may be isoelectronic but have different sizes. Choice (E) is incorrect since an atom with a filled p orbital is expected to have a large first ionization energy.

8. D

Notice that there is a large increase in ionization energy between the second and third ionizations. This implies that the atom has obtained noble gas configuration for the M^{2+} species. In other words, the atom in question must be choice (D), Ca, which lies in group II of the periodic table. The electron configuration of Ca^{2+} is $1s^2 2s^2 2p^6 3s^2 3p^6$. Choice (A) is incorrect since we would expect Ne, a noble gas, to have a significant first ionization energy. Choice (B) is incorrect since Li has a noble gas configuration in the Li^+ state and would therefore be expected to have a large second ionization energy. Choice (C) is incorrect since I is a halogen and therefore will obtain a filled p orbital by adding electrons, not subtracting them. Choice (E) is incorrect since Al lies in group III and therefore has a noble gas configuration as Al^{3+}. We would therefore expect Al to have a significant fourth ionization energy.

9. E

Choice (E) is correct since atomic radius decreases across a period but increases as you move down a group. Therefore, Rb will have the largest atomic radius of the atoms listed.

10. B

Choice (B) is correct. Since Ar is a noble gas with a filled p orbital in its outer shell, it will have special stability and therefore the highest first ionization energy of those atoms listed.

FREE-RESPONSE QUESTION

1. (a) The ground state electron configuration of Al is $1s^2 2s^2 2p^6 3s^2 3p^1$. This can also be written as [Ne] $3s^2 3p^1$. The ground-state electron configuration of Al^{3+} is [Ne] or $1s^2 2s^2 2p^6$.

   (b) The ground-state electron configuration of Br is [Ar] $4s^2 3d^{10} 4p^5$ (also acceptable: [Ar] $3d^{10} 4s^2 4p^5$ or $1s^2 2s^2 2p^6 3s^2 3p^6 3d^{10} 4s^2 4p^5$). The ground-state electron configuration of Br^- is [Kr].

   (c) Al^{3+} and Br^- are not isoelectronic since they do not have the same number of electrons. However, both have a noble gas configuration and a completely filled p orbital in the outer shell.

   (d) $Br = X$ and $Al = Y$

   We expect the first ionization energy of Br to be higher than that of Al since bromine has an electron configuration closer to that of a noble gas. This is consistent with the trends in the ionization energies of X and Y. In the ionization energies of Y, there is a large jump between I_3 and I_4, which we would expect to see for Al since I_4 corresponds to removing an electron from Al^{3+}. This is because Al^{3+} is isoelectronic with the noble gas neon (Ne), and it is especially hard to remove electrons from this type of electron configuration. We would expect the ionization energies of Br to steadily increase, like those of X, since none of the ions involved are isoelectronic with any noble gas.

   (e) We must first determine which reactant is the limiting reagent by converting to moles:

   $$2.0 \text{ g Al} \times \left(\frac{1 \text{ mol Al}}{26.98 \text{ g Al}} \right) = 0.074 \text{ mol Al}$$

   $$3.7 \text{ g Br}_2 \times \left(\frac{1 \text{ mol Br}_2}{159.8 \text{ g Br}_2} \right) = 0.023 \text{ mol Br}_2$$

   Since we would need three moles of Br_2 for every two moles of Al, Br_2 is the limiting reactant in this case. To calculate the number of moles of Br^- produced, we need to use the coefficients in the balanced chemical equation:

   $$0.023 \text{ mol Br}_2 \times \left(\frac{6 \text{ mol Br}^-}{3 \text{ mol Br}_2} \right) = 0.046 \text{ moles Br}^-$$

   The concentration of Br^- in M is given by:

   $$\frac{(0.046 \text{ mol Br}^-)}{(0.100 \text{ L})} = 0.46 \text{ M}$$

CHAPTER 6: CHEMICAL COMPOUNDS

IF YOU LEARN ONLY SIX THINGS IN THIS CHAPTER . . .

1. Ionic bonding occurs between atoms with a large electronegativity difference, typically a metal and nonmetal. Atoms are held together by an electrostatic force defined by Coulomb's law:

$$E = 2.31 \times 10^{-19}\, J \cdot nm \left(\frac{Q_1 Q_2}{r} \right)$$

2. Covalent bonding takes place between atoms with a small electronegativity difference and corresponds to an interaction in which electrons are shared. Lewis dot structures, in which the octet rule is generally obeyed, are used to represent covalent molecules.

3. Polarity refers to covalent molecules in which electrons are shared unequally, resulting in a net dipole moment when the molecule is placed in an electrical field. You must know the three-dimensional geometry of a molecule in order to predict polarity.

4. The VSEPR model is useful in predicting molecular geometry. This model states that the groups surrounding a central atom in a molecule generally want to be positioned as far apart as possible.

5. Using the VSEPR model, it's possible to predict the overall molecular geometry as well as hybridization by looking at how many steric groups surround the central atom of a molecule.

6. The empirical formula of compounds can be determined by looking at the percent composition of their various elements. To determine the molecular formula (a multiple of the empirical formula), we will need to know the molar mass.

Chemists are obsessed with structure, and the writers of the AP Chemistry exam are no exception. In this section, we will describe how atoms combine to form chemical compounds, including ionic and covalent substances. We will review ways of representing chemical structure, as well as simple methods to predict molecular geometry. On test day, you should be able to draw correct Lewis dot structures and conceptualize the three-dimensional structure of molecules as quickly as possible. Practice is key. We will provide you with the tools to get the test questions right.

In the previous section, we talked about the personality of atoms, which can be predicted using the periodic table. We did not, however, discuss the implications for chemical bonding, or how atoms combine to form larger structures. Here we will focus on two important ways atoms combine in nature: **ionic** and **covalent bonding**.

We will begin by discussing ionic bonding, which occurs between atoms with a large difference in electronegativity and results in chemical compounds that are exceptionally stable. In ionic compounds, electrons are bartered between atoms that tend to lose electrons and atoms that tend to take them. The result is an electrostatic force that holds together ions of opposite charges.

The majority of this chapter will be devoted to covalent bonding. Covalent bonding occurs when electrons are shared between atoms with similar electronegativity. We will review Lewis dot structures, which are the most common way to represent covalent molecules. The polarity of molecules, which reflects atomic greed for electrons, will also be discussed. After reading this section, you will be able to predict molecular geometry by looking at chemical formulas and Lewis structures.

Once we have discussed how simple chemical compounds are put together, we will address how chemical formulas are determined. If the molar mass of a compound is known, we can predict its molecular formula using its elemental composition.

A **chemical bond** can be defined as the forces that hold atoms together. The **bond energy** refers to the energy required to break the bond and gives us information on the strength of the bonding interaction. The **bond length** is the distance between atoms in a chemical bond and is the point at which the sum of the **attractive** and **repulsive**

AP EXPERT TIP

Never answer a question about bond strength by citing the size or mass of an atom. Answer in terms of the number of electrons on the atom, which relates to its *polarizability*, or ability to form charge separations.

energy is at a minimum. The following figure shows the potential energy versus internuclear distance as two atoms of H come together to form the molecular compound H_2.

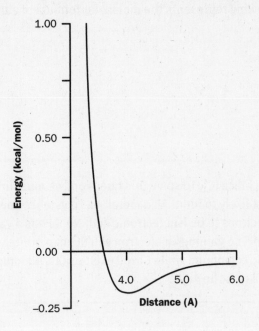

IONIC BONDING

An ionic bond is a very strong bond that forms when an atom with a tendency to lose electrons (usually from the left side of the periodic table) reacts with an atom with a tendency to gain electrons (usually from the right side of the periodic table). Thus, ionic bonding usually occurs between a **metal** and a **nonmetal**. The result is an **ionic compound** in which the attractive force between atoms is **electrostatic** and obeys **Coulomb's law**:

$$E = 2.31 \times 10^{-19} \, J \cdot nm \left(\frac{Q_1 Q_2}{r} \right)$$

where E is units of joules, r is the distance between the atoms in nanometers, and Q_1 and Q_2 are the ion charges. In ionic compounds, there is a large **electronegativity difference** between the atoms.

It's relatively easy to figure out what the ionic formula of an ionic compound will be. Just remember what we have already learned—all atoms want to be like the noble gases and have a completely filled p orbital in their valence shell. Have a look at the following example.

REVIEW QUESTION

Which of the following represents the molecular formula of aluminum chloride?

(A) AlCl

(B) $AlCl_2$

(C) $AlCl_3$

(D) Al_2Cl

(E) Al_3Cl_2

SOLUTION

The correct answer is choice (C). To solve this question, we must first determine what ions Al and Cl are most likely to form. Al, a metal, has a valence configuration of $3s^2 3p^1$ and can lose three electrons to be isoelectronic with Ne. Cl has a valence electron configuration of $3s^2 3p^5$, so gaining an electron will fill its p orbital. Therefore, Al will exist as Al^{3+} in an ionic compound while Cl will exist as Cl^-. An ionic compound will be **neutral**, so the formula must be $AlCl_3$.

MOLECULAR BONDING

Molecular bonds form between two atoms with no electronegativity difference (in **nonpolar** covalent bonding) or a relatively small electronegativity difference (in **polar** covalent bonding). A molecule such as HCl is polar since there is an electronegativity difference between the two atoms. HCl therefore has an associated **dipole moment**, which we can represent as an arrow, or vector, traveling from the less electronegative element, H, toward the more electronegative element, Cl. A more complicated molecule (such as $CHCl_3$) will have a dipole moment if the dipole moments throughout the structure do not cancel each other out. We will discuss this concept further when we look at the three-dimensional structure of molecules.

Molecular structures are represented by **Lewis dot structures**. For practice with writing Lewis dot structures and use of the **octet rule**, consult your general chemistry textbook. Here are three rules to get your structures right:

1. Elements that obey the octet rule have a total of four bonds/electron lone pairs/unpaired electrons surrounding them, for a total of eight electrons. So if your structure has an atom with one lone pair and two bonds (for a total of three), it is incorrect.

2. The number of bonds an atom usually forms can be determined by counting the number of columns the atom is away from the noble gases. In this case, the atom will have no formal charge. Therefore, N would be expected to form three bonds and have one lone pair.

3. If an atom that follows the octet rule forms more bonds than expected, it will have a positive formal charge. If an atom forms fewer bonds than expected, it will have a negative formal charge. The sum of the formal charges in a molecule must equal the net charge for that molecule.

Let's practice drawing a few Lewis structures:

REVIEW QUESTION

Draw the Lewis dot structure representations for NO_3^- and CO.

SOLUTION

The two structures are shown below. For NO_3^-, N will have a +1 formal charge (it forms four bonds instead of the usual three), and two of the oxygens will have a −1 formal charge. For CO, oxygen will have a +1 formal charge (it forms three bonds instead of two), and carbon will have a −1 formal charge (it forms three bonds instead of four). Notice in both cases that all atoms are surrounded by a total of *four* bonds and lone-pairs, or eight electrons.

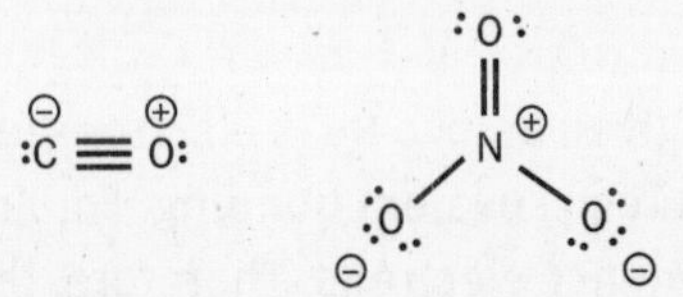

For NO_3^-, notice that we could have drawn the structure in three different ways, as shown below:

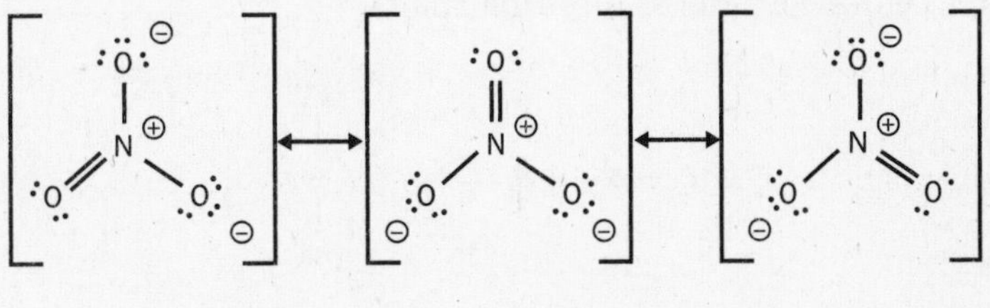

The three structures shown are known as **resonance structures**. Remember that the molecule does not alternate between the different structures but instead exists as a **resonance hybrid** of the contributing resonance structures. So for the bonds in NO_3^-, we can consider nitrogen to be forming three $1\frac{1}{3}$ bonds instead of two single bonds and a double bond.

There are some important exceptions to the octet rule. The first set of atoms that can violate the octet rule are electron-deficient, for example B and Be. B is usually represented as forming only three bonds, and Be only forms two. The second set of atoms that can violate the octet rule are electron-rich and can participate in more than the expected number of bonds by promoting electrons to unfilled *d* orbitals. To write Lewis structures for compounds in which the octet rule is exceeded, do the following:

1. Draw single bonds between bound atoms. The central atom will be the one with the expanded octet.

2. Add up the valence electrons for the molecule.

3. Make sure that the octet rule is obeyed for all atoms excluding the central atom.

4. Add the extra valence electrons to the central atom as lone pairs.

REVIEW QUESTION

Draw the Lewis dot electron configurations for $BeBr_2$ and $XeCl_2$.

SOLUTION

The Lewis dot structures are given below. Be is electron-deficient and has only two valence electrons to participate in covalent bonding. For $XeCl_2$, Xe has eight valence electrons and Cl has seven valence electrons. Therefore, there must be a total of $(7 \times 2) + 8 = 22$ valence electrons. Once the octet rule is satisfied for the two chlorine atoms and there are two bonds drawn, we have accounted for 16 of the valence electrons. Therefore, there must be $22 - 16 = 6$ remaining electrons, so these electrons are added to the central Xe atom as three lone pairs.

A third situation in which the octet rule might be violated is a molecule with an odd total number of electrons, such as NO_2. In that case, draw the Lewis structure as if the single electron were a lone pair, and make sure that the single electron is on the less electronegative atom.

MOLECULAR STRUCTURE AND THE VSEPR MODEL

This is an extremely important topic on the AP exam, and multiple questions will ask you to draw conclusions about a molecule based on its three-dimensional structure. To pick up all these points on test day, you need to remember a few essential things.

The **VSEPR model** may seem complicated, but it simply states the following two things:

1. The bonding and nonbonding pairs around an atom (also known as **steric groups**) will be positioned as far apart as possible to minimize electron-electron repulsion. This will predict the geometry of the molecule: planar, trigonal planar, tetrahedral, etc.

2. There may be subtle differences (a few degrees) in structure due to the following: repulsion between unshared pairs > repulsion between bonding/unshared pairs > repulsion between bonding pairs.

That's about all you need to know with regard to the VSEPR model. But what does this mean for the most likely structures you'll encounter on the exam? When trying to determine molecular geometry, the first thing to do is to figure out how many groups (atoms and lone pairs) surround the atom in question (these groups are called **steric groups**). For NO_3^-, we figured out that this number is three. For the electron-deficient Be, this number is two. For CO, if we are considering the carbon atom, this number is also two since carbon is surrounded by oxygen and one lone pair. For CH_4 (methane) or H_2O, this number is four. For $XeCl_2$, the number of steric groups is five (two chlorine atoms and three lone pairs). Another example of a compound with an expanded octet, SF_6, has six groups surrounding the central sulfur atom.

Next, you can immediately predict the geometry (and determine **hybridizations**) by looking at the following table:

# steric groups	fundamental geometry	hybridization	fundamental structure	possible structures
2	Linear	sp	A — X — B	$C \equiv O$ Linear
3	Trigonal Planar	sp^2	A — X (B, C)	H — B (H, H) Trigonal Planar O — N (O) Bent
4	Tetrahedral	sp^3	A — X (B, C, D)	H — C (H, H, H) Tetrahedral H — N (H, H) Pyramidal H — O (H) Bent
5	Trigonal Bipyramidal	sp^3d	A — X (D, B, C, E)	Many
6	Octahedral	sp^3d^2		Many

Often on the AP exam, you will be expected to relate the geometry of a structure to its **polarity**. Remember that if the structure is highly **symmetric**, the dipole moments associated with individual bonds will cancel out, and there will be a net dipole moment of zero. However, in **bent** structures the dipole moments will not cancel, and the result will be a molecule with a **net dipole moment**.

To see what this looks like, let's compare CO_2 and H_2O. Oxygen has a higher electronegativity than carbon, so each oxygen atom in a CO_2 molecule draws the electrons more strongly than the carbon, resulting in a partial negative charge on the oxygen atoms and a partial positive charge on the carbon. Since CO_2 is a linear compound, and the two dipoles extend outward in opposite directions toward the oxygen atoms, they cancel each other out.

Turning now to H_2O, oxygen also has a higher electronegativity than hydrogen, so, again, the oxygen atom pulls the electron pairs in the bonds more strongly than the hydrogen atoms. As a result, the two dipole moments both point toward the central oxygen atom. Because H_2O is a bent rather than linear molecule, the two dipoles do not cancel each other out as in CO_2. Instead, they add together, creating a net dipole moment in the compound. Therefore water is polar, while carbon dioxide is nonpolar.

PERCENT COMPOSITION AND EMPIRICAL FORMULAE

Now that we have figured out what a molecular formula actually means as far as structure goes, let's turn to some calculations you'll see on the AP exam.

The first concept you'll need to understand is that of the **mole**. A mole is 6.022×10^{23} units of anything. You need to be able to convert number of moles to grams using **molecular weight**, and

REVIEW QUESTION

Determine the percent composition of Al in aluminum oxide.

SOLUTION

The first thing to do here is to figure out the chemical formula of aluminum oxide. Remember that Al exists in ionic compounds as Al^{3+}, giving it a full valence shell. Oxygen, with a ground-state electron configuration of $2s^2 2p^4$, typically picks up two electrons to exist as O^{2-}. Therefore, aluminum oxide must have a chemical formula of Al_2O_3.

Let's assume that we have a mole of Al_2O_3. As the molecular formula implies, for every mole of Al_2O_3, we will have two moles of Al and three moles of O. Therefore:

$$1 \text{ mol } Al_2O_3 \times \left(\frac{101.96 \text{ g } Al_2O_3}{1 \text{ mol } Al_2O_3} \right) = 101.96 \text{ g}$$

$$2 \text{ mol } Al \times \left(\frac{26.98 \text{ g } Al}{1 \text{ mol } Al} \right) = 53.96 \text{ g}$$

Therefore, the percent composition of Al is given by:

$$\left(\frac{\text{Mass of Al in sample}}{\text{Total mass of sample}} \right) \times 100 = \left(\frac{53.96 \text{ g}}{101.96 \text{ g}} \right) \times 100 = 52.92\%$$

vice versa. Most importantly here, the chemical formula states the relationship between molar quantities of the bonding atoms. For more on this, see chapter 11. For now, check out some sample questions to see if you get the hang of it:

REVIEW QUESTION

A sample of gas containing only hydrogen and carbon is determined to be 79.9% carbon by mass.

a)   Determine the empirical formula of the gas.

b)   If the molecular weight of the gas is 30.068, what is the molecular formula of the gas?

SOLUTION

a) This time it pays to assume a 100 g sample of the gas. Since the gas is 79.9% carbon by mass, we know that there will be 79.9 g of carbon in the sample and 20.1 g of hydrogen. We need to convert these quantities to moles to determine the molar relationship of the two elements:

$$79.9 \text{ g carbon} \times \left(\frac{1 \text{ mol carbon}}{12.01 \text{ g carbon}} \right) = 6.65 \text{ mol C}$$

$$20.1 \text{ g hydrogen} \times \left(\frac{1 \text{ mol hydrogen}}{1.008 \text{ g hydrogen}} \right) = 19.9 \text{ mol H}$$

The ratio between carbon and hydrogen is clearly 1:3, so the empirical formula is $C_{1x}H_{3x}$, where x is an integer.

b) If the molecular formula were CH_3, the molecular weight would be $12.01 + 3(1.008)$, or 15.034. However, since the observed molecular weight is 30.068, we know that the molecular formula is C_2H_6. Does the answer make sense? It certainly does—you can draw a Lewis structure for C_2H_6 (ethane) but not for CH_3.

REVIEW QUESTIONS

1. For the following bond angles:

    I. HCH angle in CH_4

    II. HOH angle in H_2O

    III. HNH angle in NH_3

    Which of the following represents the bond angles I, II, III in increasing order?

	smallest bond angle		largest bond angle
(A)	I,	II,	III
(B)	II,	III,	I
(C)	I,	III,	II
(D)	II,	I,	III
(E)	III,	II,	I

2. Which of the following concerning PCl_5 and PCl_3 is/are true?

    I. PCl_3 has a net dipole moment, whereas PCl_5 does not.

    II. The geometry of PCl_3 is trigonal planar.

    III. Both PCl_3 and PCl_5 use higher energy d orbitals for bonding.

    (A) I only

    (B) II and III only

    (C) I and III only

    (D) I and II only

    (E) I, II, and III

3. The geometry of ICl_4^-, as predicted by VSEPR, is best described by the following:

    (A) Pyramidal

    (B) Tetrahedral

    (C) Square planar

    (D) Cuboidal

    (E) Trigonal

Questions 4–6 refer to the following molecules:

    (A) C_2H_2

    (B) CH_2Cl_2

    (C) BF_3

    (D) CH_3CH_2OH

    (E) HF

4. The molecule with the largest dipole moment

5. The molecule that contains a triple bond

6. The molecule that contains an sp^2-hybridized atom

7. Which of the following oxygen-containing compounds is most ionic?

    (A) SiO_2

    (B) NO_2

    (C) Al_2O_3

    (D) CaO

    (E) Cl_2O

8. The carbon atoms of acetic acid (CH_3COOH) exhibit what type(s) of hybridization?

    I. sp

    II. sp^2

    III. sp^3

    (A) I only

    (B) II only

    (C) I and II only

    (D) II and III only

    (E) I, II, and III

9. Consider C_2H_4 and C_2H_6. Which of the following is/are true?

   I. The carbon-to-carbon bond energy in C_2H_4 is greater than it is in C_2H_6.

   II. The carbon atoms in C_2H_4 are sp^2-hybridized.

   III. Both molecules have a net zero dipole moment.

(A) I only

(B) II only

(C) I and II only

(D) II and III only

(E) I, II, and III

10. The oxygen-oxygen bond length in O_3 is greater than the oxygen-oxygen bond length in O_2. Which of the following accounts for this phenomenon?

(A) Electron-electron repulsion is greater in O_3 than in O_2.

(B) The oxygen atoms in O_3 carry a formal charge.

(C) The oxygen-oxygen bonds in O_3 are single bonds.

(D) O_3 has a net dipole moment whereas O_2 does not.

(E) The bond order in O_2 is 2 while the bond order in O_3 is 1.5.

11. Order the following, from highest to lowest, with respect to their boiling points.

   I. CF_4

   II. $CaCl_2$

   III. ICl

	highest		lowest
(A)	I,	II,	III
(B)	I,	III,	II
(C)	II,	III,	I
(D)	III,	II,	I
(E)	III,	I,	II

FREE-RESPONSE QUESTION

1. $C_2H_2Cl_2$ (1, 2 dichloroethylene) is a toxic, volatile liquid used in organic synthesis. Two isomers are commercially available, neither of which has the two Cl atoms attached to the same carbon atom.

   (a) Draw a Lewis dot structure for each geometric isomer.

   (b) With respect to these two isomers, identify:

      i. The isomer with the dipole moment

      ii. The geometry of all carbon atoms

      iii. The isomer with the higher boiling point

      Justify your answers.

   (c) Would you expect it to be possible to convert one isomer to the other under mild reaction conditions? Explain.

   (d) The combustion of ethylene chloride in air produces carbon dioxide and HCl gas.

      i. Write the balanced chemical equation for this process.

      ii. Predict the sign of $\Delta S°$ for this process. Explain.

      iii. Do you expect $\Delta H°$ for this process to be different for the two isomers? Explain.

ANSWERS AND EXPLANATIONS

1. B

The question asks you to predict bond angles based on VSEPR theory. Notice that all three molecules have tetrahedral geometry (all have four groups surrounding the central atom), and thus all bond angles will be approximately 109.5°. Remember that repulsive forces are highest between electron lone pairs and lowest between bonding pairs. In H_2O, there are two lone pairs that will be positioned as far apart as possible in order to minimize repulsion, thus pushing the other groups together. So the HOH bond angle in H_2O will be smallest (about 104.5°). By using the same logic, the one lone pair in NH_3 will push the bonding pairs closer together, so the HNH angle will be about 107°. For CH_4, there are no lone pairs and thus all bond angles will be equal (109.5°).

2. A

PCl_3 has a net dipole moment, whereas PCl_5 does not.

Net dipole   No net dipole

II is incorrect; the geometry of PCl_3 is tetrahedral. III is incorrect because only PCl_3 uses p orbitals for bonding.

3. C

The geometry of ICl_4^-, as predicted by VSEPR, is square planar.

4. E

Although CH_2Cl_2, CH_3CH_2OH, and HF all have a net dipole moment, the dipole moment of HF is largest due to the large electronegativity difference between H and F.

5. A

The only molecule containing a triple bond is C_2H_2, or ethyne.

6. C

Remember that boron often does not obey the octet rule. In this case, the boron atom is surrounded by three groups, making the geometry of BF_3 trigonal planar and the boron atom sp^2-hybridized.

7. D

The greater the electronegativity difference between atoms, the more ionic the compound. In this case, CaO is the compound with the highest electronegativity difference listed. The only other ionic compound is Al_2O_3, for which the electronegativity difference is not as great. The other compounds listed are molecular compounds.

8. D

The Lewis structure for acetic acid is given by the following:

Acetic Acid

All atoms that are surrounded by four steric groups are sp^3-hybridized, including one of the oxygen atoms and one of the carbon atoms. Those atoms surrounded by three groups (for example, the central carbon that is surrounded by two oxygen atoms and one carbon atom) are sp^2-hybridized.

9. E

All three statements are correct. The carbon-to-carbon bond energy in C_2H_4 is greater than it is in C_2H_6 since it is a double bond. The hybridization in C_2H_4 is sp^2 and in C_2H_6 is sp^3. Both molecules have a net zero dipole moment and are highly nonpolar.

10. E

O_3 (ozone) has two resonance structures, as described by the following:

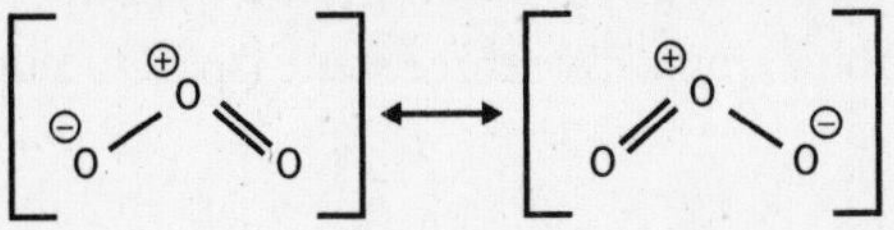

In essence, the actual structure of ozone is a "resonance hybrid" in which there are one and a half bonds between each oxygen atom. The bonding is therefore not as strong as that of O_2, which contains a double bond between the oxygen atoms. The stronger interaction in a double bond brings the O atoms closer together and decreases the bond length. Choices (B) and (D) are true, but are not responsible for determining bond length.

11. C

CF_4 is formed with covalent bonds and has no net dipole, so it should have the lowest boiling point. $CaCl_2$ is ionic and therefore has the highest boiling point. So ICl is in the middle; it has polar covalent bonds and a net dipole.

FREE-RESPONSE QUESTION

1. (a) The Lewis dot structures for both isomers are shown below. In one structure, the two chlorine atoms are opposite each other in a "trans" configuration. In the other structure, the two atoms are side by side in a "cis" configuration.

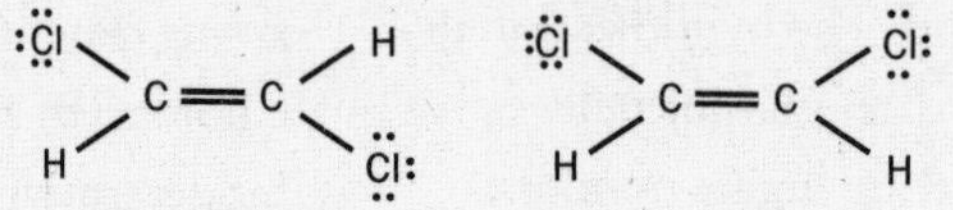

Trans   Cis

(b) i. The isomer with the dipole moment is the "cis" form since the individual dipole moments do not cancel out for the entire molecule.

ii. All carbon atoms are sp^2-hybridized and have trigonal planar geometry. The geometry is trigonal planar since all carbon atoms have three steric (space-occupying) groups surrounding them.

iii. The "cis" isomer is polar and therefore intermolecular forces will be greater. Therefore, we expect this isomer to have the higher boiling point.

(c) We would not expect to be able to convert one isomer to the other easily. Remember that there is no free rotation about a double bond (p orbitals in the bond would fail to overlap), and therefore it is not possible to convert one isomer into the other under kinetic conditions.

(d) i. The balanced chemical equation is as follows:

$$C_2H_2Cl_2 \ (l) + 2 \ O_2 \ (g) \rightarrow 2 \ CO_2 \ (g) + 2 \ HCl \ (g)$$

ii. We would expect the $\Delta S°$ to be positive, since there are more moles of gas formed on the right side of the balanced chemical equation than on the left. The degree of disorder is increasing.

iii. We would not expect $\Delta H°$ to be identical for the two isomers. The isomers have different thermodynamic stability due to the difference in structure. Since the "cis" form has a higher degree of crowding (with the Cl atoms both on the same side of the molecule), it is less stable and has a higher (more positive) energy than the "trans" form. Since the products of combustion are identical for both isomers, we expect $\Delta H°$ be more negative for the "cis" isomer.

CHAPTER 7: GASES

IF YOU LEARN ONLY NINE THINGS IN THIS CHAPTER . . .

1. The behavior of gases is described by the kinetic molecular theory, which describes how the molecules in a gas interact with each other and with the walls of their container.

2. The overall ideal gas law is represented by the following equation:

   $PV = nRT$

3. Boyle's law states that pressure and volume are inversely related, so for a sample of gas undergoing changes in pressure and volume:

   $P_1V_1 = P_2V_2$

4. Charles's law states that the temperature and volume of a gas are directly related, so for a sample of gas undergoing changes in temperature and volume:

   $$\frac{V_1}{T_1} = \frac{V_2}{T_2}$$

5. Avogadro's law states that the volume of a gas is directly related to the number of moles:

   $$\frac{V_1}{n_1} = \frac{V_2}{n_2}$$

6. Other relationships (for example the relationship between pressure and number of moles) may also be derived from the ideal gas equation. Make sure that your derived relationship makes sense with respect to the kinetic molecular theory of gases.

7. According to Dalton's law of partial pressures, the total pressure of a sample of gas is the sum of the pressures each gas exerts on its own. The mole fraction is the ratio between the number of moles of an individual gas and the total number of moles in the mixture.

$$P_{total} = P_1 + P_2 + P_3 + \cdots$$

8. Diffusion describes the mixing of gases, whereas effusion refers to the passage of gas particles through a pinhole. The relative rates of effusion for gases is given by:

$$\frac{\text{Rate of effusion of gas 1}}{\text{Rate of effusion of gas 2}} = \frac{\sqrt{M_2}}{\sqrt{M_1}}$$

9. For a real gas, discrete molecular volume as well as attractive forces must be taken into account. The van der Waals equation for a non-ideal gas takes these factors into account:

$$\left(P + \frac{n^2 a}{V^2}\right)(V - nb) = nRT$$

Gases are a hot topic on the AP exam. On test day, you need to be familiar with the calculations that you are most likely to be asked to perform, which usually involve the ideal gas laws. However, you will also need to recognize the chemical concepts at work in the calculations you make. Most important is the kinetic molecular theory, which refers to a simple model that predicts the way a sample of gas behaves. You will need to rationalize why pressure, volume, and temperature are related the way they are in both ideal and real gases.

In the previous section, we looked at the ways in which atoms come together to form chemical compounds. But how do these compounds actually exist?

A chemical compound may exist in different states in nature, such as solid, liquid, or gas. The state of a compound is determined principally by the molecular structure of the compound in question but may also be affected by environmental conditions such as temperature or pressure. For example, H_2O exists mostly as a liquid at room temperature, largely due to the hydrogen bonding between molecules. However, under other conditions of temperature and pressure, water can also exist as a solid or gas.

In general, a chemical compound exists as a gas when interactions between individual molecules are relatively weak. We will start by predicting how a gas will behave based on a simple model for molecular interactions—the kinetic molecular theory. This theory allows us to explain the ideal gas laws, which relate pressure, volume, temperature, and the number of moles of a gas. We will then turn to Dalton's law of partial pressures to consider how to treat mixtures of gases.

The kinetic molecular theory provides us with a way of describing changes in temperature, pressure, and volume. It also provides us with some important equations relating kinetic energy, temperature, and velocity. We'll discuss these equations and we'll describe effusion and diffusion, both of which describe the movement of gases.

We need to consider how a real gas behaves. The van der Waals equation takes into account nonzero molecular volume as well as the attractive forces that exist in any gas under actual conditions.

IDEAL GAS LAWS

On the AP exam, you will definitely be expected to perform calculations using the ideal gas laws of Boyle, Charles, and Avogadro. But don't worry about remembering these names. The important part is to remember the relationship between pressure, temperature, and volume in the **ideal gas equation**, which is included on the AP Chemistry exam equation sheet under "Gases, Liquids, and Solutions." The ideal gas equation states that:

$$PV = nRT$$

where P is the pressure in atm, V is the volume in L, n is the number of moles, R is the universal gas constant ($0.08206 \text{ L} \cdot \text{atm/K} \cdot \text{mol}$), and T is the absolute (Kelvin) temperature. This equation certainly comes in handy on the AP exam, but you'll also be expected to relate individual variables to each other.

- **Pressure vs. volume**—It's generally a bad idea for a submarine to travel too deep. (This concept is often reinforced in the movies—think of *The Hunt for Red October, The Abyss,* and *Das Boot.*) This is because the pressure on a submarine increases with depth. What would happen to the volume of the submarine if the walls were not extremely strong? The volume would decrease (as the submarine imploded). This leads us to the first gas law, **Boyle's law**, which states that the pressure and volume of a gas are *inversely* related. If you look at the ideal gas equation above, you can see that if the temperature is constant and number of moles of gas is constant,

$$P_1 V_1 = P_2 V_2$$

The kinetic molecular theory sounds complicated, but it's just a simple model that explains how molecules in a sample of gas behave. The model consists of some balls moving around in a container. The balls represent the molecules in a gas sample. As the volume of the container shrinks, the balls are going to collide more frequently with the sides of the container (and with each other). This results in increased force exerted on the sides of the container, which represents the pressure of the sample increasing.

- **Temperature vs. volume**—There are many ways to inflate a balloon. For example, one might add more molecules of gas into the balloon by blowing into it. This could take a while if you wanted to inflate a large balloon, such as a hot air balloon. Another way you can expand a sample of gas is to increase the temperature. This is a consequence of **Charles's law**, which states that the volume and temperature of a gas are *directly* related. If you look again at the ideal gas equation and manipulate it a bit, you will get

$$\frac{V_1}{T_1} = \frac{V_2}{T_2}$$

for a constant pressure and number of moles, where T is the temperature in Kelvin ($°C + 273$). Think about our basic moving ball model: If the temperature increases, the average speed of the molecules in a sample will increase. They will collide more frequently and exert more force on the sides of the container. As a result, the container will expand if it is allowed to do so.

- **Volume vs. number of moles**—The volume of a gas is directly related to the number of moles. This is known as **Avogadro's law**. One mole of gas will have a volume of 22.4 L at standard temperature and pressure ($0°C$ and 1 atm). Remember this number. The AP exam may give you easy multiples of this number to make calculations involving the ideal gas equation easier. This relationship makes sense according to our moving ball model. Adding more moles of gas into a container will again increase collisions between molecules, which will expand the container. At constant temperature and pressure, this can be represented as:

$$\frac{V_1}{n_1} = \frac{V_2}{n_2}$$

- **Combined gas law**—This is formed by the laws relating the temperature, pressure, and volume of a gas under two different sets of conditions. The combined gas law is:

$$\frac{P_1 V_1}{T_1} = \frac{P_2 V_2}{T_2}$$

You may be asked to manipulate the ideal gas equation to establish other relationships besides those mentioned above. Let's try a practice question to get a grasp on these calculations.

REVIEW QUESTION

Diver A is 64 feet below the surface of the water (the pressure here is approximately 3.0 atm) and has 750 mL of air in his lungs when he sees a shark. Panicking, he returns directly to the surface without breathing out. What is the volume of his lungs when he reaches the surface? How many moles of gas must he exhale on the way up to keep the volume of air in his lungs constant?

SOLUTION

The question asks us to relate the volume of a sample of gas to a change in pressure. The equation that relates these two is Boyle's law:

$$P_1 V_1 = P_2 V_2$$

Since the pressure at the surface is atmospheric pressure or 1 atm, converting to L produces:

$$V_2 = \frac{P_1 V_1}{P_2} = \frac{(3.0) \times (0.75)}{(1.0)} = 2.25 \text{ L}$$

This question demonstrates why it's so important to exhale when ascending from a dive—this diver is in serious danger of rupturing a lung.

Now we need to think about what happens as the diver ascends to the surface. Always identify which variables are changing. In this case, it's the pressure and the number of moles (the temperature and volume are being held constant). We know that these two must be directly related, so looking at the ideal gas equation:

$$\frac{P_1}{n_1} = \frac{P_2}{n_2}$$

Or $n_2 = \frac{P_1 n_1}{P_1}$

Substituting, $n_2 = \frac{(1.0) \times (0.613)}{(3.0)} = 0.0204$ moles.

This is the number of moles of gas left in the lungs, so the number of moles exhaled is given by:

$$\Delta n = 0.0613 - 0.0204 = 0.0409 \text{ moles}$$

DALTON'S LAW OF PARTIAL PRESSURE

Dalton's law of partial pressure states that for a mixture of gases, the total pressure is the sum of the pressures each gas exerts on its own. In other words:

$$P_{total} = P_1 + P_2 + P_3 + \ldots$$

for gases with partial pressures P_1, P_2, P_3, etc. This equation appears on the equation sheet for the AP exam. Remember, the partial pressure exerted by a component gas will depend on the number of moles of that gas. This is true no matter which gas is being measured. A very useful concept is that of **mole fraction**, which is the ratio of the number of moles of an individual gas to the total number of moles in the mixture. This fraction is represented using the symbol χ. If a mixture of a gas contains three moles of oxygen and one mole of nitrogen, the mole fraction of oxygen is 0.75 and of nitrogen is 0.25. To get the partial pressure of a component gas, you just need to multiply the total pressure by the mole fraction.

THE KINETIC MOLECULAR THEORY OF GASES

The kinetic molecular theory of gases sounds pretty complicated, but it really consists of a few concepts we've already discussed and a few new equations. Since these equations are given on the AP exam, it's most important that you understand which formula goes with which question.

What is **temperature**? Temperature can be thought of as the average kinetic energy of the gas particles. Kinetic energy and temperature (in Kelvin) can be represented as the following:

$$KE_{avg} = \frac{3}{2}RT$$

On the AP exam equation sheet, this appears as:

$$KE \text{ per mole} = \frac{3}{2}RT$$

All molecules at a given temperature don't have the same kinetic energy because there is always a bell-shaped distribution of molecular velocities. This distribution of kinetic energies is also known as the **Boltzmann distribution**.

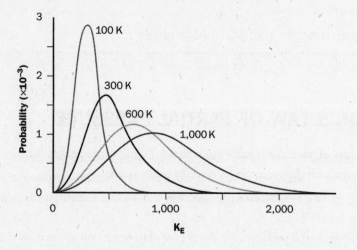

Root mean square velocity refers to the average velocity of gas particles in a sample and is given by the following equation, which appears on the AP exam equation sheet:

$$u_{rms} = \sqrt{\frac{3RT}{M}}$$

where R is 8.3145 J/K · mol, T is temperature in Kelvin, and M is the mass of a mole of gas particles in kilograms.

DIFFUSION AND EFFUSION

Diffusion describes the mixing of gases, while **effusion** refers to the passage of gas particles through a tiny orifice in a gas chamber. A question on the AP exam might ask you about the passage of gas through a pinhole. The most important thing to remember about effusion for the AP exam is that the rate of effusion of a gas is inversely proportional to the square root of its molar mass. So if we are comparing the effusion rates of two gases at the same temperature and pressure:

$$\frac{\text{Rate of effusion of gas 1}}{\text{Rate of effusion of gas 2}} = \frac{\sqrt{M_2}}{\sqrt{M_1}}$$

This equation appears on the AP exam equation sheet, but what does it mean? It implies that the smaller the molar mass, the faster the rate of effusion. Notice that the equation for root mean square velocity also reflects this (but remember how M is defined for this equation). At the same temperature, lighter gases must move more quickly than heavier particles in order to have the same kinetic energy. Since less massive gas particles move more quickly, they experience more collisions and have a greater probability of escaping through a hole in a given time. There is a common mistake students make in describing the rate of effusion on free-response questions: stating that less massive gases effuse faster because the particles are smaller (implying that they fit through the hole more easily), which is not true and won't earn credit. Gas particles are so tiny compared to even the smallest pinhole that the size of an individual molecule is irrelevant.

AP EXPERT TIP

Knowing the conditions under which the kinetic molecular theory isn't true (high pressure, low temperature, and small volume) will help you predict how non-ideal gases behave—and will help you remember what the van der Waals equation does.

THE BEHAVIOR OF REAL GASES

Real gases deviate significantly from the behavior predicted by the ideal gas equation. The AP exam will likely include a few predictable questions regarding real gases and the **van der Waals equation**, which takes into account molecular volumes and attractive forces:

$$\left(P + \frac{n^2 a}{V^2}\right)(V - nb) = nRT$$

Notice that this equation is just a modified version of $PV = nRT$. Keep in mind the following:

- The correction term for pressure results from attractive forces between gas particles. The size of this correction factor will depend on the ratio $\frac{n}{V}$. In other words, it will depend upon the concentration of gas particles in the sample. Notice that the smaller the concentration of gas particles, the smaller the magnitude of the correction factor.

- These attractive forces will contribute to making both the volume and pressure of a real gas *less* than that predicted by the ideal gas law. The attractive forces in a given gas tend to be greatest when the gas is at a low temperature, which makes sense; if you cool a gas enough, its particles clump together to form a liquid.

- The correction term for volume results from gas particles having nonzero volume. This factor tends to make both the volume and pressure of a real gas *greater* than that predicted by the ideal gas law. The volume taken up by individual gas particles tends to be greatest at high pressures—which again makes sense because squeezing more particles into a small space leaves less space between the particles.

- A real gas is most likely to behave like an ideal gas at high temperatures and low pressures. At high temperatures, the particles in a gas sample are moving so fast that the attractive forces experienced between particles is minimal. At low pressures, the volume occupied by the gas particles themselves is small compared to the overall volume. And at *low* temperature and *high* pressure, the gas would eventually condense to a liquid—a pretty large deviation from ideal gas behavior.

REVIEW QUESTIONS

1. A flask contains two moles of hydrogen, three moles of oxygen, and five moles of nitrogen. Which of the following is *NOT* true?

    (A) If the total pressure of the sample is 2 atm, the partial pressure of oxygen is 0.6 atm.

    (B) If another flask were to contain only ten moles of hydrogen (at the same temperature and volume), the total pressure would be the same.

    (C) If another flask were to contain only ten moles of hydrogen (at the same temperature and volume), the density of the gas would be the same.

    (D) The mole fraction of nitrogen is 0.5.

    (E) Changing the temperature of the flask has no effect on the mole fractions of individual gases.

2. With regard to the kinetic molecular theory of gases, which of the following is true?

    (A) The particles of gas in a sample will have no kinetic energy at 0 K.

    (B) The distribution of velocities of particles in a sample of H_2 will be identical at 25°C and 75°C.

    (C) The rate of effusion of He will be approximately 5 times faster than that of Ne at 1 atm and 25°C.

    (D) The rate of effusion of H_2 will be approximately 11 times faster than that of I_2 at 1 atm and 25°C.

    (E) The root mean square velocity for oxygen is 500 at 1 atm and 25°C. This means that the majority of particles in a sample of O_2 will have this velocity at STP.

3. Consider the three gases, H_2, Ar, and N_2 at 1 atm and 25°C. Which of the following is true?

    (A) The gas with the highest density is H_2.

    (B) The gas with the lowest average molecular speed is Ar.

    (C) The gas with the highest average molecular speed is N_2.

    (D) The gas with the lowest density is Ar.

    (E) All three gases will have equal effusion rates.

4. A gas is least likely to behave ideally under the following conditions:

    I.   High pressure
    II. Low temperature
    III. Low pressure

    (A) I only
    (B) II only
    (C) III only
    (D) I and II only
    (E) II and III only

5. A sample of N_2 gas in a flask is heated from 27°C to 150°C. If the original sample of gas is at a pressure of 1,520 torr, what is the pressure in the final sample (in atm)?

    (A) 1.4 atm
    (B) 2.8 atm
    (C) 3.2 atm
    (D) 4.3 atm
    (E) 5.6 atm

6. Consider the gases CO and N_2. Which of the following will be nearly identical at 25°C and 1 atm?

   I. Average molecular speed
   II. Rate of effusion through a pinhole
   III. Density

   (A) I only
   (B) III only
   (C) I and II only
   (D) II and III only
   (E) I, II, and III

7. Which of the following gases behaves most like an ideal gas at 25°C and 1 atm?

   (A) NO_2
   (B) NH_3
   (C) CH_4
   (D) HF
   (E) O_3

8. $2NaN_3 (s) \rightarrow 2Na (s) + 3N_2 (g)$

   The reaction above is used to generate nitrogen gas in an automobile air bag. What mass of NaN_3 is required to inflate a 20 L air bag to a pressure of 1.25 atm at 27°C?

   (A) 5.50 g
   (B) 11.0 g
   (C) 22.0 g
   (D) 33.0 g
   (E) 44.0 g

9. Consider two identical flasks, one filled with $N_2 (g)$ and one with $CO_2 (g)$ at 2.0 atm and 25°C. Assuming ideal behavior, which of the following is/are true?

   I. The N_2 molecules will have a greater average kinetic energy.
   II. The N_2 molecules will have a greater average velocity.
   III. There are an equal number of N_2 and CO_2 molecules present.

   (A) I only
   (B) II only
   (C) I and II only
   (D) II and III only
   (E) I, II, and III

10. The density of a gas at 2.0 atm and 25°C is determined to be 3.11 g/L. The identity of the gas is most likely which of the following?

    (A) CH_4
    (B) F_2
    (C) N_2O_4
    (D) O_2
    (E) CF_2Cl_2

FREE-RESPONSE QUESTION

1. A sample of sodium azide (NaN_3) is heated in a test tube and decomposed according to the following reaction:

    $$2NaN_3 \text{ (s)} \rightarrow 2Na \text{ (s)} + 3N_2 \text{ (g)}$$

    The nitrogen produced is collected by displacement of water at 27°C at a total pressure of 565 torr.

    (a) Draw the Lewis structure for N_3^-.

    (b) Label the hybridization of all atoms in N_3^-.

    (c) Do you expect $\Delta S°$ to be positive or negative for this reaction? Justify your answer.

    (d) Given that the vapor pressure of water is 26 torr at 27°C, calculate the partial pressure of N_2 in the sample in atm.

    (e) If the volume of the gas produced is 250 mL, how many moles of NaN_3 were decomposed?

ANSWERS AND EXPLANATIONS

1. C

A flask containing ten moles of hydrogen (at the same temperature and pressure) would occupy an identical volume, but due to the lower molecular weight of hydrogen compared to the other two gases, the density would be much less than that of the mixture described. Choice (A) is true since oxygen has a mole fraction of $\frac{3}{10} = 0.3$. The partial pressure of oxygen in the sample is just the total pressure multiplied by the mole fraction = (2 atm) × (0.3) = 0.6 atm. Choice (B) is true since the total pressure depends only on the sum of the partial pressures and not the identity of the gas. Choice (D) is true since nitrogen has a mole fraction of $\frac{5}{10} = 0.5$. Choice (E) is true since the number of moles of gas in the sample remains constant when temperature changes, and therefore the mole fractions will remain the same.

2. D

Since the molar mass of I_2 is 253.8 and that of H_2 is 2.016, the relative rate of effusion is the ratio of their square roots, approximately 11:1. Choice (A) is incorrect since particles at 0 K are not wholly devoid of kinetic energy and because the particles cannot be in a gas state at this temperature. Choice (B) is incorrect since we know that the root mean square velocity of particles will increase with increased temperature, thus shifting the graph for distribution of molecular velocities to the right. Choice (C) is incorrect since the ratio of molar masses is 5, not the ratio of square roots (which is approximately 2.2). Choice (E) is incorrect since the root mean square velocity represents an *average* only, and the distribution of molecular velocities is quite broad at any temperature.

3. B

Since Ar has the highest molar mass (39.95), we expect it to have the highest value for M and thus the lowest root mean square velocity.

4. D

We expect that the ideal gas laws are least likely to apply when molecular volume matters and attractive forces between molecules is greatest. As discussed earlier, this is most likely to occur at high pressure (where the volume of particles matters) and at low temperatures (since the attractive forces between gas particles is significant). At low pressures (and high volumes), the concentration of gas molecules is less and attractive forces are not as significant.

5. B

To solve the equation, we need to look at the ideal gas equation to establish a relationship between temperature and pressure.

Since $PV = nRT$

$$\frac{P_1}{T_1} = \frac{P_2}{T_2}$$

Or $P_2 = \frac{P_1 T_2}{T_1}$

What is a torr anyway? It's just another unit of pressure, identical to mm Hg. One atmosphere is 760 mm Hg or 760 torr. We should note that sometimes it's not necessary to convert to SI units. However, we would recommend always converting to standard units when in doubt.

So $1{,}520 \text{ torr} \times \left(\frac{1 \text{ atm}}{760 \text{ torr}}\right) = 2.000 \text{ atm}$.

Substituting into the equation above (remembering to convert the temperatures to Kelvin):

$$P_2 = \frac{(2.000 \text{ atm}) \times (423 \text{ K})}{(300 \text{ K})} = 2.8 \text{ atm}$$

6. E

The two gases have nearly identical molar masses (approximately 28 g/mol). At constant temperature and pressure, all of the listed quantities depend primarily on the molar mass of the gas.

7. C

We expect that a gas will behave most ideally when intermolecular interactions are minimized. We expect that intermolecular interactions will be strongest for polar molecules and weakest for non-polar molecules. Of all the molecules listed, only CH_4 (methane) has a net zero dipole moment.

8. E

To solve this problem, we must start by determining the number of moles of N_2 in the air bag using the ideal gas equation:

$$PV = nRT$$

Solving for n:

$$n = \frac{PV}{RT} = \frac{(1.25 \text{ atm})(20 \text{ L})}{(0.0821 \text{ L} \cdot \text{atm} \cdot \text{mol}^{-1}\text{K}^{-1})(300 \text{ K})}$$

$$n = 1.0 \text{ mol } N_2$$

According to the balanced chemical equation, 2 mol of NaN_3 are needed for every 3 mol of N_2 produced. So:

$$\frac{1.0 \text{ mol } N_2 \times 2 \text{ mol } NaN_3}{3 \text{ mol } N_2} = 0.67 \text{ mol } NaN_3$$

The corresponding mass of NaN_3 is given by:

$$\text{\# of grams of } NaN_3 = (0.67 \text{ mol } NaN_3) \times$$

$$\left(\frac{65 \text{ g } NaN_3}{1 \text{ mol } NaN_3}\right) = 44 \text{ g } NaN_3$$

9. D

Since the two gases are both at the same temperature, pressure, and volume, there must be an identical number of moles (and therefore molecules) of gas present in each flask. N_2 has a lower molecular weight than CO_2, and therefore has a higher average velocity (remember that average molecular velocity is inversely proportional to the square of the molar mass). However, the average kinetic energy of both gases should be the same since average kinetic energy is determined by temperature only, which is identical for both flasks.

10. B

To solve this equation, we need to manipulate the ideal gas equation to give an expression for molar mass.

We know that $PV = nRT$.

But $n = \frac{m}{M}$ where m is the mass of gas present and M is the molar mass of the gas.

Substituting, $PV = \frac{mRT}{M}$. So $M = \frac{mRT}{PV}$.

Since $\frac{m}{V}$ is just the density of the gas (d), the expression simplifies to:

$$M = \frac{dRT}{P} =$$

$$\frac{(3.11 \text{ g/L})(0.0821 \text{ L} \cdot \text{atm} \cdot \text{mol}^{-1}\text{K}^{-1})(298 \text{ K})}{(2.0 \text{ atm})} =$$

38 g/mol

Of all the molecules listed, only F_2 has a molar mass of 38 g/mol. Therefore, the gas is most likely F_2.

FREE-RESPONSE QUESTION

1. **(a)** The Lewis structure for the azide anion is as follows:

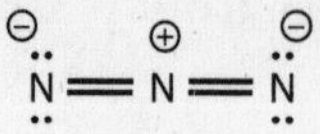

Azide

(b) The N_3^- anion has linear geometry. The central N atom has *sp* hybridization whereas the other atoms have *sp²* hybridization.

(c) The change in entropy should be positive since there is a gas produced from a solid reactant.

(d) The vapor pressure of a liquid can be thought of as that small portion of a liquid that is behaving like a gas. At equilibrium, there are equal numbers of molecules changing from liquid to gas and vice versa. The important thing here is that the vapor pressure of a liquid must be included when calculating partial pressures. In this case, we need to subtract it for this classic "gas collected over water" question.

Since $P_{TOTAL} = P_{WATER} + P_{NITROGEN}$

$P_{NITROGEN} = P_{TOTAL} - P_{WATER} = 565 - 26 = 539$ torr

$539 \text{ torr} \times \left(\dfrac{1 \text{ atm}}{760 \text{ torr}} \right) = 0.709 \text{ atm}$

(e) We need to use the ideal gas equation to calculate the number of moles of gas that were formed.

$PV = nRT$

$n = \dfrac{PV}{RT} = \dfrac{(0.71 \text{ atm}) \times (0.25 \text{ L})}{(0.08206 \text{ L} \cdot \text{atm} \cdot \text{mol}^{-1}\text{K}^{-1}) \times (300 \text{ K})} = 0.0072 \text{ mol}$

Then, by looking at the balanced chemical equation, we can see that for every three moles of N_2 produced, there were two moles of NaN_3 consumed.

$0.0072 \text{ mol } N_2 \times \left(\dfrac{2 \text{ mols } NaN_3}{3 \text{ mols } N_2} \right) = 0.0048 \text{ mols } NaN_3$

CHAPTER 8: PROPERTIES OF MATTER

IF YOU LEARN ONLY EIGHT THINGS IN THIS CHAPTER . . .

1. London dispersion forces are the relatively weak interactions that occur between nonpolar molecules, such as CH_4 and I_2. For nonpolar compounds, physical properties will depend on the total number of electrons in the molecule (more electrons = stronger interactions).

2. Dipole-dipole interactions occur between molecules with a net dipole moment. An important type of dipole-dipole interaction is hydrogen bonding, which occurs between the atoms F, O, and N and a proton that is bonded to F, O, or N.

3. Metals can be represented as an array of metal cations in a sea of electrons. This structure accounts for the malleability and electrical conductivity of metals.

4. Ionic bonding is an extremely powerful type of bonding between a metal and a nonmetal. Ionic compounds conduct electricity when in solution or melted.

5. Network solids, for example diamond, can be thought of as a single giant molecule in which atoms are covalently bonded in every direction.

6. For a given chemical formula, multiple structures (and therefore multiple physical properties) are possible. For example, SiO_2 can exist as the amorphous solid glass or as the crystalline solid quartz, while C can exist as carbon or as graphite.

7. Changes of state will occur in response to changes in temperature and/or pressure. We can observe changes of state by following a heating curve.

8. Phase diagrams describe the states of a material with respect to temperature and pressure. The triple point corresponds to the temperature and pressure at which all three phases of a material exist, whereas the critical point corresponds to the temperature above which a gas cannot be liquefied no matter what pressure is applied.

"Properties of Matter" is a pretty broad topic. In this section, we will present only high-yield topics, focusing on those questions you are most likely to encounter on test day. We will focus on two major areas. First, we will relate the properties of bulk matter to underlying chemical structure. Second, we will look at the phases of matter, paying close attention to favorite AP topics such as phase diagrams.

In the last section, we discussed gases, which are an important state of matter. Why does nitrogen (N_2) exist as a gas instead of some other state? Since nitrogen is a nonpolar covalent molecule, forces between individual nitrogen molecules are weak. Therefore, nitrogen can only exist as a liquid at low temperatures and high pressures, when molecules of a gas are brought closer together and these intermolecular forces become more important.

Here, we will expand our discussion to include other states of matter, namely solids and liquids. The most important determinant of state is the strength of bonding and/or intermolecular interactions within a material. Molecules with weak intermolecular forces (such as hydrocarbons) have low melting and boiling points. Materials in which the bonding interaction is strong (for example ionic and network solids) have exceptionally high melting and boiling points.

Once we have defined the sorts of materials that exist in nature, we need to consider how materials undergo changes in state. When a substance boils or melts, it undergoes a phase change that we can describe using heating or cooling curves. It's possible to predict the state of a substance under specific conditions using a phase diagram, which defines the state of a material if the pressure and temperature are known.

After reading this section, you will have an understanding of how structure relates to the **physical properties** of a variety of materials. You will also be more familiar with the various states matter can occupy and how this can be predicted based on environmental conditions.

We have already described intramolecular bonding, which describes the forces that keep atoms within a molecule together. In this section we will discuss the **intermolecular forces** that cause the particles within a solid or liquid to aggregate. These forces are very important in determining the physical properties of a substance, e.g., its melting and boiling point. On the AP exam, you will often be asked to compare the physical properties of two substances just by looking at their molecular structures.

MOLECULAR FORCES AND MATTER: A SURVEY

The properties of matter are determined primarily by the forces holding the constituent particles together. These forces may be intramolecular or intermolecular. Here, we will present the different types of forces that hold matter together from weakest to strongest.

London dispersion forces are relatively weak interactions that occur between all molecules. In nonpolar molecules and noble gases, London dispersion forces are the only intermolecular forces present, and thus determine the properties of those substances. We know these forces must exist since even nonpolar molecules exist as liquids and solids at certain temperatures and pressures. The basic concept here is that even a molecule with no net dipole moment can act as a **temporary dipole** depending on where its electrons are and therefore **induce a dipole** in a neighboring atom. Examples of molecules that interact with each other this way are CH_4, I_2, and BH_3, all of which have no net dipole moment. One trend of which you should be aware is that associated with atomic number. The higher the atomic number, the greater the number of electrons and thus the higher the **polarizability**, which indicates the likelihood of inducing a dipole. Thus, I_2 has a higher boiling point than F_2 (in fact, I_2 is a solid at room temperature).

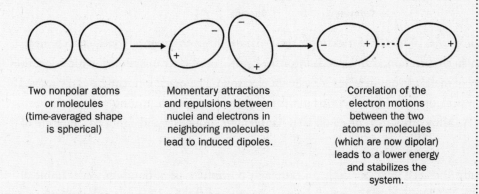

| Two nonpolar atoms or molecules (time-averaged shape is spherical) | Momentary attractions and repulsions between nuclei and electrons in neighboring molecules lead to induced dipoles. | Correlation of the electron motions between the two atoms or molecules (which are now dipolar) leads to a lower energy and stabilizes the system. |

Dipole-dipole forces are the forces that exist between molecules with a net dipole moment, which line up so that the positive and negative ends are close to each other. These forces govern the behavior of some of the most important molecules in nature, including H_2O and NH_3. Again, total electron count is important: HI has a higher boiling point than HBr, which has a higher boiling point than HCl.

Why does HF have a higher boiling point than all three of these? There is a special type of bonding in the most polar X-H bonds called **hydrogen bonding**. This type of bonding applies only to the elements F, O, and N and occurs between a lone pair on one of these elements and a hydrogen atom bonded to F, O, or N. One important example of hydrogen bonding is the interaction between base pairs in DNA, which is shown in the following figure.

AP EXPERT TIP

For "properties of water" questions, water's properties are the result of hydrogen bonding.

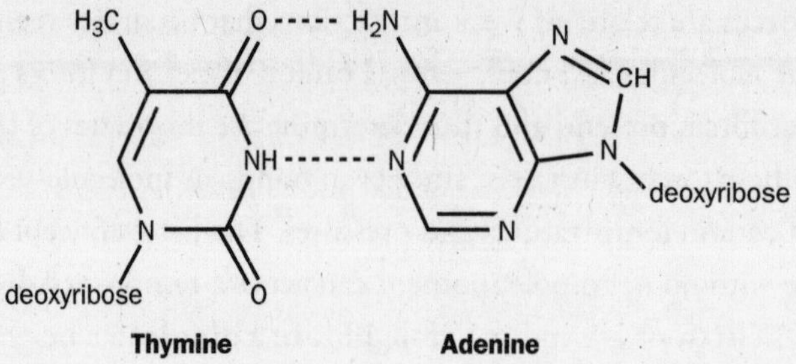

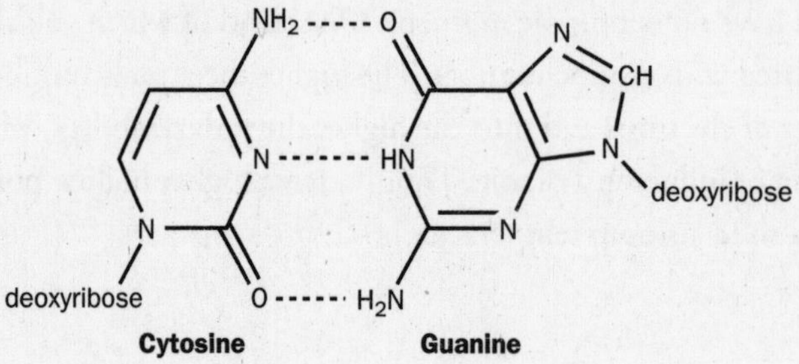

Thymine   **Adenine**

Cytosine   **Guanine**

The unique properties of **metals**, for example their malleability and superior conduction of heat and electricity in all directions, can be accounted for by the sea of electrons model, which describes a regular array of metal cations in a "sea" of valence electrons. The metal ions can be easily moved around by physical force, and the electrons in the metal are free to move around to conduct heat or electricity. An **alloy** refers to a substance that has metallic properties and contains a mixture of elements.

We have already discussed ionic bonding, an extremely powerful type of bonding. An example of an ionic solid is NaCl, in which Na and Cl atoms are arranged at the points of a regular lattice. Ionic solids have exceptionally high melting points and conduct electricity in solution or when melted to the liquid state, though they are electrical insulators in the solid state.

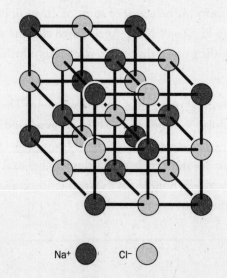

Na⁺ ⬤   Cl⁻ ◯

Finally, **network solids** are solids with exceptionally high melting points. These solids can be thought of as a giant molecule in which atoms have covalent bonds in every direction. For example, diamond is just a tetrahedral array of carbon atoms. Network solids are poor conductors of electricity (with the exception of graphite).

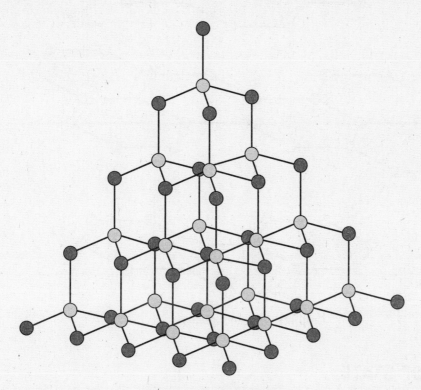

Diamond (tetrahedral)

THE IMPORTANCE OF STRUCTURE

The other key thing to remember is that different structures are possible for a given atomic formula. In this case, it will be the degree of structural order that will determine the physical properties of the material. One example is SiO_2, which may exist in **crystalline** form (as in quartz) or as an **amorphous solid** (as in glass). In this case, the two will have quite different physical properties, reflecting the increased disorder present in glass. Another example is C, which may exist as diamond or graphite. As we discussed earlier, diamond represents an extremely stable form of carbon in which all carbon atoms are bonded to each other, in tetrahedral fashion, to give a super molecule. In contrast, the structure of graphite is many stacked plates of sp^2-hybridized carbon atoms. While diamond is hard, colorless, and cannot conduct electricity, graphite is black, somewhat malleable, and a conductor.

AP EXPERT TIP

If you see diamond or Si (as in SiO_2) in a structure, the answer is "network solid." C and Si easily form network bonds.

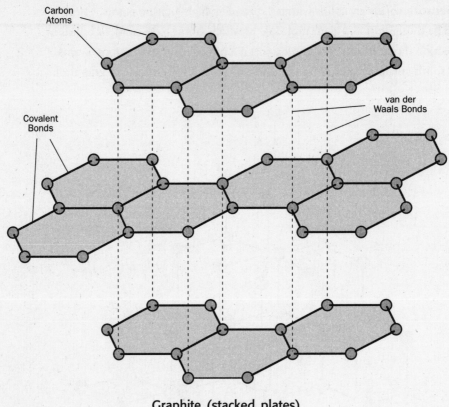

Graphite (stacked plates)

CHANGES OF STATE

Changes of state will occur in response to changes in temperature or pressure. In general, as a **solid** (such as ice) is heated, it will melt to form a **liquid**. If this liquid continues to be heated, at some point the liquid will boil to form a **gas**. This process can be described by a **heating curve**, which is just a plot of temperature versus energy added. The AP exam will expect you to understand what is going on at all points on the curve qualitatively, so here goes:

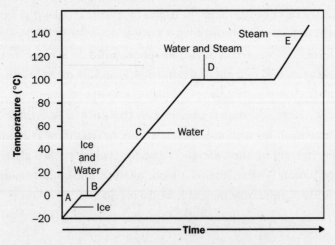

Section A represents water in solid form, which changes temperature as energy is added. The energy is enough to increase molecular vibrations and alter the temperature of the solid, but not enough to disrupt its crystalline structure.

Section B represents a **phase transition**, namely **melting**. The temperature at which this transition occurs is constant and is the **melting point**. In this section, the added energy is able to disrupt the ordered, crystalline structure of ice by breaking hydrogen bonds. Remember that in this section, both solid and liquid are present. There is an **enthalpy change** at the melting point known as the **heat of fusion**. This enthalpy is given in units of kJ/mol.

Section C represents water in liquid form. The kinetic energy of the water molecules increases as energy is added to the system.

Section D represents the second phase transition, namely **boiling**. The temperature at which this transition occurs is constant at a given pressure and is the **boiling point**. In this section, the added energy disrupts the hydrogen bonds between water molecules in the liquid phase, giving them enough energy to escape to the vapor phase. In this section, both the liquid phase and the vapor phase are present. There is an enthalpy change at the boiling point known as the **heat of vaporization**, also given in units of kJ/mol. Remember that the boiling point can also be defined as the temperature at which the vapor pressure of the liquid is equal to the atmospheric pressure, and the **normal boiling point** is defined as the temperature at which the vapor pressure is exactly one atmosphere.

Section E represents water in the **vapor** form, or steam. As energy is added, the water molecules continue to gain kinetic energy. Notice that the heating curve of water helps to explain why a steam burn can be so dangerous; the temperature of steam can exceed 100°C.

REVIEW QUESTION

With respect to the heating curve for water, as shown on p. 112, which of the following is NOT true?

(A) The slopes of sections A and C are different since the different states of water have different molar heat capacities (the energy required to raise the temperature of 1 mole of a substance by 1°C).

(B) The temperature in section B represents the freezing point of H_2O.

(C) In section C, the vapor pressure of the liquid is increasing as energy is added.

(D) As a gas condenses to form a liquid, energy is absorbed.

(E) If the enthalpy of fusion of H_2O is approximately 6.0 kJ/mol, the energy required to melt 36 grams of ice is approximately 12.0 kJ.

SOLUTION

The correct answer is choice (D). This statement is not true since energy is released as water vapor condenses to form a liquid. Choice (A) is true since the different phases of water will have different molar heat capacities. Choice (B) is true since the freezing point and melting point of any solid are the same. Choice (C) is true since increasing the temperature will increase the vapor pressure of a liquid. When this vapor pressure is equal to the atmospheric pressure, boiling will occur. Choice (E) is true since 36 grams of water (molar mass $= 18$) represents approximately 2 moles. Therefore, $\Delta H = (6.0 \text{ kJ/mol}) \times (2.0 \text{ mol}) = 12 \text{ kJ}$.

PHASE DIAGRAMS

Phase diagrams are an extremely important topic for the AP exam. Basically, these diagrams depict the phases of a substance in terms of pressure and temperature. Just knowing the temperature and pressure of a sample will allow you to predict whether it will be a solid, liquid, or gas.

The phase diagram for carbon dioxide is shown on the following page. Have a look at this diagram and note the following:

- The **triple point** corresponds to the temperature and pressure at which all three phases of carbon dioxide (solid, liquid, and gas) can exist.

- The **critical point** is defined by the **critical temperature** and **critical pressure**. The critical temperature is the temperature above which a gas cannot be liquefied no matter what pressure is applied. The critical pressure is the pressure required to produce a liquid at this temperature. For a sample of CO_2 gas at 60°C, what pressure is required to liquefy the gas? The gas cannot be liquefied at this temperature no matter how high the pressure is, since 60°C is higher than the critical temperature.

Notice that a phase diagram allows you to predict what phase change will occur under certain conditions. For example, suppose that a sample of solid CO_2 (dry ice) is removed from the freezer at −90°C and allowed to sit on a laboratory bench at room temperature. According to the phase diagram, at 1 atm a sample of solid CO_2 will transform from a solid to gas directly, or **sublimate**. What will happen when a sample of CO_2 gas at −50°C and 1 atm is pressurized slowly to 100 atm?

According to the phase diagram, the sample will liquefy, then solidify if the temperature is held constant.

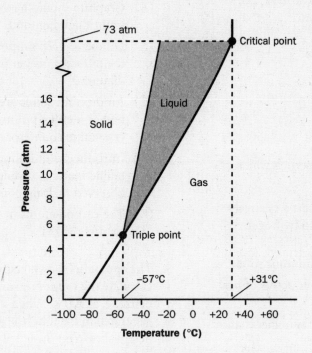

REVIEW QUESTIONS

1. Which of the following would you expect to have the highest boiling point?

   (A) H_2Se
   (B) H_2Te
   (C) H_2O
   (D) H_2S
   (E) SiH_4

2. With respect to physical properties, which of the following is *NOT* true?

   (A) CH_2O (formaldehyde) has a relatively high boiling point due to hydrogen bonding.
   (B) $BaCl_2$ is an example of an ionic solid.
   (C) Ionic solids are poor conductors of heat and electricity.
   (D) The melting point of Fe is higher than that of $C_6H_{12}O_6$ (glucose).
   (E) The melting point of diamond is higher than that of Cu (copper).

3. Put the following in order of increasing melting point: SiH_4, SiO_2, CH_4, NH_3, PH_3

   <u>lowest</u>   <u>highest</u>
   (A) NH_3, PH_3, CH_4, SiH_4, SiO_2
   (B) SiO_2, PH_3, NH_3, CH_4, SiH_4
   (C) SiH_4, CH_4, NH_3, PH_3, SiO_2
   (D) CH_4, SiH_4, PH_3, NH_3, SiO_2
   (E) CH_4, SiH_4, SiO_2, NH_3, PH_3

4. Consider diamond and graphite. Which of the following is true?

   (A) Graphite should have a higher melting point than diamond.
   (B) Differences in composition account for the different properties of graphite and diamond.
   (C) Amorphous solids are likely to have higher melting points than their crystalline counterparts.
   (D) Introducing an impurity into a crystalline material should increase the observed melting point.
   (E) The carbon atoms in diamond are sp^3-hybridized.

5. The apparatus shown below is used to calculate the vapor pressure of C_6H_6, or benzene, at 300 K. When a vacuum is generated in the tube, the mercury rises to the height shown on the left. Then, a sample of benzene is injected into the tube, which floats to the surface and causes the mercury to be displaced downward by the amount shown on the right. Which of the following represents the vapor pressure of benzene in torr?

   (A) 0 torr
   (B) 104 torr
   (C) 656 torr
   (D) 760 torr
   (E) 864 torr

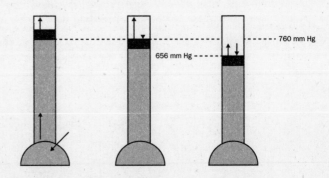

6. Which of the following will be true on the top of Mt. Everest (29,028 feet above sea level)?

   (A) The boiling point of water will be higher than that at sea level.

   (B) The vapor pressure of water at 60°C will be the same as that at sea level at an identical temperature.

   (C) An egg placed in boiling water at the summit will cook faster than one placed in boiling water at sea level.

   (D) Atmospheric pressure will be approximately the same as that at sea level.

   (E) The environment at the top will be similar to that of Honolulu, HI.

Questions 7–9 refer to the following substances:

   (A) H_2
   (B) CH_3OH
   (C) CH_2Cl_2
   (D) KCl
   (E) CO

7. It is a solid at 25°C and 1 atm.

8. It is capable of hydrogen bonding.

9. Its properties are governed by London (dispersion) forces.

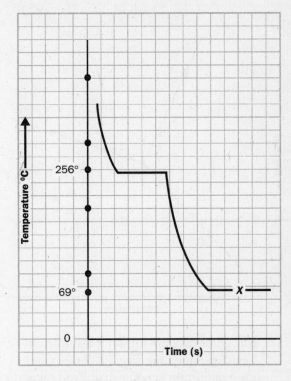

10. Liquid biphenyl ($C_{12}H_{10}$) at 85°C was cooled to 40°C, following the trends represented in the curve above. At point X, which of the following is true?

   I. The temperature at point X corresponds to the melting point of biphenyl.

   II. Both liquid and solid phases are present.

   III. No heat is being transferred to the environment.

   (A) I only
   (B) II only
   (C) I and II only
   (D) II and III only
   (E) I, II, and III

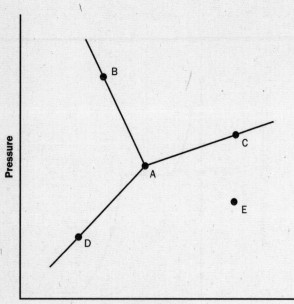

11. The phase diagram for water is shown above. Which point on the diagram corresponds to the equilibrium between liquid and gas at the normal boiling point?

(A) A
(B) B
(C) C
(D) D
(E) E

FREE-RESPONSE QUESTIONS

1. You are given samples of two pure compounds, one of which is a gas and one of which is a liquid at room temperature. By elemental analysis, both compounds are found to have the following composition (by mass):

   52% C

   13% H

   35% O

   (a) Determine the empirical formula of both compounds.

   (b) The molar mass of both compounds is determined to be 46.0 g/mol. Name one method for calculating the molar mass of a pure gas and one method for calculating the molar mass of a pure liquid.

   (c) Some properties of the two compounds are listed below:

	Compound X	Compound Y
Melting point	−114°C	−138°C
Boiling point	78°C	−22°C
Net dipole?	Yes	Yes
Functional group	Alcohol	Ether

      i. Draw plausible Lewis structures of compounds X and Y.

      ii. In which compound are intramolecular forces stronger? Justify your answer.

      iii. Is either compound capable of hydrogen bonding? If so, draw a schematic representation of a typical hydrogen bond in this substance.

      iv. Which compound is likely to be more soluble in water? Justify your answer.

   (d) Given the information in the table above, draw a schematic representation of a phase diagram for the alcohol.

ANSWERS AND EXPLANATIONS

1. C

The intermolecular attraction in each case results from a dipole-dipole interaction, but only H_2O can participate in hydrogen bonding. Therefore, H_2O will have the highest boiling point.

2. A

It is not true that CH_2O (formaldehyde) has a relatively high boiling point due to hydrogen bonding. Formaldehyde is made up of oxygen and hydrogen atoms. However, a hydrogen bond must occur between a lone pair on an electronegative atom (F, O, N) and a hydrogen atom bonded to an electronegative atom. In formaldehyde, the hydrogen atoms are bonded to carbon. Choice (B) is true since $BaCl_2$ represents a compound formed between a metallic atom and nonmetallic atom with a large electronegativity difference. Choice (C) is true since ionic solids only conduct electricity in solution, not as solids. Choice (D) is true since glucose is a molecular solid in which intramolecular interaction occurs by hydrogen bonding, whereas Fe is a metallic solid. Metallic solids generally have higher melting points than molecular solids. Choice (E) is true since diamond is a network solid and therefore has an exceptionally high melting point (3,500°C).

3. D

This is a pretty tough question. First, identify the types of intermolecular bonding at work in all cases. CH_4 and SiH_4 both have a net zero dipole moment and therefore interact by weak London dispersion cases; SiH_4 has more electrons and therefore stronger dispersion forces. For both NH_3 and PH_3, the intermolecular forces are dipole-dipole, but only NH_3 can participate in hydrogen bonding. Finally, SiO_2 is an example of a network solid and should

have a much higher melting point than all the other substances considered.

4. E

The carbon atoms in diamond exist in a tetrahedral arrangement and are sp^3-hybridized. Choice (A) is not true since diamond has a greater degree of structural order than graphite and has an extremely high melting point. Choice (B) is not true since diamond and graphite have an identical composition with differing structure. Choice (C) is not true since amorphous solids have a higher degree of disorder in their structure and therefore have lower melting points than their related crystalline solids. Choice (D) is not true since the presence of an impurity in an otherwise crystalline sample lowers the observed melting point.

5. B

The mercury is pushed downward due to the vapor pressure exerted by the benzene. The higher the vapor pressure, the further down the mercury will be displaced. In general, liquids with weak intermolecular forces will have higher vapor pressures. Thus, benzene, a nonpolar molecule, would be expected to have a relatively high vapor pressure. To solve this question, we use the following equation:

$$P_{atm} = P_{vapor} + P_{Hg\ column}$$

or
$$P_{vapor} = P_{atm} - P_{Hg\ column} =$$
$$760\ mmHg - 656\ mmHg =$$
$$104\ mmHg = 104\ torr$$

6. B

The vapor pressure of water depends on the temperature, not on the atmospheric pressure. The important thing here is that the atmospheric pressure at the summit will be significantly less than that at sea level. Therefore, since water boils when its vapor pressure equals atmospheric pressure, it will boil at a lower temperature at high altitudes. This will also cause an egg to take longer to cook.

7. D

Only KCl is a solid under standard conditions. KCl is an ionic compound and should therefore have a high melting point.

8. B

Of the substances listed, only CH_3OH, or methanol, is capable of hydrogen bonding. The requirement for a hydrogen bond is an H atom bonded to a highly electronegative atom: F, O, or N.

9. A

The only nonpolar molecular compound listed is H_2. This material is a gas at room temperature due to relatively weak London forces.

10. C

At point X, a phase transition is occurring. As the biphenyl molecules slow down, intermolecular interactions result in crystallization to form a solid. At this point, both solid and liquid phases are present and in equilibrium. While the temperature in this plateau region of the curve is constant, heat is still being transferred from the system to the environment, corresponding to the latent heat of fusion.

11. C

Since this point lies on the boundary between the liquid and gas phases of water, it corresponds to the boiling point of water at a given pressure.

FREE-RESPONSE QUESTION

1. **(a)** Assuming 100 g of the compound(s), we have 52 g of carbon, 13 g of hydrogen, and 35 g of oxygen. Converting these to moles:

    $$52 \text{ g C} \times \left(\frac{1 \text{ mol}}{12 \text{ g}} \right) = 4.3 \text{ mol C}$$

    $$13 \text{ g H} \times \left(\frac{1 \text{ mol}}{1 \text{ g}} \right) = 13 \text{ mol H}$$

    $$35 \text{ g O} \times \left(\frac{1 \text{ mol}}{16 \text{ g}} \right) = 2.2 \text{ mol O}$$

    From these values, it is apparent that the ratio of C:H:O is 2:6:1. Therefore, the empirical formula is C_2H_6O.

    (b) The molar mass of a gas can be calculated using the ideal gas law. The molar mass of a liquid can be calculated by vaporization, then by applying the ideal gas law (the Dumas method).

    (c)   i. We know that the molecular formula of both compounds is also C_2H_6O. Using the data provided, we can see that compound X is an alcohol with a much higher boiling point than compound Y, which is an ether. The Lewis structures are given below:

   ii. The intramolecular forces must be stronger in compound X since this compound has such a high boiling point. Individual molecules are less able to escape their intermolecular interactions with neighboring solvent molecules and escape into the gas phase.

   iii. Compound X is capable of hydrogen bonding as follows:

   iv. Compound X is much more polar (and also capable of hydrogen bonding) and therefore is much more soluble in polar water. This is evidenced by 151-proof rum, which is greater than 75% Compound X, more commonly known as ethanol.

(d) The correct phase diagram for ethanol (EtOH) would include correctly labeled axes (pressure on vertical, temperature on horizontal), as well as the correct progression of solid to liquid to gas as solid EtOH is warmed at 1 atm, with the corresponding melting and boiling points.

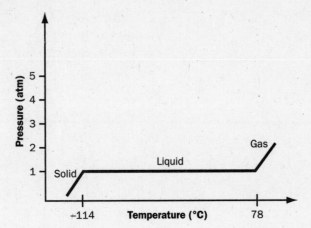

CHAPTER 9: SOLUTIONS

IF YOU LEARN ONLY EIGHT THINGS IN THIS CHAPTER . . .

1. You must be able to define molarity, molality, mass percent, and mole fraction for any solution. These definitions are important since the behavior of solutions is often defined with respect to one of these.

2. The basic rule when predicting solubility is "like dissolves like." This statement means that highly polar molecules (such as glucose) are likely to dissolve in polar solvents (like water), whereas nonpolar molecules will be insoluble in aqueous media.

3. The solubility of a gas in a liquid is proportional to the ambient pressure, as predicted by Henry's law:

   $P = kC$

4. The solubility of a gas in a liquid is expected to decrease as the temperature is raised.

5. Colligative properties you will need to know for solutions include boiling point elevation, freezing point depression, and vapor pressure depression.

6. Boiling point elevation and freezing point depression are both directly proportional to the molality of the solution, as described by the following equations:

   $\Delta T = iK_b m_{solute}$

   $\Delta T = iK_f m_{solute}$

   Both may be used to calculate the molar mass of a solid.

7. For vapor pressure depression, remember that the vapor pressure of a solvent at a given temperature is given by Raoult's law:

$$P_{solution} = \chi_{solvent} P^0_{solvent}$$

8. Osmotic pressure may be defined as the pressure required to stop osmosis (the movement of water) across a semipermeable membrane. This phenomenon is useful in calculating the molar mass of large molecules, such as proteins in solution. The osmotic pressure is given by:

$$\pi = MRT$$

Liquids we encounter in everyday life are rarely pure. Instead, they are usually solutions with a number of dissolved components. Even the clear tap water we drink contains significant amounts of fluoride and calcium carbonate, as well as other ions. Solutions are equally ubiquitous on the AP exam, where you will be asked a number of questions regarding solutions and their properties. But if you understand the different ways of describing solutions, the properties of molecules that allow solution formation, and the so-called colligative properties of solutions, you will be in great shape for this section of the exam.

In this section, we will start by clarifying the different ways of representing solutions, including molarity, molality, and mass percent. How does one go about making a solution? Most people do so every day, whether it's by adding sugar to coffee or salt to a pot of boiling water. However, it's not the case that a given solute will dissolve in just any solvent. Similarity of chemical structure forms the basis for the rule that "like dissolves like."

What does the chemical structure on the next page have to do with solutions? This chemical structure corresponds to a chemical known as perflubron, a fluorocarbon with potential use in liquid breathing. Liquid breathing is more than just a fantasy propagated by movies such as *The Abyss*. Actually, experiments on the subject using rats began in the 1960s, and more recently perflubron has been used in studies looking at the treatment of lung disorders in newborns. The high solubility of oxygen and carbon dioxide in perflubron make it a potential way to allow liquid "respiration" in patients with severe lung disease. The solubility of gases in liquids will be addressed in this chapter, with special attention to the effects of pressure and temperature.

Perflubron

Finally, we will look at how the physical properties of solutions are different than those of pure liquids. Why is salt poured on the road in a snowstorm? Why do people cooking at high altitudes add salt to boiling water? We will examine these questions, and we will address the colligative properties of solutions as a method for determining molar mass.

THE COMPOSITION OF SOLUTIONS

A **solution** is generally described as a homogeneous mixture in which some quantity of **solute** (for example, NaCl) is dissolved, or stabilized, by a **solvent** (for example, H_2O). Before we get started, we need to define a few ways of describing solution composition. Some of the equations we will encounter later will use different variables for **solution concentration**.

Molarity is the most common way chemists define solution composition and is defined as:

$$\text{Molarity} = M = \frac{\text{Moles of solute}}{\text{Liters of solution}}$$

Molality is defined as:

$$\text{Molality} = m = \frac{\text{Moles of solute}}{\text{Kg of solvent}}$$

Mole fraction is defined as the ratio of the number of moles of a given component to the number of moles of solution. The mole fraction of A, χ_A in a solution consisting of A, B, C ... is calculated by:

$$\chi_A = \frac{\text{Moles of } A}{\text{Moles of } A + \text{moles of } B + \text{moles of } C \ldots}$$

Finally, **mass percent** is the percent by mass of the solute in the solution. It may be given by the following:

$$\text{Mass percent} = \frac{\text{Grams of solute}}{\text{Grams of solvent}} \times 100$$

Molarity is expressed in units of mol/L that are often referred to as **molar** or abbreviated as M. Molality is expressed in units of mol/kg that are often referred to as **molal** or abbreviated as m. Mole fraction and mass percent are unitless values. Let's try a few review questions to make sure that you understand these ideas and know how to convert one expression for concentration into another.

REVIEW QUESTION

The Dead Sea contains approximately 332 g of salt per kg of seawater. Assume that this salt is all sodium chloride (NaCl). Given that the density of the Dead Sea is approximately 1.20 kg/L, calculate the following:

(a) The mass percent of NaCl

(b) The mole fraction of NaCl

(c) The molarity of the the Dead Sea in M

SOLUTION

This question asks us to be familiar with the various ways of representing the concentration of a solution.

For (a), we use the equation on the previous page for mass percent. The question tells us that the solution contains 0.332 kg of salt per kg of solution. Therefore, the mass percent is given by $\left(\frac{0.332}{1.0}\right) \times 100 = 33.2\%$.

For (b), we need to determine the number of moles of both NaCl and H_2O present in a given sample of seawater. Assume that we have 1 kg of seawater. This means we must have 332 g of NaCl and 668 g of water in the sample (by conservation of mass). Therefore:

$$\text{\# of moles of NaCl} = 332 \text{ g NaCl} \times \left(\frac{1 \text{ mol NaCl}}{58.44 \text{ g NaCl}}\right) = 5.68 \text{ mol NaCl}$$

$$\text{\# of moles of } H_2O = 668 \text{ g } H_2O \times \left(\frac{1 \text{ mol } H_2O}{18.02 \text{ g } H_2O}\right) = 37.1 \text{ mol } H_2O$$

The mole fraction of NaCl is given by $\chi_{Na} = \left(\frac{5.68}{5.68 + 37.1}\right) = 0.133$.

For (c), we need to express the concentration in terms of moles of NaCl per liter. Again, let's assume that we have 1 kg of solution. We have already calculated the number of moles of NaCl present in this sample, but what is its volume? For this we need the density.

So 1 kg solution $\times \left(\frac{1 L \text{ solution}}{1.20 \text{ kg solution}}\right) = 0.833$ L solution.

Therefore, the molarity is given by $\frac{5.68 \text{ mol NaCl}}{0.833 \text{ L solution}} = 6.82$ M.

REVIEW QUESTION

Suppose you have a 5.0 M aqueous solution of HCl. You want to make 500 mL of a 0.25 M solution. Describe how you would prepare this solution.

SOLUTION

This is a straightforward dilution problem that could very well appear on your AP exam. Just remember that the whole point of molarity is that it allows you to find the number of moles of solute present just by multiplying by the volume. The number of moles of HCl present in the final sample will be equal to the number of moles of HCl we add with the concentrated or "stock" solution. So, the initial volume (V_1) times the initial concentration (C_1) will equal the final volume (V_2) times the final concentration (C_2).

$$V_1 C_1 = V_2 C_2$$

$$\left(\frac{5.0 \text{ mol}}{L}\right) V_1 = \left(\frac{0.25 \text{ mol}}{L}\right)(0.5 \text{ L})$$

$$V_1 = \frac{(0.25 \text{ mol/L})(0.5 \text{ L})}{(5.0 \text{ mol/L})} = 0.025 \text{ L}$$

That's half the problem, but the prompt specifies that you must describe the process: measure 0.025 L of 5.0 M HCl into an appropriate vessel (for a precise dilution, you would use a volumetric flask) and dilute to the 500-mL mark with water. It's critical to describe the process this way to avoid losing those points on test day; many students needlessly lose points by failing to provide a description at all or by saying something like "dilute 25 mL of 5.0 M HCl with 475 mL water." This answer is incorrect because volumes are not additive; you must say "fill to the mark" to get full points.

MOLECULAR STRUCTURE AND SOLUBILITY

What are the factors affecting solubility of a solute in some solvent? The general rule of thumb is that "like dissolves like." This statement usually applies to the polarities of the solute and solvent. What types of things would we expect to dissolve in water? We know that water is a relatively polar solvent because its molecules participate in hydrogen bonding. We know that **ionic solids** often dissolve well in water, due to the ability of the polar solvent to interact with positively and negatively charged ions. We would also expect molecules with many polar bonds, such as ethanol and glucose, shown in the following figure, to dissolve well in water. Molecules that are less polar, which are sometimes described as **fat-soluble** or **hydrophobic**, are less soluble in water.

Glucose   Ethanol

THE DEPENDENCE OF GAS SOLUBILITY ON PRESSURE

The solubility of a gas depends on both pressure and temperature. The important thing to remember is that the amount of a gas dissolved in a solution is directly proportional to the pressure of the gas on the surface of the solution. The equation describing this is **Henry's law:**

$$P = kC$$

where P is the gas pressure (atm), C is concentration (mol/L), and k is Henry's law constant (in units of L · atm/mol). Henry's law constant is dependent on the solute, solvent, and temperature.

THE DEPENDENCE OF GAS SOLUBILITY ON TEMPERATURE

The solubility of a gas in a liquid will decrease as the temperature is increased. Try opening a warm bottle of seltzer water to prove this to yourself. In contrast, we should mention that the solubility of solids in liquids may either increase or decrease as the temperature is raised. This is difficult to predict and is best determined by experiment, so you probably will not be asked about this on the AP exam.

COLLIGATIVE PROPERTIES

If you can remember the terms **boiling point elevation, freezing point depression,** and **vapor pressure depression,** you are already halfway there in terms of understanding **colligative properties**. If you add a solute to a solvent (for example, NaCl to H_2O), the boiling point will be higher, the freezing point will be lower, and the vapor pressure will be lower than that of the pure solvent. Further, the greater the concentration of solute added, the greater the effect on the boiling point, freezing point, and vapor pressure.

For **boiling point elevation**, you can calculate what the increase in boiling point will be using the following equation:

$$\Delta T = i K_b m_{solute}$$

where ΔT is the change in boiling point in °C, K_b is a constant characteristic of the solvent, m_{solute} represents the molality of the solute, and i represents the total number of particles that result from dissolving the solute in the solvent. For example, if we dissolve 1 mole of Na_2SO_4 in 1 kg of water, we expect that 1 mole of Na_2SO_4 will dissociate in solution to form 2 moles of Na^+ ions and 1 mole of SO_4^{2-} ions. Therefore, i for this solution is 3. For a molecule that does not dissociate in solution, such as glucose, i would just be 1.

For **freezing point depression**, you can calculate how much the freezing point will decrease using a very similar equation:

$$\Delta T = iK_f m_{solute}$$

where ΔT is the change in freezing point in °C, K_f is a constant characteristic of the solvent, m_{solute} represents the molality of the solute, and i is the total number of particles that result from dissolving the solute in the solvent.

For **vapor pressure depression**, remember that the vapor pressure of a solvent at a given temperature is given by **Raoult's law**:

$$P_{solution} = \chi_{solvent} P^0_{solvent}$$

where $P_{solution}$ is the observed vapor pressure of the solution, $\chi_{solvent}$ is the mole fraction of the solvent, and $P^0_{solvent}$ is the vapor pressure of the pure solvent.

OSMOTIC PRESSURE

Osmotic pressure, another colligative property, can be defined as the pressure required to stop osmosis (the movement of water) across a **semipermeable membrane**. A semipermeable membrane allows **solvent** but not solute molecules to pass through it. The pure solvent (let's say water) wants to flow from left to right across the semipermeable membrane down its concentration gradient. The flow will stop when the applied pressure "pushes back" enough on the column of solution. The osmotic pressure is given by:

$$\pi = MRT$$

where π is the osmotic pressure in atm, M is the molarity of the solution, R is the gas law constant (0.08206 L · atm/mol · K), and T is the temperature in Kelvin.

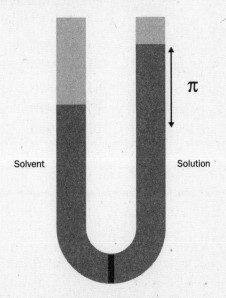

Solvent π Solution

Membrane permeable only to solute

REVIEW QUESTIONS

1. 100 mL of a 1 M HCl and 100 mL of a 5 M NaCl solution are mixed. What is the final molarity of the chloride (Cl^- ion)?

   (A) 0.5 M

   (B) 2 M

   (C) 3 M

   (D) 4 M

   (E) 5 M

2. The structures of vitamins C and E are shown. Which of the following is correct?

Vitamin C

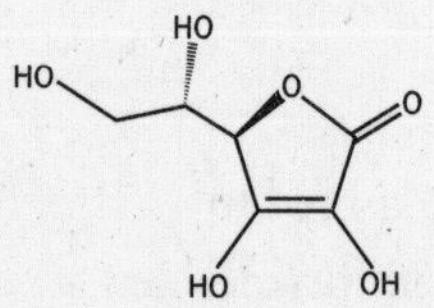

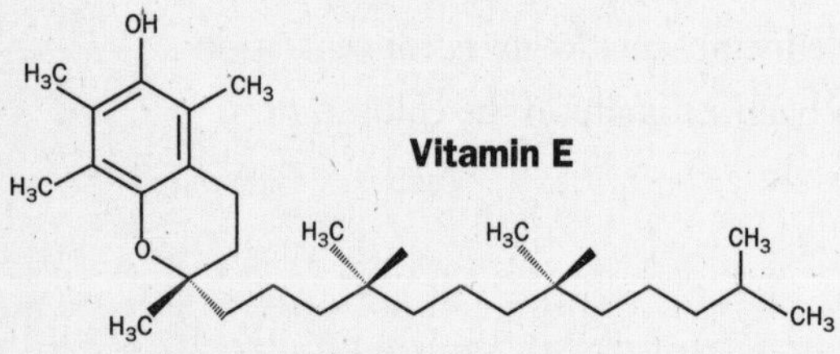

Vitamin E

   (A) Vitamin E contains more polar bonds than vitamin C.

   (B) The melting point of vitamin E is likely higher than that of vitamin C.

   (C) Vitamin C should have excellent solubility in hexane.

   (D) Vitamin C should have a higher solubility in water than vitamin E.

   (E) Vitamin C can be described as a fat-soluble vitamin.

3. A 75-gram sample of an unknown solid is dissolved in 1 kg of water. The resulting solution is found to have a freezing point of $-0.93°C$. If the freezing point depression constant for water (K_f) is $1.86°C/m$ and the unknown solid does not dissociate in water, which of the following is most likely the solid?

   (A) Citric acid (MW = 60 g/mol)

   (B) Ribose (MW = 150 g/mol)

   (C) Ascorbic acid (MW = 176 g/mol)

   (D) Dextrose (MW = 180 g/mol)

   (E) Maltose (MW = 360 g/mol)

4. Which of the following aqueous solutions will have the highest boiling point?

   (A) 0.5 m NaBr

   (B) 0.75 m $C_2H_6O_2$ (ethylene glycol)

   (C) 1.25 m $C_6H_{12}O_6$ (glucose)

   (D) 1.0 m $KMnO_4$ (potassium permanganate)

   (E) 0.75 m LiCl

5. 50.0 grams of all trans retinol (vitamin A, molar mass = 286.5 g/mol) are dissolved in 100.0 mL of ethanol (molar mass = 46.07). The density of ethanol is 0.787 g/mL. If the vapor pressure of pure ethanol at 20°C is 44.4 torr, what is the observed vapor pressure of the retinol solution at this temperature?

   (A) 35.4 torr

   (B) 36.9 torr

   (C) 38.2 torr

   (D) 40.4 torr

   (E) 46.3 torr

6. What mass of $FeSO_4 \cdot 6H_2O$ ($260 \text{ g} \cdot \text{mol}^{-1}$) is required to produce 500 mL of a 0.10 M iron (II) sulfate solution?

   (A) 9 g

   (B) 13 g

   (C) 36 g

   (D) 72 g

   (E) 144 g

7. Sodium carbonate (Na_2CO_3) is least soluble in which of the following liquids?

   (A) CH_3OH

   (B) CF_3COOH

   (C) H_2O

   (D) $CH_3(CH_2)_4CH_3$

   (E) $CHCl_3$

8. An aqueous solution of KCl is heated from 15°C to 85°C in a closed container. Which of the following properties of the solution remains the same?

   I. Molality

   II. Molarity

   III. Density

   (A) I only

   (B) III only

   (C) I and II only

   (D) II and III only

   (E) I, II, and III

9. Under which of the following sets of conditions would the solubility of CO_2 (g) in water be lowest? In each case, the given pressure is the pressure of CO_2 (g) above the solution.

   (A) 5.0 atm and 75°C

   (B) 1.0 atm and 75°C

   (C) 5.0 atm and 25°C

   (D) 1.0 atm and 25°C

   (E) 3.0 atm and 25°C

FREE-RESPONSE QUESTION

1. A scuba diver is swimming at 64 feet below the surface, where the pressure is approximately 3.0 atm. The Henry's law constant for N_2 at 25°C is 1,600 L · atm/mol, and the mole fraction of N_2 in the atmosphere (and in the diver's compressed air) is 0.79.

   (a) Calculate the concentration of dissolved N_2 in the diver's bloodstream at this depth.

   (b) The diver is forced to ascend suddenly to the surface (where the pressure is 1 atm). Calculate the concentration of dissolved N_2 in the diver's bloodstream once it reaches equilibrium.

   (c) If the total blood volume of the diver is about 5 L, calculate the number of moles of N_2 dissolved in the diver's blood at 3 atm and at the surface.

   (d) What is the volume of N_2 gas theoretically released in the diver's bloodstream as he ascends?

ANSWERS AND EXPLANATIONS

1. C

Both of the solutions we are mixing contain Cl^-, so we need to figure out the total number of moles of Cl^- in the final solution.

For the HCl solution: $0.1 \text{ L} \times \left(\dfrac{1 \text{ mol } Cl^-}{\text{L sol}} \right) =$ $0.1 \text{ mol } Cl^-$

For the NaCl solution: $0.1 \text{ L} \times \left(\dfrac{5 \text{ mol } Cl^-}{\text{L sol}} \right) =$ $0.5 \text{ mol } Cl^-$

The total number of moles of Cl^- is 0.6, so the molarity of the final solution is $\dfrac{0.6 \text{ mol}}{0.2 \text{ L}} = 3 \text{ M}$.

2. D

Since vitamin C has a significant number of polar O-H bonds (in the alcohol groups), we expect it to have a much higher solubility in water than vitamin E. Choice (A) is incorrect since there are significantly more polar bonds in vitamin C than in vitamin E. Choice (B) is incorrect since vitamin E is a much less polar molecule than vitamin C and therefore will have weaker intermolecular interactions and a lower melting point. Choice (C) is incorrect since hexane is a nonpolar solvent in which ionic and polar compounds are poorly soluble. Choice (E) is incorrect since the term *fat-soluble* refers to less polar, more hydrophobic chemical compounds.

3. B

According to the equation for freezing point depression:

$$\Delta T = i K_f \, m_{\text{solute}}$$

so $\quad m_{\text{solute}} = \dfrac{\Delta T}{i K_f} = \dfrac{(0.93°C)}{(1)(1.86°C/m)} = 0.50 \text{ m}$

$$= \dfrac{0.50 \text{ mol solute}}{\text{kg water}}$$

Therefore, the molality of the solution must be 0.50. Since the solute is dissolved in 1 kg of water, this means that there must be 0.50 moles of the mystery solid dissolved in the solution.

Finally, we can calculate that the molar mass $=$ $\dfrac{75 \text{ g}}{0.50 \text{ mol}} = \dfrac{150 \text{ g}}{\text{mol}}$. The mystery solid is most likely ribose.

4. D

The solution that will have the highest boiling point is $KMnO_4$ (potassium permanganate). We need to account for the total molality of particles present in solution. $KMnO_4$ (potassium permanganate) will dissociate in solution to form K^+ and MnO_4^-. Therefore, the effective molality ($i \times m$) of this solution is 2.0. $C_6H_6O_6$ and $C_2H_6O_2$ are molecular compounds, so their effective molalities are the same as the given molality.

5. D

In order to solve this question, we use Raoult's law:

$$P_{\text{solution}} = \chi_{\text{solvent}} P^0{}_{\text{solvent}}$$

We are given the vapor pressure of pure ethanol, but we need to calculate the mole fraction of ethanol in this solution.

$100.0 \text{ mL ethanol} \times \left(\dfrac{0.787 \text{ g ethanol}}{1 \text{ mL ethanol}} \right) \times$ $\left(\dfrac{1 \text{ mol ethanol}}{46.07 \text{ g ethanol}} \right) = 1.71 \text{ mol ethanol}$

$50.00 \text{ g retinol} \times \left(\dfrac{1 \text{ mol retinol}}{286.5 \text{ g retinol}} \right) = 0.17 \text{ mol}$ retinol

Therefore, the mole fraction of ethanol $\chi = \dfrac{(1.71)}{(1.71 + 0.17)} = 0.910$.

The observed vapor pressure $= (0.910) \times (44.4 \text{ torr}) =$ 40.4 torr.

6. B

We must first calculate the number of moles of iron (II) sulfate present in our final solution. In order to do this, we use the following expression:

$$\text{\# of moles} = (\text{concentration in M}) \times (\text{volume in L}) = \left(0.10\frac{mol}{L}\right)(0.500\ L) = 0.050\ mol$$

Each mole of hydrate dissolved produces 1 mole of $FeSO_4$, so we must dissolve 0.05 moles of the hydrate to produce the desired solution. The mass of hydrate needed is given by:

$$\text{\# of grams of hydrate} = (\text{\# of moles of hydrate}) \times (\text{molar mass hydrate}) = (0.05\ mol)\left(260\frac{g}{mol}\right) = 13.0\ g$$

7. D

Sodium carbonate (Na_2CO_3) is least soluble in hexane ($CH_3(CH_2)_4CH_3$). The first thing is to recognize that Na_2CO_3 is an ionic compound. As you know, "like dissolves like," so the more polar, the better. Sodium carbonate should be highly soluble in water and somewhat soluble in methanol (CH_3OH), trifluoroacetic acid (CF_3COOH), and chloroform ($CHCl_3$). However, since hexane ($CH_3(CH_2)_4CH_3$) is a nonpolar solvent, we would expect that sodium carbonate would not dissolve well in it.

8. A

When an aqueous solution of KCl is heated from 15°C to 85°C, the molality of the solution remains the same. As the solution is heated, increased molecular velocities result in expansion. Since the mass of the solution does not change, we expect that the density of the solution (mass/volume) will decrease. The molarity of the solution (moles solute/L solution) will also decrease, since the number of moles of solute present in the solution will remain the same while the volume of the solution will increase. The molality of the solution (moles solute/kg solvent) is constant since the mass of solvent remains the same.

9. B

According to Henry's law, the solubility of a gas in a liquid will be proportional to the partial pressure of that gas above the solution. In addition, the solubility of a gas in a liquid is inversely proportional to the temperature. Therefore, the solubility of CO_2 in water should be lowest at low partial pressures of CO_2 and high temperatures.

FREE-RESPONSE QUESTION

1. This is the classic "baby's got the bends" question that uses Henry's law to predict what sort of shape this unfortunate diver will be in after rising to the surface.

   (a) We need to use Henry's law to calculate the concentration of N_2 dissolved in the diver's bloodstream at 3 atm. So:

   $$P_{N_2} = k_{N_2} C_{N_2}$$

   $$C_{N_2} = \frac{P_{N_2}}{k_{N_2}}$$

   To calculate the partial pressure of nitrogen at 3 atm, we just need to multiply by the mole fraction, so $P_{nitrogen} = (0.79)(3.0 \text{ atm}) = 2.37 \text{ atm}$.

   Therefore, the concentration of dissolved nitrogen is given by $\frac{2.37 \text{ atm}}{1,600 \text{ L} \cdot \text{atm/mol}} = 1.5 \times 10^{-3} \text{ mol/L}$.

   (b) The same equation applies—the concentration of nitrogen is $\frac{0.79 \text{ atm}}{1,600 \text{ L} \cdot \text{atm/mol}} = 4.9 \times 10^{-4} \text{ mol/L}$.

   (c) We need to calculate how many total moles of N_2 are dissolved in 5 L of blood at both pressures.

   At 3 atm: # of moles of $N_2 = (1.5 \times 10^{-3} \text{ mol/L}) \times (5 \text{ L}) = 7.5 \times 10^{-3} \text{ mol } N_2$

   At 1 atm: # of moles of $N_2 = (4.9 \times 10^{-4} \text{ mol/L}) \times (5 \text{ L}) = 2.5 \times 10^{-3} \text{ mol } N_2$

   (d) We saw that there were 7.5×10^{-3} moles of N_2 dissolved when the diver was at 3 atm but only 2.5×10^{-3} moles of N_2 dissolved at the surface. Where did the remaining 5.0×10^{-3} moles of N_2 go? Since this extra N_2 was no longer soluble in the bloodstream at the lower pressure, it became a gas. How much gas? We know that a mole of gas at SATP (Standard Ambient Temperature, 25°C, and Pressure, 1 atm) occupies 22.4 L.

   Therefore, the volume of gas released $= (5.0 \times 10^{-3} \text{ mol } N_2) \times (22.4 \text{ L } N_2/\text{mol } N_2) = 0.11 \text{ L} = 110 \text{ mL}$. This is actually quite a bit and can put the diver at risk of life-threatening conditions such as pulmonary embolism.

CHAPTER 10: THERMODYNAMICS

IF YOU LEARN ONLY SIX THINGS IN THIS CHAPTER . . .

1. Gibbs free energy can be used to predict the spontaneity of a chemical reaction and is given by the following:

   $$\Delta G^\circ = \Delta H^\circ - T\Delta S^\circ$$

   If ΔG° is negative, the reaction will be spontaneous under the conditions specified. If ΔG° is positive, then the reaction will not occur.

2. Entropy may be defined as the degree of disorder in a system or may be interpreted in terms of degrees of freedom. Since a gas has many more degrees of freedom than a liquid or solid:

   $$S_{solid} < S_{liquid} < S_{gas}$$

3. The equilibrium constant K for a reaction is defined by ΔG° as demonstrated by the following equation:

   $$\Delta G^\circ = -RT \ln K$$

4. Remember that ΔG° has no relationship to the rate of a chemical reaction. Furthermore, the presence of a catalyst will not alter ΔG° or K, but rather the rate at which equilibrium is reached.

5. Since entropy, enthalpy, and Gibbs free energy are all state functions, they are subject to Hess's law:

$$\Delta H^\circ = \Sigma n_p \Delta H^\circ_F(\text{products}) - \Sigma n_r \Delta H^\circ_F(\text{reactants})$$

$$\Delta S^\circ = \Sigma n_p \Delta S^\circ_F(\text{products}) - \Sigma n_r \Delta S^\circ_F(\text{reactants})$$

$$\Delta G^\circ = \Sigma n_p \Delta G^\circ_F(\text{products}) - \Sigma n_r \Delta G^\circ_F(\text{reactants})$$

6. The Born-Haber cycle for predicting lattice energy represents an important application of Hess's law.

Thermodynamics is a favorite topic on the AP Chemistry exam. For many of the questions, you will need to apply your understanding of enthalpy, entropy, and Gibbs free energy. Questions about thermodynamics can apply to almost any topic in chemistry. On test day, your understanding of a few basic concepts will be very important, but we will also cover the calculations that you are most likely to be asked to perform.

Energy may be kinetic or potential in nature. The energy stored in chemical bonds represents a sort of potential energy, and reactions that alter chemical structure are almost always associated with a change in energy. The two most important components we use in quantifying these energetic changes are **enthalpy** and **entropy**. Enthalpy describes the heat released or absorbed during a chemical reaction while entropy refers to the degree of disorder in a system.

Together, enthalpy and entropy define a quantity called Gibbs free energy, which is useful in predicting whether or not a reaction will be spontaneous under specified conditions and in defining the equilibrium constant K for a system at equilibrium.

Enthalpy, entropy, and Gibbs free energy are all examples of state functions. What is the meaning of "state function" anyway? Let's consider a skiing analogy where the top of the mountain corresponds to a high energy state and the lodge below corresponds to a low energy state, both in terms of potential energy. Skier A takes two black diamond runs to the bottom whereas Skier B manages to find six bunny slopes. Bragging rights aside, the two skiers have not done anything energetically different. The energetic state reached is independent of the path taken to get there.

At the end of this section, we will address a few useful applications of treating energy as a state function that takes advantage of the additivity of energies. We will use Hess's law in a variety of ways to predict and calculate unknown thermodynamic parameters.

BASIC LAWS AND GIBBS FREE ENERGY

The **first law of thermodynamics** states that energy is neither created nor destroyed, while the **second law of thermodynamics** states that the entropy in the universe is always increasing. The **third law of thermodynamics** states that the entropy of a perfect crystal at 0 K is zero. By far the most important thing to know for the AP exam is the equation for **Gibbs free energy**:

$$\Delta G = \Delta H - T\Delta S$$

where G is Gibbs free energy (in joules), H is the enthalpy, T is the temperature in K, and S is entropy. Often, these thermodynamic quantities are shown as $\Delta G°$, $\Delta H°$, and $\Delta S°$ to indicate the **standard state** at 1 atm and $25°C$. The key points about this equation are the following:

- If ΔG is negative, then the reaction is **spontaneous** under the conditions specified. If ΔG is positive, then the reaction will not occur under those same conditions.

- Looking at the equation above, we can see that ΔG is more likely to be negative for a reaction for which the ΔH is negative (an **exothermic** reaction).

- We can also predict that ΔG is more likely to be negative for a reaction in which the entropy is positive.

- Changing the temperature may have an impact on whether or not a reaction will have a negative Gibbs free energy.

Remember: if ΔG is negative for a reaction, this doesn't mean that it will take place under the conditions specified. There is also an activation energy (E_a) that must be overcome in order for the reaction to occur. So ΔG does not tell you anything about the rate of the reaction.

THE MEANING OF ENTROPY

Entropy can be viewed as a measurement of disorder or randomness in a system. In the context of molecules, it is often interpreted in terms of **degrees of freedom**. Molecules in a crystalline solid have less entropy than those in a gas. In the gaseous state, molecules are relatively far apart and are free to move in all directions. In fact:

$$S_{\text{solid}} < S_{\text{liquid}} < S_{\text{gas}}$$

You will be asked on the AP exam to predict the entropy of a reaction based on this concept.

For example:

REVIEW QUESTION

Which of the following reactions is predicted to have a positive value for $\Delta S°$?

(A) $N_2 \ (g) + 3H_2 \ (g) \rightarrow 2NH_3 \ (g)$

(B) $CO \ (g) + 2H_2 \ (g) \rightarrow CH_3OH \ (l)$

(C) $2SO_2 \ (g) + O_2 \ (g) \rightarrow 2SO_3 \ (g)$

(D) $CaCO_3 \ (s) \rightarrow CaO \ (s) + CO_2 \ (g)$

(E) $NH_3 \ (g) \rightarrow NH_3 \ (l)$

SOLUTION

The correct answer is choice (D). In choice (D), there are no gaseous molecules on the left-hand side of the equation and one molecule of gas on the right-hand side. Therefore, the degree of disorder in the system is increasing, and the entropy is predicted to be positive. In every other case, there is positional entropy lost since there are fewer molecules of gas on the right-hand side of the equation.

EQUILIBRIUM AND GIBBS FREE ENERGY

Another often-tested topic is the relationship of free energy to chemical equilibrium. Chemical equilibrium occurs when the rate of the forward reaction equals the rate of the reverse reaction, maintaining the concentrations of reactants and products. We will discuss other aspects of chemical equilibrium later, but for now:

$$\Delta G° = -RT \ln K$$

where $G°$ is Gibbs free energy (in joules), R is the gas constant (8.3145 J/ K · mol), T is the temperature (in Kelvin), and K is the equilibrium constant.

This relationship implies that the spontaneity of a reaction will be related to K. In other words, spontaneity will be related to whether or not the reactants or products are favored at equilibrium. Remember the following rules:

If $K > 1$, then $\Delta G° < 0$ and the reaction will be spontaneous.

If $K < 1$, then $\Delta G° > 0$ and the reaction will not be spontaneous.

Note: These rules are only true for $\Delta G°$ for reactions at chemical equilibrium.

HESS'S LAW

How do we go about calculating $\Delta G°$, $\Delta S°$, and $\Delta H°$ from known thermodynamic data? We can use **Hess's law**, which states that:

$$\Delta H° = \Sigma n_p \Delta H°_F(\text{products}) - \Sigma n_r \Delta H°_F(\text{reactants})$$

$$\Delta S° = \Sigma n_p \Delta S°_F(\text{products}) - \Sigma n_r \Delta S°_F(\text{reactants})$$

$$\Delta G° = \Sigma n_p \Delta G°_F(\text{products}) - \Sigma n_r \Delta G°_F(\text{reactants})$$

where $\Delta H°_F$, $\Delta S°_F$, and $\Delta G°_F$ are the standard enthalpies, entropies, and Gibbs free energies of formation, respectively (usually expressed in kJ/mol), and where n_p and n_r are the stoichiometric coefficients of the products and reactions, respectively, from the balanced reaction.

This method for calculating $\Delta G°$, $\Delta S°$, or $\Delta H°$ takes advantage of the fact that these are all **state functions**. In addition to applying Hess's law, you will have other opportunities to add and subtract state functions on the AP Chemistry exam. Often, we can sum two separate equations for which thermodynamic data is known to determine thermodynamic parameters for a third reaction.

REVIEW QUESTION

From the thermodynamic data below, calculate $\Delta H°$ for the following reaction:

$2C\ (s) + O_2\ (g) \rightarrow 2CO\ (g)$

given: $C\ (s) + O_2\ (g) \rightarrow CO_2\ (g)$ $\Delta H° = -393.7$ kJ

$CO\ (g) + \frac{1}{2}\ O_2\ (g) \rightarrow CO_2\ (g)$ $\Delta H° = -283.3$ kJ

SOLUTION

Notice that in order to obtain the equation we want, we must reverse the second expression and multiply both by 2. Then, when we add the two equations, CO_2 will cancel:

$2C\ (s) + 2O_2\ (g) \rightarrow 2CO_2\ (g)$ $\Delta H° = -787.4$ kJ

$2CO_2\ (g) \rightarrow O_2\ (g) + 2CO\ (g)$ $\Delta H° = 566.6$ kJ

_____

$2C\ (s) + O_2\ (g) \rightarrow 2CO\ (g)$ $\Delta H° = -220.8$ kJ

We can do the exact same thing for $\Delta G°$ and $\Delta S°$.

What does it mean to be a state function anyway? It refers to a property of the system that depends only on its present state. In other words, it doesn't matter how you get there. Energetically speaking, there can be many paths you can take to arrive at the same destination or state.

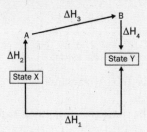

We can tell that ΔH_1 must be equal to $\Delta H_2 + \Delta H_3 + \Delta H_4$ since we arrived at the same place, or state, for both paths we took. This is what is meant by state function.

THE BORN-HABER CYCLE

The **Born-Haber cycle** is an important application of this thinking. It is a method of calculating the **lattice energy** for a crystal from other parameters that can be measured directly. The lattice energy is defined as the change in energy that takes place when gaseous ions are packed together to form an ionic solid. This is represented by the following equation:

$$M^+ (g) + X^- (g) \rightarrow MX (s)$$

As written, this is specific to monovalent ions. While most texts define lattice energy this way (always negative), some define it for the opposite reaction (always positive, in terms of the amount of energy required to break the lattice rather than the amount released in forming the lattice).

Since a crystal represents a highly stable, ordered state, we expect that a tremendous amount of energy will be released when an ionic solid forms from its ions. This implies that the reaction will be **exothermic** and have a negative sign.

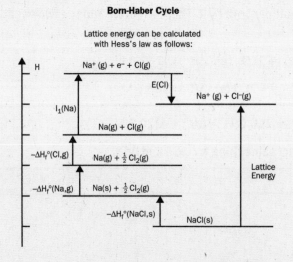

Born-Haber Cycle

We see that if we start with LiF (s) that we can arrive at Li^+ (g) + F^- (g) by two different paths. We could go directly there taking path A, which corresponds to a change in energy equal in magnitude but opposite in sign to the lattice energy. Or, alternatively, we could go there taking path B, which involves the following steps:

(A) LiF (s) $\rightarrow$ Li (s) + $\frac{1}{2}$ F_2 (g) $\hspace{2cm}$ $\Delta H = -\Delta H_F$ (LiF)

(B) Li (s) + $\frac{1}{2}$ F_2 (g) $\rightarrow$ Li (g) + $\frac{1}{2}$ F_2 (g) $\hspace{1cm}$ $\Delta H = \Delta H_{SUB}$ (Li)

(C) Li (g) + $\frac{1}{2}$ F_2 (g) $\rightarrow$ Li^+ (g) + $\frac{1}{2}$$F_2$ (g) $\hspace{0.6cm}$ $\Delta H = \Delta H_{ION}$ (Li)

(D) Li^+ (g) + $\frac{1}{2}$ F_2 (g) $\rightarrow$ Li^+ (g) + F (g) $\hspace{0.8cm}$ $\Delta H = \frac{1}{2}\Delta H_{DISS}$ (F_2)

(E) Li^+ (g) + F (g) $\rightarrow$ Li^+ (g) + F^- (g) $\hspace{1.2cm}$ $\Delta H = \Delta H_{EA}$ (F)

$\hspace{1cm}$ LiF (s) $\rightarrow$ Li^+ (g) + F^- (g) $\hspace{0.6cm}$ $\Delta H = -\Delta H_F$ (LiF) + ΔH_{SUB} (Li) + ΔH_{ION} (Li) + $\frac{1}{2}\Delta H_{DISS}$ (F_2) + ΔH_{EA}(F)

Notice that step A is just the opposite of an equation that represents the standard enthalpy of formation of LiF. Step B is just the energy required to turn Li (s) into a gas—in other words, the enthalpy of sublimation. Step C represents the ionization energy for Li—in other words, the energy required to remove an electron from the atom. Step D represents the breaking apart of F_2 to form its constituent atoms, or the bond dissociation energy for F_2. In this case, the equation we need is for the dissociation of $\frac{1}{2}$ mole of F_2, so the energy required is half of the bond dissociation energy. Step E corresponds to the energy released when an electron is added to F in its gaseous state, or the electron affinity of fluorine. We can see that by adding steps A–E together, we get an expression corresponding to the lattice energy. Remember how the lattice energy is defined as the energy released when an ionic solid forms from its ions. Therefore, the energy we just calculated for

$\hspace{1cm}$ LiF (s) $\rightarrow$ Li^+ (g) + F^- (g)

will be equal in magnitude, but opposite in sign, to the lattice energy. In other words:

$$-\Delta H_{LATT} (LiF) = -\Delta H_F (LiF) + \Delta H_{SUB} (Li) + \Delta H_{ION} (Li) + \frac{1}{2}\Delta H_{DISS} (F_2) + \Delta H_{EA} (F)$$

On your AP exam, it's unlikely that you will be asked a long and complicated question about the Born-Haber cycle. Most importantly, just remember that since energy is a state function, you can use different routes for getting to the same place. The change in energy will be the same no matter what route you choose, and the result is the application of Hess's law, which allows us to predict enthalpies based on known thermodynamic data. Enthalpy, entropy, and free energy are all examples of state functions.

REVIEW QUESTIONS

1. Which of the following regarding reaction spontaneity is true?

   (A) A reaction with a positive $\Delta S°$ will always be spontaneous.

   (B) A reaction with a negative $\Delta H°$ will always be spontaneous.

   (C) A reaction with a positive $\Delta S°$ and a negative $\Delta H°$ will always be spontaneous.

   (D) A reaction with a negative $\Delta S°$ and a negative $\Delta H°$ will never be spontaneous.

   (E) A reaction with a positive $\Delta S°$ and a positive $\Delta H°$ will never be spontaneous.

2. When a solid sample of $NaNO_3$ is added to a cup of water, the temperature of the resulting solution decreases. Which of the following MUST be true?

   I. The entropy of the solution must be positive.

   II. The entropy of the solution must be negative.

   III. The enthalpy of the solution must be negative.

   (A) I only

   (B) II only

   (C) I and II only

   (D) II and III only

   (E) I and III only

3. Consider the following reaction at equilibrium at 25°C:

   $$N_2 (g) + 3H_2 (g) \rightleftharpoons 2 NH_3 (g)$$
   $$\Delta G° = -33.3 \text{ kJ}$$

   Which of the following is NOT true?

   (A) The quantity $\dfrac{[NH_3]^2}{[N_2][H_2]^3}$ is >1 at equilibrium.

   (B) $\Delta H°$ must be negative.

   (C) The reaction is spontaneous under the conditions specified.

   (D) $\Delta G°$ is independent of temperature.

   (E) For the equilibrium $2 NH_3 (g) \rightleftharpoons N_2 (g) + 3H_2 (g)$, we expect that $\Delta G° = 33.3$ kJ.

4. At what temperature is the following process spontaneous at 1 atm? Assume that $\Delta H°$ and $\Delta S°$ do not change with temperature.

   $$CHCl_3 (l) \rightarrow CHCl_3 (g)$$

   $\Delta H° = 31.4$ kJ/mol and $\Delta S° = 94.0$ J/mol $\cdot$ K

   (A) 61°C

   (B) 58°C

   (C) 43°C

   (D) 25°C

   (E) 14°C

5. Which of the following has a standard enthalpy of formation equal to zero?

   (A) $S_8 (g)$

   (B) $CO_2 (g)$

   (C) $HCl (g)$

   (D) $Cl_2 (g)$

   (E) $O_3 (g)$

Questions 6–8 refer to the following thermodynamic quantities:

   (A) ΔG

   (B) ΔH

   (C) ΔS

   (D) E_a

   (E) K

6. Is decreased by addition of a catalyst

7. Is always negative for a spontaneous process

8. Has units of J/mol · K

9. For an ideal gas, which of the following quantities will be pressure-dependent?

   I. Enthalpy

   II. Entropy

   III. Gibbs free energy

   (A) I only

   (B) II only

   (C) I and II only

   (D) II and III only

   (E) I, II, and III

10. $C_3H_8\ (g) + 5\ O_2\ (g) \rightarrow 3\ CO_2\ (g) + 4\ H_2O\ (l)$

For the combustion of propane represented above, $\Delta H° = -2,220$ kJ. What would the value of $\Delta H°$ be if the reaction produced H_2O (g) instead of liquid water? (For the phase change $H_2O\ (l) \rightarrow H_2O\ (g)$, $\Delta H° = +44$ kJ/mol.)

   (A) $-2,044$ kJ

   (B) $-2,176$ kJ

   (C) $-2,264$ kJ

   (D) $-2,396$ kJ

   (E) $-2,440$ kJ

FREE-RESPONSE QUESTION

1. Given the following equilibrium:

    $$H_2 \text{ (g)} + Br_2 \text{ (g)} \rightleftarrows 2HBr \text{ (g)}$$

    (a) Can you predict the sign of $\Delta S°$? Explain.

    (b) Given the following set of enthalpies/entropies of formation:

    $$\Delta H°_F \text{ (HBr)} = -36.3 \text{ kJ/mol and } \Delta S°_F \text{ (HBr)} = 198.7 \text{ J/mol} \cdot \text{K}$$

    $$\Delta S°_F \text{ (Br}_2) = 245.5 \text{ J/mol} \cdot \text{K}$$

    $$\Delta S°_F \text{ (H}_2) = 130.7 \text{ J/mol} \cdot \text{K}$$

    calculate $\Delta G°$ for the reaction at 25°C.

    (c) What is the equilibrium constant K for the reaction at 25°C?

    (d) Is there any temperature at which we would expect the reaction to be non-spontaneous?

ANSWERS AND EXPLANATIONS

1. C

Looking at the Gibbs free energy equation, we predict that if $\Delta S°$ is positive and $\Delta H°$ is negative, then $\Delta G°$ will be negative at all temperatures. Choice (A) and choice (B) are not true since you must know the value of both $\Delta S°$ and $\Delta H°$ in order to predict spontaneity. Choice (D) and choice (E) are not true since it's always possible for a favorable entropy to counteract an unfavorable enthalpy and vice versa.

2. A

If the solution cools, this implies that heat is absorbed, and therefore the enthalpy of the solution ($\Delta H°$) is positive. If this is the case, then the entropy ($\Delta S°$) must be positive in order for the reaction to occur.

3. D

We expect $\Delta G°$ to be dependent on temperature according to the expression $\Delta G° = \Delta H° - T\Delta S°$. Choice (A) is true since this is an expression for K, and we know that $K>1$ for a reaction with a negative $\Delta G°$. Choice (B) is true since the reaction has a negative entropy since there are more molecules of gas on the left side of the equation than the right. Therefore, the reaction must have a negative $\Delta H°$ for $\Delta G°$ to be negative. Choice (C) is true since the reaction has a negative Gibbs free energy and therefore is spontaneous. Choice (E) is true since this equation represents the reverse reaction, which we would expect to have a $\Delta G°$ that is equal in magnitude but opposite in sign.

4. A

Since $\Delta H°$ is positive, we know that the reaction will not be spontaneous at all temperatures. Instead, we need to increase T until the entropy part of Gibbs free energy dominates, giving a net negative value for $\Delta G°$. Looking at the equation for free energy:

$$\Delta G° = \Delta H° - T\Delta S°$$

If we can find the temperature for which $\Delta G° = 0$, then we know that at all temperatures above this the reaction should be spontaneous.

$$0 = \Delta H° - T\Delta S°$$

solving for T: $\quad T = \dfrac{\Delta H°}{\Delta S°} = \dfrac{31,400 \text{ J/mol}}{94.0 \text{ J/mol·K}}$

$$= 334 \text{ K}$$
$$= 61.0°C$$

Therefore, we predict that at all temperatures above 61°C the reaction will be spontaneous. Thus, 61°C is the normal boiling point for $CHCl_3$ (chloroform).

5. D

The standard enthalpy of formation ($\Delta H_F°$) is defined as the change in enthalpy associated with the formation of a mole of a compound from its elements, with all elements in their standard state. Therefore, $\Delta H_F°$ is zero when the substance in question is already in its standard state. Since Cl exists as Cl_2 (g) at 1 atm and 25°C, Cl_2 (g) is in its standard state and has an enthalpy of formation equal to zero.

6. D

Addition of a catalyst will affect only the activation energy, which will increase the rate of the reaction but not change any of the other thermodynamic parameters listed.

7. A

The principal requirement for a spontaneous process is that Gibbs free energy is negative. The change in entropy or enthalpy may be positive for a spontaneous process.

8. C

Only entropy may have units of J/mol · K.

9. D

Enthalpy is independent of pressure. However, entropy depends on the volume of a gas and therefore its pressure. At higher volumes (corresponding to lower pressures), the molecules of gas can occupy more positions than at smaller volumes. Therefore, the positional entropy is higher at lower pressures. Since the entropy of a gas is pressure-dependent, and $\Delta G° = \Delta H° - T\Delta S°$, this implies that Gibbs free energy must also be pressure-dependent.

10. A

We can use Hess's law and sum the two reactions as follows:

$$C_3H_8 \text{ (g)} + 5\,O_2 \text{ (g)} \rightarrow 3\,CO_2 \text{ (g)} + 4\,H_2O \text{ (l)}$$

$$\Delta H° = -2{,}220 \text{ kJ}$$

$$4\,H_2O \text{ (l)} \rightarrow 4\,H_2O \text{ (g)} \qquad \Delta H° = 176 \text{ kJ}$$

$$C_3H_8 \text{ (g)} + 5\,O_2 \text{ (g)} \rightarrow 3\,CO_2 \text{ (g)} + 4\,H_2O \text{ (g)}$$

$$\Delta H° = -2{,}044 \text{ kJ}$$

FREE-RESPONSE QUESTION

1. **(a)** No. For a reaction with the same number of molecules of gaseous products as gaseous starting materials, it is not possible to predict $\Delta S°$ without further information, such as $\Delta S°_F$ values.

   (b) First, we must calculate $\Delta H°$ and $\Delta S°$ for the reaction:

   $$\Delta H° = \Sigma n_p \Delta H°_F(\text{products}) - \Sigma n_r \Delta H°_F(\text{reactants})$$

   $$\Delta H° = 2\ (-36.3\ \text{kJ/mol}) - 0\ (\text{since the } \Delta H°_F \text{ of both } Br_2 \text{ and } H_2 \text{ are 0})$$

   $$\Delta H° = -72.6\ \text{kJ/mol}$$

   $$\Delta S° = \Sigma n_p \Delta S°_F(\text{products}) - \Sigma n_r \Delta S°_F(\text{reactants})$$

   $$\Delta S° = 2\ (198.7\ \text{J/mol} \cdot \text{K}) - 245.5\ \text{J/mol} \cdot \text{K} - 130.7\ \text{J/mol} \cdot \text{K}$$

   $$\Delta S° = 21.2\ \text{J/mol} \cdot \text{K} = 0.0212\ \text{kJ/mol} \cdot \text{K}$$

   recalling that $\Delta G = \Delta H - T\Delta S$,

   $$\Delta G° = -72.6\ \text{kJ/mol} - (298\ \text{K})\ (0.0212\ \text{kJ/mol} \cdot \text{K})$$

   $$\Delta G° = -78.9\ \text{kJ/mol}$$

   The reaction is spontaneous at this temperature. Pay careful attention to units in this final equation; the temperature must be in Kelvin and the $T\Delta S$ term must be converted from J to kJ.

   (c) Recall that $\Delta G° = -RT \ln K$. Rearranging this equation gives $\ln K = -\Delta G°/RT$, and substituting the $\Delta G°$ from above (converted from kJ/mol to J/mol so the J will cancel), the temperature in Kelvin, and $R = 8.31\ \text{J/mol} \cdot \text{K}$ gives:

   $$\ln K = \frac{-(-78.900\ \text{J/mol})}{(8.31\ \text{J/mol} \cdot \text{K})(298\ \text{K})}$$

   $$\ln K = +31.9$$

   $$K = 6.9 \times 10^{13}$$

   This answer makes sense; the reaction is spontaneous, so we would expect to see a large value for the equilibrium constant.

   (d) Since the $\Delta H°$ is negative and the $\Delta S°$ is positive, this reaction is spontaneous at all temperatures.

CHAPTER 11: KINETICS

IF YOU LEARN ONLY NINE THINGS IN THIS CHAPTER . . .

1. The spontaneity of a chemical reaction has no influence on reaction rate.

2. A reaction rate is usually defined in terms of rate of disappearance of reactants or appearance of products. However, it is not this definition, but rather the rate law that is usually used in solving kinetics problems.

3. For a given chemical reaction, the rate law includes both the rate constant k as well as the dependence of rate on reactant concentrations:

    $$\text{Rate} = k[A]^n[B]^m \ldots$$

    where the overall reaction order is given by $n + m + \cdots$

4. The integrated rate law predicts how the concentration of a reactant or product will vary over time. You must be familiar with the integrated rate laws for zero-order, first-order, and second-order reactions.

5. First-order kinetics are most likely to appear on the AP Chemistry exam. For a first-order process:

$A \rightarrow B$	General reaction form
$\text{Rate} = k[A]$	Rate law
$\ln[A]_t = -kt + \ln[A]_0$	Integrated rate law
$t_{\frac{1}{2}} = \ln[2]/k$	Half-life

6. The graph of ln[A] versus t yields a graph with slope = −k.

7. A reaction mechanism describes the elementary reactions that constitute an overall chemical reaction. The overall rate depends only on the slow or rate-limiting step of the reaction.

8. The activation energy E_a refers to the energetic barrier that needs to be overcome in order for a reaction to occur and is the principal determinant of reaction rate. The reaction rate will vary with temperature according to the Arrhenius equation:

$$k = Ae - E_a/RT$$

This equation can be used to calculate E_a if the rate constants are known for a reaction at two or more different temperatures.

9. The activation energy is lowered by the presence of a catalyst. A catalyst may appear in the overall rate law for a reaction and may be homogeneous or heterogeneous.

For a reaction to be useful, either in the laboratory or in nature, it must occur at a reasonable rate. In this section, we will turn to chemical kinetics, which helps us to describe how a reaction proceeds. Since reaction rate is so important in the laboratory, you will certainly be expected to understand the determinants of rate and to predict kinetic behavior under different reaction conditions.

In the last section, we looked at the energetics of chemical reactions, with special emphasis on reaction spontaneity. However, in nature it's more often favorable kinetics that determine whether or not a reaction actually occurs. In the human body, biochemical pathways are regulated by sets of enzymes, proteins that are able to enhance the rate of reaction by factors as high as 10^8. The tight control of enzymes determines how molecules are transformed and shuttled throughout the human body.

There are two important points here: First, chemical transformations in the human body (for example, the phosphorylation of glucose required to perform glycolysis) must be thermodynamically favorable in order to occur. Second, they will not proceed at a reasonable rate without the presence of enzymatic catalysis. So any reaction in the human body is under the dual control of thermodynamics and kinetics.

In this section, we will start by examining ways to represent the rate of a reaction. We will then discuss the way reactions actually occur in nature for a variety of chemical systems. Finally, we will look again at the ways in which thermodynamics and kinetics intersect by looking at catalysis and defining the energetic quantity known as activation energy.

KINETICS VS. THERMODYNAMICS

One of your most important tasks on the AP Chemistry exam will be to distinguish **kinetics**, which deals with reaction rates, from the thermodynamic concepts discussed in the previous section. Remember that *spontaneous* does not mean fast. Combustion, for instance, often has a large negative Gibbs free energy. Let's look at the combustion of ethanol, or CH_3CH_2OH, as an example. Remember that in order to represent a combustion reaction, we just add oxygen to the substance in question, list water and carbon dioxide as the products, and balance the equation:

$$CH_3CH_2OH\ (l) + 3O_2\ (g) \rightarrow 3H_2O\ (g) + 2CO_2\ (g)$$

What is the Gibbs free energy for this reaction?

Remember that we can use Hess's law and the standard Gibbs free energies of formation:

$$\Delta G_{rxn}° = [3\Delta G°_F(H_2O) + 2\Delta G°_F(CO_2)] - [\Delta G°_F(CH_3CH_2OH) + 3\Delta G°_F(O_2)]$$

$$= [3(-228.6\ kJ/mol) + 2(-394.4\ kJ/mol)] - [(-174.8\ kJ/mol) + 3(0)]$$

$$= -1,299.8\ kJ/mol$$

The impressive result is that there is a large negative Gibbs free energy associated with this reaction, meaning that it is spontaneous. However, we also know that ethanol does not spontaneously combust under standard conditions. This illustrates that the thermodynamics of a reaction say nothing about reaction speed. The combustion reaction above does not have a fast rate simply because there is a large, favorable $\Delta G°$ associated with it.

DEFINITION OF REACTION RATE

How is the **reaction rate** defined for a reaction? Let's take a look at a reaction corresponding to the formation of HI (hydrogen iodide gas) from hydrogen and iodine:

$$H_2\ (g) + I_2\ (g) \rightarrow 2HI\ (g)$$

When H_2 and I_2 are placed in a flask, those molecules with sufficient kinetic energy will collide and undergo a chemical reaction, with formation of HI. As time passes, the concentration of substrates H_2 and I_2 in the flask will decrease, and the concentration of HI will increase. The reaction rate can be defined as the change in concentration of a reactant or product per unit time. The rate for the above reaction can be given by:

$$Rate = \frac{\Delta[H_2]}{\Delta t}$$

Notice that since the concentration of H_2 in the flask is decreasing, using the expression above for rate will give us a negative number. Since we want to describe reaction rate using a positive number, we can use the negative of the above to express rate. Also, the rate of disappearance of products should equal the rate of appearance of products, corrected for the number of moles of HI formed according to the balanced chemical equation. So the rate of this reaction is better represented as:

$$\text{Rate} = \frac{-\Delta\,[\text{H}_2]}{\Delta t} = \frac{-\Delta\,[\text{I}_2]}{\Delta t} = \frac{\frac{1}{2}\,\Delta[\text{HI}]}{\Delta t}$$

If the concentration of H_2 were plotted against time, we would end up with a graph similar to that shown below. The reaction rate could be determined at any point by looking at the slope of the curve—in other words, the instantaneous change in the concentration of H_2 at a given point in time. Notice that this rate is not constant but decreases with time. The **initial rate** of the reaction is the instantaneous rate determined just after $t = 0$, where the curve is usually somewhat linear. This value is used to determine the rate law for the reaction, as we will see later.

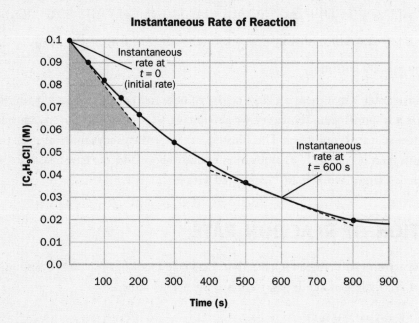

RATE LAWS

Another way to represent the rate of a chemical reaction is using a **rate law**, which shows how the reaction rate depends on the concentrations of the reactants. Remember, the rate law for a chemical reaction must be determined experimentally. That is, we cannot write the rate law for a reaction just by looking at the balanced chemical equation.

Let's suppose that we conduct an experiment to determine how the rate of the previous reaction varies with changing the initial concentrations of H_2 and I_2:

Experiment	Initial Concentration of I_2	Initial Concentration of H_2	Initial Rate (mol/L $\cdot$ s)
A	0.100 M	0.50 M	5.2×10^{-6}
B	0.100 M	1.00 M	1.04×10^{-5}
C	0.400 M	0.50 M	2.08×10^{-5}

Looking at the data, we see that as we doubled the initial concentration of H_2 in experiment B, the initial reaction rate also doubled. In other words, the initial reaction rate was directly proportional to the initial concentration of H_2. This is another way of saying that the order of the reaction with respect to H_2 is one. Experiment C shows that as we changed the initial concentration of I_2 by a factor of four, the initial reaction rate also increased by a factor of four. Again, the increase in rate was directly proportional to the increase in the initial concentration, so the order with respect to I_2 is also one. We can represent the rate of this reaction as the following:

$$\text{Rate} = k[I_2][H_2]$$

where k is a proportionality constant known as the **rate constant**. This reaction is first order in I_2, first order in H_2, and second order overall.

For any given chemical reaction, the rate law has the general form:

$$\text{Rate} = k[A]^n[B]^m \ldots$$

where A, B, ... are reaction components on which the rate depends, and the overall order is given by $n + m + \ldots$. The components A, B, ... might be reactants (most common), products, or catalysts. The numbers n, m, etc. are usually positive integers but may be fractions or negative numbers, and they are always determined experimentally.

ZERO-, FIRST-, AND SECOND-ORDER KINETICS

What do these rate laws say about how the concentration of reactants and products will vary with time? On the AP Chemistry exam, it's unlikely that you'll be asked to perform any long and complicated calculations. However, you will be expected to understand how reaction order relates to different kinetic behavior. In particular, you will need to be able to recognize **zero-order, first-order, and second-order kinetics**. Let's look at the basic things you'll need to know, starting with **zero-order** and working our way up.

A **zero-order reaction** is a reaction in which the rate does not depend on the concentrations of any of the reaction components. If we have the following reaction:

$$A \rightarrow B$$

the rate of disappearance of A (or appearance of B) will not change as the reaction proceeds. Therefore the rate is given by:

$$\text{Rate} = k[A]_0 = k$$

The **integrated rate law**, which reflects how the concentrations of reactants will vary according to time, is a linear equation:

$$[A]_t = -kt + [A]_0$$

If you see a graph of reactant (or product) concentrations vs. time that is **linear**, it corresponds to a zero-order reaction.

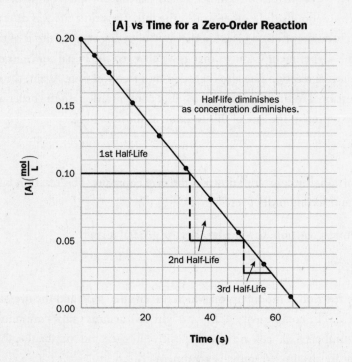

[A] vs Time for a Zero-Order Reaction

In a **first-order reaction,** the rate is proportional to the concentration of the reactant. The decomposition of hydrogen peroxide is an example of first-order kinetics:

$$2H_2O_2 \, (aq) \rightarrow 2H_2O \, (l) + O_2 \, (g)$$

for which the rate law is given by:

$$\text{Rate} = k[H_2O_2]$$

We might be tempted to predict that the reaction is second-order by looking at the balanced chemical equation. However, the order must be determined experimentally.

The reaction rate will be highest when the concentration of H_2O_2 is high but will decrease as the reaction proceeds and the hydrogen peroxide disappears. Radioactive decay is another important example of first-order kinetics. If you see a question on the AP exam relating to radioactive decay (α, β particles, etc.), you are dealing with first-order kinetics.

For a first-order reaction of the general form:

$$A \rightarrow B$$

the rate law is given by:

$$\text{Rate} = k[A]$$

And the integrated rate law is given by:

$$\ln[A]_t = -kt + \ln[A]_0$$

Nobody likes to deal with equations like this in the heat of the moment on test day, but remember that there are only a handful of calculations you may be asked to perform. In addition, the integrated rate law is included on the Section II equation sheet. Most likely, you'll be given some information and will be asked to solve for the missing variable in the equation above.

Also notice that the integrated rate law shown above:

$$\ln[A]_t = -kt + \ln[A]_0$$

has the general form:

$$y = mx + b$$

Therefore, a graph of $\ln[A]$ (on the vertical axis) versus t (on the horizontal axis) should yield a straight line with slope $= -k$.

[A] vs Time for a First-Order Reaction

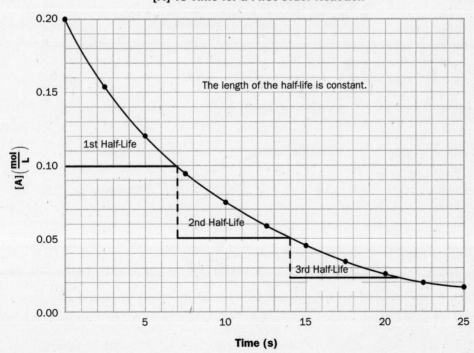

A **second-order reaction** is a reaction for which the rate is proportional to the square of a reactant. Remember: it's possible for a reaction to be first-order in two different reactants and second-order overall, a case that is complicated and unlikely to appear on the AP exam. For a second-order reaction:

$$A \rightarrow B$$

the rate law is given by:

$$\text{Rate} = k[A]^2$$

And the integrated rate law, derived using calculus is:

$$\frac{1}{[A]_t} = \frac{kt + 1}{[A]_0}$$

This equation is also in the form:

$$y = mx + b$$

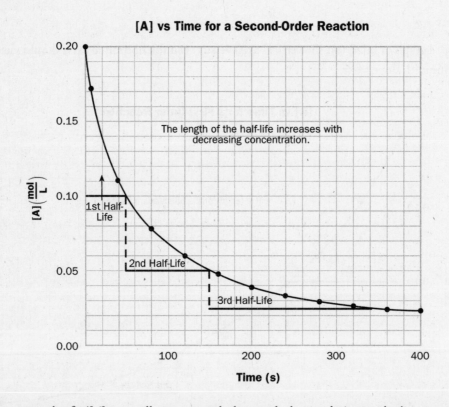

[A] vs Time for a Second-Order Reaction

The length of the half-life increases with decreasing concentration.

Therefore, a graph of $1/[A]$ vs. t will give a straight line with slope $= k$. Again, don't get too excited about this for the AP exam. Understanding first-order kinetics will lead to the highest yield on test day.

To summarize the different types of kinetics we discussed here:

Order	Rate Law	Integrated Law	Straight Line	Slope of Plot
0	Rate = k	$[A]_t = -kt + [A]_0$	$[A]$ vs. t	$-k$
1	Rate = $k[A]$	$\ln[A]_t = -kt + \ln[A]_0$	$\ln[A]$ vs. t	$-k$
2	Rate = $k[A]^2$	$\dfrac{1}{[A]_t} = \dfrac{kt+1}{[A]_0}$	$\dfrac{1}{[A]}$ vs. t	k

REACTION MECHANISMS

We've provided a pretty simple description of how reactions occur, but remember that they don't typically occur in a single step. Usually for a given overall reaction, there is a sequence of elementary reactions that describes the **reaction mechanism**. Studying the kinetics of a reaction is one of the principal ways the mechanism of a reaction is described. For example, let's look at the reaction of nitrogen dioxide with carbon monoxide to form nitric oxide and carbon dioxide:

$$NO_2 + CO \rightarrow NO + CO_2$$

One might be tempted to look at the balanced chemical equation and simply write the rate law:

$$\text{Rate} = k[NO_2][CO]$$

But this would be incorrect, since the rate law (and reaction mechanism) can only be determined experimentally. This would imply that the reaction takes place via a simple bimolecular mechanism without the presence of any intermediates or **catalysts**. By experiment, the rate law is determined to be:

$$\text{Rate} = k[NO_2]^2$$

The observed rate is consistent with a slightly more complicated mechanism:

Step I $NO_2 + NO_2 \rightarrow NO_3 + NO$ (slow)

Step II $NO_3 + CO \rightarrow NO_2 + CO_2$ (fast)

Remember the following important points regarding any reaction mechanism:

- The sum of the individual steps should give the balanced chemical equation.

- The mechanism must agree with the experimental data.

- A mechanism usually includes intermediates—chemical compounds seen in the elementary reactions but not in the balanced chemical reaction. In the mechanism above, NO_3 is an intermediate since it appears as the product of step 1 and a reactant in step 2.

AP EXPERT TIP

Questions about reactions have shown up in recent years as final parts to kinetics questions. They are easy to understand and can be worth a few points.

- A mechanism may include a catalyst, or chemical component that makes the reaction go faster. A catalyst is usually consumed in the first step and produced by a later step. The mechanism shown previously is without an obvious catalyst.

- The rate-limiting step is the elementary reaction in the mechanism on which the entire rate depends. It is the slowest step of the mechanism. If the mechanism of the reaction is known, you can write the overall rate law by just looking at the rate-limiting step.

REACTION ENERGY DIAGRAMS AND ACTIVATION ENERGY

Reaction energy diagrams are where kinetics meets thermodynamics and are a potential source of confusion on the AP exam. As we stated at the beginning of this section, *spontaneous* does not mean fast. In other words, a reaction (such as combustion) can be highly thermodynamically favorable and still not occur at any significant rate. For any chemical reaction, molecules must collide with sufficient **kinetic energy** to react, as well as collide in the proper **orientation** for a reaction to occur, which is also rare. Thus, there is an energetic barrier to the reaction. The energy required to overcome this energetic barrier is known as the **activation energy**, or E_a. The rate of a reaction will depend on the activation energy rather than on the overall change in energy between reactants and products. The higher the activation energy, the slower the reaction will occur.

The energetics of chemical reactions can be described using a reaction energy diagram. This diagram summarizes energetic changes that occur during the course of a chemical reaction, with **potential energy** on the y-axis and **reaction coordinate** or **progress** on the x-axis.

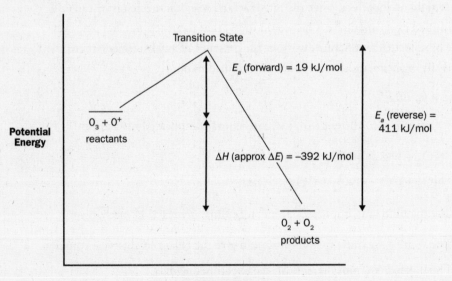

Notice that the reaction is exothermic but that there is an activation energy that must be overcome in order for the reaction to occur. The structure at the top of an energy diagram like this one is called the **transition state** or **activated complex**. It describes the exact intermediate geometry required in order for the reaction to occur.

If reaction rate depends principally on activation energy, what can we do to increase reaction rate? First, we can lower the activation energy. This is accomplished via use of a catalyst, which may be either chemical or biological. Enzymes, or proteins that facilitate chemical reactions in the human body, are the oldest and most important form of catalysts. However, they are less likely to show up on the AP exam. We can also increase the temperature to increase reaction rate. In doing so, please note that we are not decreasing the activation energy of the reaction. Instead, we are increasing the percentage of reactant molecules that possess enough kinetic energy for a chemical reaction to occur when they collide.

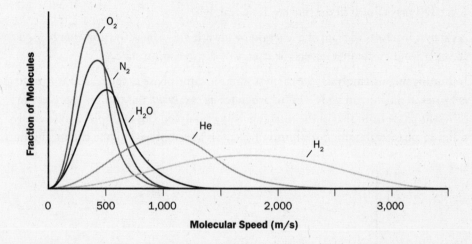

How do we determine the activation energy for a reaction? We can use the **Arrhenius equation**, which relates activation energy to the reaction rate constant. As we just stated, a reaction rate depends not only on concentrations of reactants, but also on temperature. The Arrhenius equation is given by:

$$k = Ae^{\frac{-E_a}{RT}}$$

where A is a frequency factor that depends on how often molecules collide in the proper orientation, E_a is the activation energy in joules, R is the gas-law constant (8.314 J/mol • K), and T is the temperature in Kelvin.

On the AP Chemistry exam, you are likely to be asked to determine E_a by looking at the reaction rate constant k at two different temperatures. The equation you can use to calculate E_a in this case is:

$$\ln\left(\frac{k_2}{k_1}\right) = -\frac{E_a}{R}\left(\frac{1}{T_2} - \frac{1}{T_1}\right)$$

CATALYSIS

Catalysis is an extremely important phenomenon used by the chemical industry and in the human body, which employs enzymes as catalysts to carry out important biological tasks. We have already mentioned several of the most important features of catalysts, but here is a summary of what you will need to know for the AP exam:

- In a reaction mechanism, the catalyst is often consumed in an early step and regenerated by a later elementary reaction. The catalyst doesn't appear in the overall reaction, though it is present in both the reactant mixture and the product mixture. (Contrast this with an intermediate, which is generated in an early step and consumed in a later step, and is therefore absent from both the reactant and product mixtures.)

- A catalyst may appear in the rate law for a reaction.

- A catalyst increases the rate of a reaction by lowering E_a, the activation energy for a reaction. In other words, a catalyst lowers the energy of the transition state.

- In **homogeneous catalysis**, the catalyst is in the same phase as the reactants (for example, enzymes in the human body). In **heterogeneous catalysis**, the catalyst is present in a different phase from that of the reactants being catalyzed (for example, a solid catalyst for a solution-phase reaction). Both forms of catalysis are employed by the chemical industry.

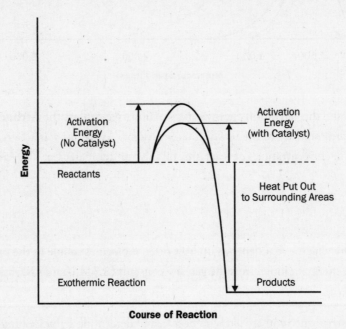

REVIEW QUESTIONS

1. Nitrogen monoxide and oxygen were combined in a flask at 25°C and allowed to react as follows:

$$2NO \ (g) + O_2 \ (g) \rightarrow 2NO_2 \ (g)$$

The concentrations of reactants were varied as follows, and initial rates were calculated. Which of the following is the rate law for the reaction?

Experiment	Initial Concentration of O_2	Initial Concentration of NO	Initial Rate (mol/L · s)
A	0.020 M	0.050 M	0.038
B	0.020 M	0.100 M	0.152
C	0.080 M	0.100 M	0.608

(A) Rate = $k[O_2]^2[NO]^2$
(B) Rate = $k[NO]^2$
(C) Rate = $k[O_2][NO]^2$
(D) Rate = $k[O_2][NO]$
(E) Rate = $k[O_2]$

2. Chlorofluorocarbons (CFCs) released by hairspray and other aerosol propellants are thought to play a major role in the destruction of the ozone layer. CFCs can generate chlorine radicals in the atmosphere. For example, $CF_2Cl_2 \ (g)$ is a CFC that undergoes the following decomposition when it absorbs a high-energy photon of light:

$$CF_2Cl_2 \ (g) \rightarrow CF_2Cl + Cl$$

The following represents a proposed mechanism for ozone destruction:

Step I $O_3 + Cl \rightarrow O_2 + ClO$ (slow)

Step II $ClO + O \rightarrow Cl + O_2$ (fast)

Which of the following is NOT true with respect to this mechanism?

(A) Cl is a catalyst for the destruction of ozone.
(B) For this reaction, rate = $k[O_3][Cl]$.
(C) Step I is the rate-limiting step for this reaction.
(D) Increasing ClO will increase the reaction rate.
(E) The balanced chemical reaction for the destruction of ozone by this mechanism is given by: $O_3 + O \rightarrow 2 \ O_2$.

3. Which of the following is correct?

(A) The rate of a reaction is always proportional to the concentration of the reactants.
(B) The activation energy E_a is constant with respect to temperature.
(C) The rate constant k is constant with respect to temperature.
(D) A reaction with a large negative $\Delta G°$ will occur at a fast rate.
(E) For the second-order reaction $2A \rightarrow B + C$, we can calculate the value of k using a plot of $\ln[A]$ versus t.

4. Which of the following statements is/are correct? Assume that in each case, $\Delta H°$ closely approximates the overall energy change.

   I. For an endothermic reaction, E_a is always greater than $\Delta H°$.
   II. For an exothermic reaction, the magnitude of E_a is always greater than that of $\Delta H°$.
   III. For an exothermic reaction, adding a catalyst will decrease the magnitude of $\Delta H°$.

   (A) I only
   (B) II only
   (C) I and II only
   (D) II and III only
   (E) I, II, and III

5. A sample of the unstable isotope ^{20}Na is generated in the laboratory. Initally, there are 2.40 mmol of the isotope detected. After 0.30 seconds, there are 0.30 mmol of ^{20}Na detected. Which of the following is the half-life for this reaction?

   (A) 0.10 s
   (B) 0.15 s
   (C) 0.20 s
   (D) 0.25 s
   (E) 0.30 s

6. $2\,NO\,(g) + O_2\,(g) \rightarrow 2\,NO_2\,(g)$

   The reaction between nitrogen monoxide and oxygen occurs as above. One proposed mechanism is the following:

   Step I $NO\,(g) + O_2\,(g) \rightarrow NO_2\,(g) + O\,(g)$ (slow)

   Step II $NO\,(g) + O\,(g) \rightarrow NO_2\,(g)$ (fast)

   Which of the following rate expressions agrees best with this possible mechanism?

   (A) Rate $= k\,[NO][O]$
   (B) Rate $= k\,[NO][O_2]$
   (C) Rate $= k\,[NO]^2[O_2]$
   (D) Rate $= k\,[NO]/[O_2]$
   (E) Rate $= k\,[NO]/[O]$

7. $C_6H_6\,(l) + \frac{15}{2}\,O_2\,(g) \rightarrow 6\,CO_2\,(g) + 3\,H_2O\,(l)$

   The above reaction represents the combustion of benzene. If a sample of benzene is burning at 0.25 mol/L · s, which of the following is the rate at which $CO_2\,(g)$ is being produced?

   (A) 0.25 mol/L · s
   (B) 0.75 mol/L · s
   (C) 1.5 mol/L · s
   (D) 2.5 mol/L · s
   (E) 3.0 mol/L · s

Questions 8–9 refer to the following reaction mechanism:

Step I $O_3 (g) + NO (g) \rightarrow NO_2 (g) + O_2 (g)$
       (slow)

Step II $NO_2 (g) + O (g) \rightarrow NO (g) + O_2 (g)$
       (fast)

The overall reaction is given by:

$O_3 (g) + O (g) \rightarrow 2\,O_2 (g)$

(A) O_3
(B)   NO
(C) O_2
(D)   O
(E) NO_2

8. Identify the catalyst.

9. Identify the reaction intermediate.

FREE-RESPONSE QUESTION

1. $2 H_2 (g) + 2 NO (g) \rightarrow N_2 (g) + 2 H_2O (g)$

    The reaction represented above is conducted at two different temperatures with the following results:

Experiment	Initial Concentration of H_2	Initial Concentration of NO	Initial Rate (mol/L·s) @ 25°C	Initial Rate (mol/L·s) @ 75°C
A	0.025 M	0.010 M	0.065	0.75
B	0.025 M	0.020 M	0.26	3.0
C	0.100 M	0.020 M	1.04	12.0

    (a) Write down the rate law for the reaction.

    (b) Calculate the value of k at both temperatures. Include units in your answer.

    (c) Calculate E_a, the activation energy of the reaction.

    (d) The following mechanisms have been proposed for the reaction:

    Mechanism #1:

    $H_2 (g) + NO (g) \rightarrow H_2O (g) + N (g)$ (slow)

    $N (g) + NO (g) \rightarrow N_2 (g) + O (g)$ (fast)

    $H_2 (g) + O (g) \rightarrow H_2O (g)$ (fast)

    Mechanism #2:

    $2 H_2 (g) + 2 NO (g) \rightarrow N_2 (g) + 2 H_2O (g)$

    Mechanism #3:

    $H_2 (g) + 2 NO (g) \rightarrow N_2O (g) + H_2O (g)$ (slow)

    $N_2O (g) + H_2 (g) \rightarrow N_2 (g) + H_2O (g)$ (fast)

    i. Which mechanism is most consistent with the observed rate law? Justify your answer.

    ii. For Mechanism #1, diagram the geometry required for a productive collision in the rate-determining step.

    iii. Which mechanism is, in general, the least likely to happen as written? Justify your answer.

ANSWERS AND EXPLANATIONS

1. C

To solve this question, we must determine the order of the reaction with respect to different components. Let's first look at experiments A and B. In experiment B, we are keeping the initial concentration of O_2 the same, but we are doubling the concentration of NO, from 0.050 M to 0.100 M. The result is a fourfold increase (2^2) in the initial reaction rate. Therefore, the reaction order with respect to NO is two. Looking at experiments B and C, we see that increasing the initial concentration of O_2 by a factor of four also results in a fourfold increase (4^1) in the reaction rate. Therefore, the order of the reaction with respect to O_2 is one. The reaction rate is given by:

$$\text{Rate} = k[O_2][NO]^2$$

and the overall order of the reaction is three.

2. D

Since ClO is an intermediate that does not appear in the rate law, changing its concentration should not affect the reaction rate. Choice (A) is true since Cl appears as a reactant in the first step, a product in a later step, and does not appear in the overall balanced chemical reaction. Choice (B) is true since Step I is the slow or rate-determining step of the reaction. Since Step I represents an elementary (in this case bimolecular) reaction, we can determine the rate law just by looking at the chemical equation. Choice (C) is therefore also correct. Choice (E) is true since you arrive at this overall equation by summing the two elementary steps.

3. B

E_a is constant with respect to temperature. Remember that increasing the temperature increases the percentage of reactant molecules with kinetic

energy $> E_a$, but does not change E_a itself. Choice (A) is incorrect since this statement only holds true for first-order reactions, but not for zero-order (for which the rate does not depend at all on the concentration of reactants) or second-order (for which the rate depends on the square of the concentration of reactants). Choice (C) is incorrect since k will vary according to temperature; this is how we calculate E_a using the Arrhenius equation. Choice (D) is incorrect since reaction rate depends only on E_a, not on $\Delta G°$. Remember the difference between kinetics and thermodynamics. Choice (E) is incorrect since for a second-order reaction, we would determine k by drawing a graph of $\frac{1}{[A]}$ vs. t (not ln $[A]$ as for a first-order reaction).

4. A

E_a refers to the energetic barrier that must be overcome for a reaction to occur, which is larger than $\Delta H°$ for an endothermic reaction. The other statements are untrue. Remember that addition of a catalyst will lower E_a but will never change $\Delta G°$ or $\Delta H°$.

5. A

When you encounter first-order kinetics on the AP exam, it's more than likely that the question will relate to half-life. Notice that in this review question, the concentration of ^{20}Na decreased by a factor of eight over the course of 0.30 seconds. A factor of eight (which is 2^3) corresponds to three half-lives.

Time	Half-Lives	Amount Present (mmol)
0	0	2.40
0.10	1	1.20
0.20	2	0.60
0.30	3	0.30

6. B

According to the given mechanism, step I is the slow or rate-determining step. Since a reaction mechanism includes only elementary reactions, you can write down the rate expression just by looking at this step. Choice (C) is incorrect since this rate expression is derived from the overall reaction and not the rate-determining step.

7. C

The rate of production of CO_2 is given by the stoichiometry of the balanced chemical equation. Since the rate of disappearance of C_6H_6 is 0.25 mol/L·s, the rate of appearance of CO_2 is six times as fast.

8. B

NO is the catalyst since it is consumed in an early step, regenerated in a later step, and does not appear in the overall reaction. Notice in this case that the first step is limiting the rate = $k[O_3][NO]$, demonstrating that the catalyst can appear in the rate law of a reaction.

9. E

NO_2 is the intermediate since it is generated in an early step, consumed in a later step, and does not appear in the overall reaction.

FREE-RESPONSE QUESTION

1. (a) The rate law is given by:

    $$\text{Rate} = k\,[NO]^2[H_2]$$

    We can see that doubling the concentration of NO produces a fourfold (2^2) increase in the reaction rate, and quadrupling the concentration of H_2 produces a fourfold increase in the reaction rate (4^1).

    (b) We can determine the value of k in both cases by substituting values for rate at the two different temperatures:

    At 25°C, rate = $k_1\,[NO]^2[H_2]$

    $1.04\ M/s = k_1\,[0.020\ M]^2[0.100\ M]$ $k_1 = 26{,}000\ M^{-2}\,s^{-1}$

    At 75°C, rate = $k_2\,[NO]^2[H_2]$

    $12.0\ M/s = k_2\,[0.020\ M]^2[0.100\ M]$ $k_2 = 300{,}000\ M^{-2}s^{-1}$

    Notice that the units of k must be $M^{-2}s^{-1}$ in order to yield the desired rate units.

    (c) In order to calculate activation energy, we need to use the Arrhenius equation:

    $$\ln\left(\frac{k_2}{k_1}\right) = -\frac{E_a}{R}\left(\frac{1}{T_2} - \frac{1}{T_1}\right) \text{ where } T \text{ is in Kelvin}$$

    Substituting, $\ln\left(\dfrac{300{,}000}{26{,}000}\right) = -\dfrac{E_a}{8.31\ \text{J/mol}\cdot K}\left(\dfrac{1}{348\ K} - \dfrac{1}{298\ K}\right)$

    $E_a = 1.42 \times 10^{-4}\ \text{J/mol}$

    (d) i. Mechanism #3 is most consistent with the observed reaction rate, since the rate-determining (or slow) step consists of two molecules of NO combining with one molecule of H_2.

    ii. For Mechanism #1, the collision must occur with a hydrogen atom of the H-H molecule striking the O atom of NO to release the nitrogen atom.

    $$\text{H–H} \cdots \rightarrow \begin{array}{c} O \\ \| \\ N \end{array}$$

    iii. Mechanism #2 is particularly unlikely to be correct since it requires a simultaneous collision of four molecules in correct orientation for the reaction to occur, a highly unlikely event in an elementary step.

CHAPTER 12: CHEMICAL REACTIONS AND STOICHIOMETRY

IF YOU LEARN ONLY FOUR THINGS IN THIS CHAPTER . . .

1. Balancing chemical equations takes practice. The essential outcome is that the number of atoms of each element is the same on both sides of a chemical equation.

2. A balanced chemical equation can be understood in terms of molecules or moles. Moles provide a useful way of relating products and reactants to each other. Molar mass, grams, and moles are related by the following equation:

   (# of grams) = (# of moles) × (molar mass)

3. The limiting reagent is the reactant that is completely consumed during the course of a chemical reaction and limits the amount of product that may be formed.

4. No chemist is perfect, so the quantity percent yield is given by the following:

   $$\text{Percent yield} = \left(\frac{\text{Actual yield}}{\text{Theoretical yield}}\right) \times 100$$

In this section, we will focus on some nuts and bolts. Stoichiometric calculations are certain to make an appearance on your AP Chemistry exam, so be sure to practice the basic skills of balancing chemical equations, predicting yields, and identifying limiting reagents.

BALANCING CHEMICAL EQUATIONS

Balancing chemical equations is a skill that will come up again and again on your AP exam. There are two types of questions that you may be asked. In the first case, you will be presented with a complete chemical reaction that is unbalanced, i.e., missing the whole number coefficients that should be in front of individual molecules. In the second case, you may be asked to write down the chemical reaction and then balance it. Be sure to distinguish between balancing by inspection and redox reactions; in the latter, charge must be balanced as well.

Trial and error is the best way to learn how to balance equations. Practice is key. Don't get hung up on balancing chemical equations on test day. A general method to learning this skill would include the following:

1. Write down the unbalanced equation containing the correct molecular formula of all reactants and products.

2. Determine whether or not the chemical equation is already balanced.

3. If the equation is not balanced, start by balancing the element that occurs in the fewest number of reactant and product molecules. Start with carbon or any atom with a high molar mass.

4. Balance the remaining elements in order of number of molecules and/or molar mass.

5. Make sure that the number of atoms of each element is balanced.

Alternatively, you may choose a method such as the following:

1. Write down the unbalanced equation containing the correct molecular formula of all reactants and products.

2. Determine whether or not the chemical equation is already balanced.

3. If the equation is not balanced, begin by setting the coefficient for the most complex component to 1.

4. Solve for the remaining reactants using decimals as necessary.

5. If decimals now exist, multiply to make all coefficients whole numbers.

REVIEW QUESTION

Which of the following represents the coefficients A, B, C, and D in the balanced chemical equation for the following reaction of iron with water?

[A] Fe + [B] H_2O → [C] Fe_3O_4 + [D] H_2

(A) 3, 4, 1, 4

(B) 2, 3, 4, 1

(C) 1, 2, 1, 4

(D) 3, 4, 2, 2

(E) 2, 4, 1, 1

SOLUTION

The correct answer is choice (A). We can see by looking at the unbalanced reaction that the equation is not already balanced. We start by balancing Fe, since it is a high molar-mass atom and only appears once on the reactant and product sides. We see that there are three Fe atoms on the product side, so we can balance Fe by adding a coefficient of 3 to Fe on the reactant side. So:

$$3\ Fe + [B]\ H_2O \rightarrow 1\ Fe_3O_4 + [D]\ H_2$$

Having balanced Fe, we can then turn to oxygen, since it occurs with Fe in Fe_3O_4 (and we have already balanced Fe). We notice that there are four atoms of oxygen on the product side, so we can balance oxygen by placing a coefficient of 4 in front of H_2O on the reactant side:

$$3\ Fe + 4\ H_2O \rightarrow 1\ Fe_3O_4 + [D]\ H_2$$

Finally, notice that all we have left to do is balance hydrogen. In this case, we now have eight H atoms on the left side, so we need to place a coefficient of 4 in front of H_2 to balance the equation:

$$3\ Fe + 4\ H_2O \rightarrow 1\ Fe_3O_4 + 4\ H_2$$

Another type of question may require you to balance a **combustion reaction**. Balancing a combustion reaction can sometimes be tricky because you are usually not given a complete chemical equation. For a combustion reaction, you need to add oxygen to the reactant side. The products of the reaction will be water and carbon dioxide.

REVIEW QUESTION

Write the balanced chemical reaction for the combustion of glucose, $C_6H_{12}O_6$, in air.

SOLUTION

To solve this question, we start by showing the unbalanced reaction:

$$C_6H_{12}O_6 + O_2 \rightarrow CO_2 + H_2O$$

We can start by balancing carbon. There are six C atoms on the left-hand side, so we can place a coefficient of 6 in front of CO_2 on the right-hand side. This gives us:

$$C_6H_{12}O_6 + O_2 \rightarrow 6\ CO_2 + H_2O$$

Now we can move to hydrogen. There are 12 H atoms on the left, so we can add a coefficient of 6 in front of H_2O:

$$C_6H_{12}O_6 + O_2 \rightarrow 6\,CO_2 + 6\,H_2O$$

Finally, we need to balance oxygen. We now have a total of $12 + 6$ or 18 O atoms on the right-hand side, and $C_6H_{12}O_6$ contributes six on the left. Therefore, we need to make sure that there are 12 more oxygen atoms on the left, which we can accomplish by adding a coefficient of 6 in front of O_2:

$$C_6H_{12}O_6 + 6\,O_2 \rightarrow 6\,CO_2 + 6\,H_2O$$

REACTION STOICHIOMETRY

Now that we have learned how to balance a chemical equation, let's have a look at what a balanced chemical equation actually means. For the combustion of glucose:

$$C_6H_{12}O_6 + 6\,O_2 \rightarrow 6\,CO_2 + 6\,H_2O$$

we could summarize the reaction in two ways:

1. One **molecule** of glucose reacts with six **molecules** of oxygen to form six **molecules** of carbon dioxide and six **molecules** of water.

2. One **mole** of glucose reacts with six **moles** of oxygen to form six **moles** of carbon dioxide and six **moles** of water.

Therefore, the balanced chemical reaction allows us to predict in what ratios the reaction will occur. In other words, if for the above reaction you started with 2 moles of $C_6H_{12}O_6$ and unlimited oxygen, you would predict that 12 moles of CO_2 and 12 moles of H_2O could be produced. The concept of mole ratio relates the number of moles of reactants and products to each other.

Also remember the following two important features of working with moles:

1. One mole of a molecule (or anything for that matter) is 6.022×10^{23} units.

2. The relationship between grams and moles is as follows:

$$(\text{\# of grams}) = (\text{\# of moles}) \times (\text{molar mass})$$

So in order to convert between grams and moles, look at the periodic table and calculate the molar mass of a given molecule or atom.

Let's try an example:

REVIEW QUESTION

The oxidation of Fe to Fe_2O_3 is described by the following balanced chemical equation:

$$4\ Fe\ (s) + 3\ O_2\ (g) \rightarrow 2\ Fe_2O_3\ (s)$$

If 2.03 g of iron is allowed to oxidize in the presence of excess oxygen, what is the mass of Fe_2O_3 formed when the reaction is complete?

(A) 1.23 g

(B) 1.48 g

(C) 2.29 g

(D) 2.56 g

(E) 2.90 g

SOLUTION

The correct answer is choice (E). We start by converting to moles:

$$2.03\ g\ Fe \times \left(\frac{1\ mol\ Fe}{55.8\ g\ Fe} \right) = 0.0364\ mol\ Fe$$

We can then use the mole fraction defined by the balanced chemical reaction above to determine how many moles of Fe_2O_3 are formed:

$$0.0364\ mol\ Fe \times \left(\frac{2\ mol\ Fe_2O_3}{4\ mol\ Fe} \right) = 0.0182\ mol\ Fe_2O_3$$

Finally, we need to figure out what mass of Fe_2O_3 is formed by multiplying by the molar mass of Fe_2O_3. The molar mass is given by $2(55.8) + 3(16.0) = 159.6$ g/mol, so:

$$0.0182\ mol\ Fe_2O_3 \times \left(\frac{159.6\ g\ Fe_2O_3}{1\ mol\ Fe_2O_3} \right) = 2.90\ g\ Fe_2O_3$$

COMBUSTION ANALYSIS STOICHIOMETRY

With a simple understanding of combustion stoichiometry, we can predict the chemical outcome of a reaction and the empirical formula of organic molecules (those containing only carbon, hydrogen, and oxygen). A combustion reaction is run in the following way. First, the water and carbon dioxide collection chambers are preweighed (this is called the *tare weight*). A known amount (mass) of organic sample is admitted into the furnace, burned completely in excess oxygen, and then the collection chambers are reweighed. The mass increase in each collector is simply the mass of the corresponding reaction product. The reaction is performed with excess oxygen so that the product is limited by the amount of carbon and hydrogen present in the organic reagent. The combustion reaction stoichiometry is then a useful way of identifying at least the constituency of an unknown.

REVIEW QUESTION

A 0.25 g dry cleaning solvent containing C, H, and Cl was submitted for combustion analysis, and the following data were obtained:

Mass of CO_2 absorbed after combustion = 89.76222 g

Mass of CO_2 absorbed before combustion = 89.31122 g

Mass of H_2O absorbed after combustion = 36.6039 g

Mass of H_2O absorbed before combustion = 36.5422 g

What is the molecular formula of the solvent?

(A) C_3H_2Cl

(B) C_6H_5Cl

(C) $C_6H_2Cl_4$

(D) $C_6H_4Cl_2$

(E) $C_6H_6Cl_2$

SOLUTION

The correct answer is choice (D). We can determine the mass (g) of CO_2 from the mass of the CO_2 absorbed before and after combustion. Mass of CO_2 = 89.76222 g − 89.31122 g = 0.451 g. Similarly, the mass of H_2O = 36.6039 g − 36.5422 g = 0.0617 g.

We find the mass of C and H from the mass fraction of CO_2 and H_2O:

$$\text{Mass fraction of C in } CO_2 = \frac{\text{mol C} \times \dfrac{12.01\text{g C}}{\text{mol C}}}{44.01 \text{ g } CO_2} = \frac{0.2729 \text{ g C}}{1\text{g } CO_2}$$

$$\text{Similarly, the mass fraction of H in } H_2O = \frac{1.008 \text{ g H}}{18.02 \text{ g } H_2O}$$

$$= \frac{0.1119 \text{ g H}}{1 \text{ g } H_2O}$$

The mass of C in 0.451 g of CO_2 is the product of the mass of CO_2 produced by the combustion of C to CO_2 and the mass fraction of C in CO_2:

Mass of C = 0.451 g × 0.2729 = 0.123 g C.

The mass of H in this sample is likewise the product of the mass of H_2O produced and the mass fraction of H in H_2O:

Mass of H $= 0.1119 \times 0.0617$ g $= 0.00690$ g H

Since the sample contains only C, H, and Cl, we can calculate the weight of Cl from the difference between the weight of the sample (0.25 g) and the mass of H and C:

Mass of Cl $= 0.25$ g $- 0.2729$ g $- 0.0069$ g $= 0.120$ g

We are now ready to determine the empirical formula by calculating the moles for each of the elements:

C:H:Cl $= 0.0102$ mol : 0.00685 mol : 0.00339 mol

The empirical formula is C_3H_2Cl, and the multiple $= 2$.

This is choice (A) but is not the molecular formula. Remember to multiply the empirical formula by the multiple two. The answer is choice (D). The molecular formula is $C_6H_4Cl_2$. Note that choice (E) can be discounted by virtue of the fact that it is not a correct molecular formula.

LIMITING REAGENT

Sometimes in a chemical reaction, there is not enough of one reactant to go around. In such cases, this reactant, or **limiting reagent**, will determine how much product can be formed. During the course of a reaction, the limiting reagent is the reactant that is completely consumed.

To clarify the concept of limiting reagent, consider your laundry habits. Most people try to put off doing laundry as long as possible but cannot avoid doing so any longer when the supply of clean underwear runs out. This implies that clean underwear is the limiting reagent with respect to avoiding laundry. There is considerable debate on this topic, with dissenting groups claiming that socks are in fact the limiting reagent.

In any case, let's try a sample question involving the concept of limiting reagents. We follow a simple plan:

1. Determine the number of moles of each reactant.
2. Use the mole fraction, defined by the balanced chemical equation, to determine which reactant is the limiting reagent.
3. Use the limiting reagent to determine how much product can be formed.

REVIEW QUESTION

Urea ($(NH_2)_2CO$) is a compound used in fertilizer, since it can react with water to form ammonia, which can be a source of nitrogen for plants. The reaction proceeds as follows:

$$(NH_2)_2CO \ (s) + H_2O \ (l) \rightarrow 2 \ NH_3 \ (aq) + CO_2 \ (g)$$

If 5.00 g of urea (molar mass = 60.0 g/mol) reacts with 1.35 g of H_2O, what is the volume of CO_2 (g) that can be formed at 1.00 atm and 0°C?

(A) 0.34 L

(B) 0.84 L

(C) 1.68 L

(D) 2.45 L

(E) 3.26 L

SOLUTION

The correct answer is choice (C). In order to determine which reactant is the limiting reagent, we need to determine how many moles of $(NH_2)_2CO$ and H_2O are present:

$$5.00 \text{ g urea} \times \left(\frac{1 \text{ mol urea}}{60.0 \text{ g urea}} \right) = 0.0833 \text{ mol } (NH_2)_2CO$$

$$1.35 \text{ g water} \times \left(\frac{1 \text{ mol water}}{18.0 \text{ g water}} \right) = 0.0750 \text{ mol } H_2O$$

According to the balanced equation, we know that the mole ratio of the reactants (mol H_2O/mol $(NH_2)_2CO$) = 1, so since there are fewer moles of H_2O present, it must be the limiting reagent (notice, though, that this is only true because of the 1:1 relationship of the reactants). Therefore, we use only the amount of H_2O to determine how much product can be formed. According to the balanced chemical equation:

$$0.0750 \text{ mol } H_2O \times \left(\frac{1 \text{ mol } CO_2}{1 \text{ mol } H_2O} \right) = 0.075 \text{ mol } CO_2$$

What is the volume of CO_2 formed? We know that at standard temperature and pressure (1.00 atm and 0°C) that one mole of any gas occupies 22.4 L, so:

$$0.075 \text{ mol } CO_2 \times \left(\frac{22.4 \text{ L } CO_2}{1 \text{ mol } CO_2} \right) = 1.68 \text{ L } CO_2$$

PERCENT YIELD

The final concept we need to discuss in this section is **percent yield**. As you have no doubt figured out in the laboratory, no chemist is perfect. The percent yield reflects the ratio between how much product was actually obtained and the theoretical yield of the reaction. Or:

$$\text{Percent yield} = \frac{\text{(Actual yield)}}{\text{(Theoretical yield)}} \times 100\%$$

Let's try an example that incorporates the concept of limiting reagent:

REVIEW QUESTION

The reaction of lithium hydroxide and carbon dioxide takes place according to the following balanced chemical equation:

$$2\ LiOH\ (s) + CO_2\ (g) \rightarrow Li_2CO_3\ (s) + H_2O$$

A 1.65 g sample of LiOH is allowed to react with 0.150 mol of CO_2, and 2.08 g of lithium carbonate are formed. Which of the following represents the percent yield for this synthesis?

(A) 73.3%

(B) 81.9%

(C) 89.5%

(D) 93.4%

(E) 97.2%

SOLUTION

The correct answer is choice (B). In this question, since neither reactant is present in excess, we must figure out which one is the limiting reagent. Since the ratio of reactants is not 1:1, it's less obvious which will be limiting. So let's add an easy step and calculate the number of moles of product that could theoretically be made from each reactant:

$$1.65\ g\ LiOH \times \left(\frac{1\ mol\ LiOH}{23.94\ g\ LiOH}\right) \times \left(\frac{1\ mol\ Li_2CO_3}{2\ mol\ LiOH}\right) = 0.0345\ mol\ Li_2CO_3$$

$$0.150\ mol\ CO_2 \times \left(\frac{1\ mol\ Li_2CO_3}{1\ mol\ CO_2}\right) = 0.150\ mol\ Li_2CO_3$$

Clearly, LiOH is limiting, and only 0.0345 mol Li_2CO_3 can be made. The theoretical yield of Li_2CO_3 in grams is:

$$0.0345\ mol\ Li_2CO_3 \times \left(\frac{73.88\ g\ Li_2CO_3}{1\ mol\ Li_2CO_3}\right) = 2.54\ g\ Li_2CO_3$$

$$\text{Percent yield} = \left(\frac{\text{Actual yield}}{\text{Theoretical yield}}\right) \times 100 = \left(\frac{2.08\ g}{2.54\ g}\right) \times 100 = 81.9\%$$

REVIEW QUESTIONS

1. $C_6H_{10}O_4$ (l) $+ 2 NH_3$ (g) $+ 4 H_2$ (g) $\rightarrow$
   $C_6H_{16}N_2$ (l) $+ 4 H_2O$ (l)

   According to the reaction above, how many moles of hexamethylenediamine ($C_6H_{16}N_2$) can be produced when three moles of $C_6H_{10}O_4$ react with four moles of NH_3 and four moles of H_2 in a flask?

   (A) 1
   (B) 2
   (C) 3
   (D) 4
   (E) 5

2. Aluminum reacts with sulfur gas to form aluminum sulfide (Al_2S_3). If 23.2 g of aluminum are reacted in excess sulfur, what is the theoretical yield of aluminum sulfide (in g)?

   (A) 21.9 g
   (B) 64.6 g
   (C) 87.6 g
   (D) 112.5 g
   (E) 134.3 g

3. Zn (s) $+ 2 HCl$ (aq) $\rightarrow ZnCl_2$ (aq) $+ H_2$ (g)

   Zinc reacts with hydrochloric acid in aqueous solution according to the reaction above. How many mL of a 0.50 M solution of HCl must be added to Zn to produce 5.6 L of gas at STP?

   (A) 250 mL
   (B) 500 mL
   (C) 750 mL
   (D) 1,000 mL
   (E) 2,000 mL

ANSWERS AND EXPLANATIONS

1. A

The first step is to identify the limiting reagent. In this case, we are already given the number of moles of each reactant. Let's suppose that $C_6H_{10}O_4$ is the limiting reagent. Since there are three moles of this compound present, we would have to have 6 moles of NH_3 and 12 moles of H_2 to react it completely. This is not the case, so $C_6H_{10}O_4$ is not the limiting reagent. If NH_3 were the limiting reagent, we would need twice the number of moles of H_2 to react it completely. We have four moles of NH_3 but only four moles of H_2. Therefore, H_2 must be the limiting reagent. Since we can form one mole of $C_6H_{16}N_2$ for every four moles of H_2, we can form one mole of this product.

2. B

It turns out we don't even need a balanced equation for this, so the formula of sulfur doesn't matter. The formula of aluminum sulfide is Al_2S_3. We know that aluminum metal, Al (s), is limiting, and there are no other aluminum-containing species besides Al (s) and Al_2S_3 (s), so the ratio must be $\dfrac{1 \text{ mol } Al_2S_3}{2 \text{ mol } Al}$ (in the balanced equation, it could be 1:2 or 2:4 or 3:6, etc., but that doesn't matter here).

$$23.2 \text{ g Al} \times \left(\frac{1 \text{ mol Al}}{26.98 \text{ g Al}} \right) = 0.860 \text{ mol Al}$$

$$0.860 \text{ mol Al} \times \left(\frac{1 \text{ mol } Al_2S_3}{2 \text{ mol Al}} \right) =$$
$$0.430 \text{ mol } Al_2S_3$$

$$0.430 \text{ mol } Al_2S_3 \times \left(\frac{150.14 \text{ g } Al_2S_3}{1 \text{ mol } Al_2S_3} \right) =$$
$$64.6 \text{ g } Al_2S_3$$

3. D

Remember that one mole of gas at STP has a volume of 22.4 L. Therefore:

$$5.6 \text{ L } H_2 \times \left(\frac{1 \text{ mol } H_2}{22.4 \text{ L } H_2} \right) = 0.25 \text{ mol of } H_2$$

According to the balanced chemical equation, we need two moles of HCl per mole of H_2 produced. Therefore, we need 0.5 mol of HCl.

0.5 mol HCl = (x L) (0.50 M) where x is the volume of solution.

$x = 1.0$ L

Therefore, 1.0 L, or 1,000 mL, of the 0.5 M solution is needed to produce the desired quantity of H_2 (g).

CHAPTER 13: EQUILIBRIA

IF YOU LEARN ONLY SEVEN THINGS IN THIS CHAPTER . . .

1. Chemical equilibrium is a dynamic process. At equilibrium, the rate of the forward reaction is equal to the rate of the reverse reaction. The equilibrium position is independent of the starting conditions or the direction of approach.

   For the general expression:

   $$aA + bB \rightleftarrows cC + dD$$

   The equilibrium constant K is given by:

   $$K = \frac{[C]^c [D]^d}{[A]^a [B]^b}$$

   a. The equilibrium constant (K) of a reaction that has been multiplied by a number, $n[aA + bB \rightleftarrows cC + dD]$, is the equilibrium constant raised to a power equal to that number (K^n).

   b. The equilibrium constant (K') of a reaction in the reverse direction, $cC + dD \rightleftarrows aA + bB$, is the inverse of the equilibrium constant of the reaction in the forward direction ($1/K$).

   c. The equilibrium constant for a net reaction made of two or more steps is the product of the equilibrium constants of the individual steps.

2. Le Chatelier's principle states that if any change is applied to a system at equilibrium, the equilibrium will shift so as to reduce that change. The system will obey this principle in response to changes in reagent concentration, temperature, and pressure.

3. If we increase the concentration of a component in a system at equilibrium, the equilibrium will shift away from that component. K will not be affected.

4. In response to an increase in pressure brought about by compressing or expanding the container, an equilibrium involving gases will shift to the side of the reaction producing the fewer number of moles of gaseous components, and K will not be affected. As long as there is no volume change, adding an inert gas will not change the partial pressures or the concentrations of the equilibrium species, so the reaction will not shift.

5. In response to an increase in temperature, the equilibrium will shift in the direction of the endothermic process, since energy is consumed in an endothermic reaction. K will be affected.

6. The solubility product constant K_{sp} describes to what extent a compound will dissociate in solution. You must be able to relate molar solubility to K_{sp}.

7. The presence of a common ion in solution will substantially diminish molar solubility, and forms the basis for selective precipitation.

Reactions in nature often do not go to completion. Instead, reactions are usually reversible, reaching chemical equilibrium when the rate of the forward reaction slows to the extent that this rate equals the rate of the opposite, or reverse, reaction. In this section, we will address the important concept of chemical equilibrium, with an emphasis on some topics that are often tested on the AP exam, such as calculations using the equilibrium constant K and applications of Le Chatelier's principle.

In the last section, we discussed the stoichiometry of reactions, which defines the molar ratio between reactants and products in a chemical reaction. Let's suppose we conducted an experiment in the laboratory in which we precipitated the solid ZnF_2 by addition of F^- ions to a solution containing Zn^{2+}:

$$Zn^{2+} (aq) + 2 F^- (aq) \rightarrow ZnF_2 (s)$$

If you looked at the above expression stoichiometrically, you would assume that for every one mole of Zn^{2+} (aq) present in solution, you could produce one mole of ZnF_2. Notice that this assumption implies that the reaction goes to completion.

If you collected the precipitate, dried it, and failed to get a 100 percent yield, would this imply that you were a bad chemist? No, it wouldn't. The above reaction does not go to completion, but instead exists in chemical equilibrium with the reverse process. It is impossible to precipitate all of the Zn^{2+} present in this solution.

In this section, we will discuss the meaning of chemical equilibrium and review some of the most likely calculations you will be asked to perform on test day. We will pay particular attention to the effect of temperature and pressure on chemical equilibrium, which is predicted using Le Chatelier's principle. Le Chatelier's principle essentially states that if we disturb any process in nature, nature will respond so as to minimize the impact of this disturbance.

THE MEANING OF CHEMICAL EQUILIBRIUM

When we considered reaction kinetics, we treated chemical reactions as though they occurred in one direction and one direction only. For example, consider the reaction between nitrogen dioxide and carbon monoxide:

NO_2 (g) + CO (g) $\rightarrow$ NO (g) + CO_2 (g) (forward reaction)

One might assume from looking at this expression that if equimolar amounts of NO_2 and CO were mixed, that every last molecule of each reagent would be used up. In the real world, reactions don't usually occur this way. Instead, reactions in nature usually occur until **chemical equilibrium** is reached, the state where the concentration of all products and reactants remain constant with time. In the case above, the reverse reaction also occurs, i.e.:

NO (g) + CO_2 (g) $\rightarrow$ NO_2 (g) + CO (g) (reverse reaction)

If we mixed one mole of NO_2 and one mole of CO together in a flask, initially the rate of the forward reaction would be quite fast. However, as we discussed in the kinetics section, the forward reaction would proceed more and more slowly. Eventually, the rate of the forward reaction would equal the rate of the reverse reaction. Once this is the case, the reaction has reached equilibrium. Remember: equilibrium is a dynamic process. The forward and reverse reactions do not simply stop once equilibrium is reached. The reaction above is better represented as:

NO (g) + CO_2 (g) $\rightleftarrows$ NO_2 (g) + CO (g)

To emphasize that equilibrium involves a balance between opposite reactions, the equilibrium expression is usually represented with a double arrow ($\rightleftarrows$). Remember the following points regarding chemical equilibrium:

- The equilibrium may be **product-favored** (products predominate over reactants) or **reactant-favored** (reactants predominate over products).

- The position of equilibrium is independent of the starting conditions or direction of approach. For the above reaction, the final relative concentrations of reactants and products would be the same whether we reacted one mole of NO_2 and one mole of CO or we reacted one mole of NO with one mole of CO_2.

- Equilibrium concentrations are not affected by the presence of a catalyst. Instead, a catalyst will allow the reaction to reach equilibrium more quickly.

THE EQULIBRIUM CONSTANT *K*

Most of the questions you will be asked on the AP exam regarding equilibrium will relate to the **equilibrium constant**. The general equilibrium expression is:

$$aA + bB \rightleftarrows cC + dD$$

The equilibrium constant *K* is given by:

$$K = \frac{[C]^c [D]^d}{[A]^a [B]^b}$$

In this expression, the square brackets indicate the concentration of chemical species at equilibrium. Note that lowercase *k* is a rate constant, as outlined in the last chapter, while the equilibrium constant is denoted by a capital *K*. The equation sheet has equations that include both, and you will need to be able to choose the right one.

For example, consider the synthesis of ammonia from elemental nitrogen and hydrogen:

$$N_2 \text{ (g)} + 3\,H_2 \text{ (g)} \rightleftarrows 2\,NH_3 \text{ (g)}$$

The equilibrium expression for this synthesis is given by the following:

$$K = \frac{[NH_3]^2}{[N_2][H_2]^3}$$

where the concentrations of all gases are expressed in mol/L. Notice that the units for the equilibrium constant will vary according to the number of reactants and products. In most textbooks, the units on *K* are omitted altogether and it is treated as a dimensionless number.

The equilibrium constant *K* for this reaction is 6.0×10^{-2} L^2/mol^2 at 500°C. What does this imply about the final concentrations of N_2, H_2, and NH_3 in any equilibrium situation? Notice that there are essentially infinite sets of equilibrium concentrations, or **equilibrium positions**, that could yield this value for *K*.

Remember: when using the equilibrium constant *K*, *pure solids and pure liquids are not included in the equilibrium constant* when there are components present in different phases. For example, for the decomposition of water into hydrogen and oxygen:

$$2H_2O \text{ (l)} \rightleftarrows 2H_2\text{(g)} + O_2 \text{ (g)}$$
$$K = [H_2]^2[O_2]$$

This reaction represents a **heterogeneous equilibrium**. Experimental results indicate that the position of a heterogeneous equilibrium does not depend on the amounts of pure solids or liquids present.

How do we determine if a system is at equilibrium? If we use actual concentrations of reactants and products present instead of equilibrium concentrations, we could use an identical expression to calculate the reaction quotient Q instead of equilibrium constant K. For example, let's take another look at the reaction:

$$N_2 \text{ (g)} + 3 H_2 \text{ (g)} \rightleftarrows 2 NH_3 \text{ (g)} \qquad K = 6.0 \times 10^{-2} \text{ L}^2/\text{mol}^2 \text{ at } 500°C$$

Suppose that at a given instant, $[NH_3] = 0.76$ M, $[H_2] = 0.12$ M, $[N_2] = 0.23$ M. The reaction quotient Q will be:

$$Q = \frac{[NH_3]^2}{[N_2][H_2]^3} = \frac{[0.76]^2}{[0.23][0.12]^3} = 1.5 \times 10^3 \text{ L}^2/\text{mol}^2$$

Notice that Q is much greater than K. This implies that the reaction is not at equilibrium. In fact, the large value of Q suggests that there is a preponderance of products present at the time the concentrations were measured. However, we know that at equilibrium, the reaction is reactant-favored since K is only 6.0×10^{-2} L^2/mol^2. This implies that from this starting point, the system will shift to the left in order to generate more reactants. The following general rules apply:

- If Q is equal to K, the reaction is already at equilibrium.

- If Q is less than K, the concentration of products is too small. The system must *shift to the right* to form products and consume reactants.

- If Q is greater than K, the concentration of products is too large. The system must *shift to the left* to form reactants and consume products.

CALCULATIONS USING *K*

Calculations involving K can sometimes be confusing. Because the AP Chemistry exam is designed to test your knowledge of chemistry (rather than your knowledge of mathematics), you will probably not be asked to solve any particularly difficult equations. Let's have a look at the two types of questions you may be asked.

The first type of question will be fairly straightforward. You may be asked to use the expression for equilibrium constant to solve for K and/or a missing concentration.

REVIEW QUESTION

Iodine and bromine liquids are combined at high pressure and high temperature and react as follows:

$$I_2 + Br_2 \rightleftarrows 2\ IBr$$

The equilibrium constant for the above reaction is 280 at 250°C. If the equilibrium concentrations of IBr and Br_2 are 1.3 M and 0.23 M respectively, what is the equilibrium concentration of I_2 at 250°C?

(A) 1.3×10^{-3} M

(B) 8.4×10^{-3} M

(C) 1.2×10^{-2} M

(D) 2.6×10^{-2} M

(E) 6.7×10^{-1} M

SOLUTION

The correct answer is choice (D). In order to solve this question, we need to know how to represent the equilibrium constant K for the reaction.

$$K = \frac{[IBr]^2}{[I_2][Br_2]}$$

Rearranging and solving for $[I_2]$, we get:

$$[I_2] = \frac{[IBr]^2}{K[Br_2]}$$

$$[I_2] = \frac{[1.3]^2}{280[0.23]} = 2.6 \times 10^{-2}\ M$$

The second type of question you might be asked is slightly more complicated, and you must look for an opportunity to simplify your calculations.

REVIEW QUESTION

At 35°C, $K = 1.6 \times 10^{-5}$ mol/L for the reaction:

$$2NOCl\,(g) \rightleftarrows 2NO\,(g) + Cl_2\,(g)$$

0.15 mol of pure NOCl is placed in a 1.0 L flask. Calculate the concentrations of all species when equilibrium is reached.

SOLUTION

Notice that K for this reaction is extremely small. So we can predict that at equilibrium, the concentration of NOCl will be almost identical to its initial concentration. We must start this problem by making the following table:

Initial Concentration (mol/L)	Change (mol/L)	Equilibrium Concentration (mol/L)
$[NOCl]_0 = 0.15$	$-2x$	$[NOCl] = 0.15 - 2x$
$[NO]_0 = 0$	$+2x$	$[NO] = 2x$
$[Cl_2]_0 = 0$	$+x$	$[Cl_2] = x$

We know that the initial concentration of NOCl is 0.15 M. To represent the change that occurs as the reaction approaches equilibrium, we can start by calling the increase in concentration of $Cl_2 = x$. By looking at the balanced chemical equation for this reaction, we can see that if $[Cl_2]$ increases by x, then $[NO]$ will increase by $2x$ and $[NOCl]$ will decrease by $2x$. Therefore, the concentrations at equilibrium are as represented on the right-hand side of the table. So K is given by:

$$K = \frac{[Cl_2][NO]^2}{[NOCl]^2} = \frac{[x][2x]^2}{[0.15 - 2x]^2} = 1.6 \times 10^{-5}$$

The algebra in this equation seems pretty complicated, but since you are not taking a math exam, there must be a way to simplify this equation. Because the K for this reaction is so small, we can ignore the change in concentration of NOCl; we just don't have enough decimal places in the original concentration to be able to see the change. The equation simplifies to:

$$\frac{[x][2x]^2}{[0.15 - 2x]^2} = \frac{[x][2x]^2}{[0.15]^2} = 1.6 \times 10^{-5}$$

$$4x^3 = 1.2 \times 10^{-6}$$

$$x = 0.3 \times 10^{-2}$$

Remember that we have only calculated x, the change in concentration of Cl_2. Therefore:

$[Cl_2] = x = 0.3 \times 10^{-2}\,M$

$[NO] = 2x = 0.6 \times 10^{-2}\,M$

$[NOCl] = 0.15 - 2[0.3 \times 10^{-4}] = 0.15\,M$

Let's check whether or not we get the correct value of K using these values:

$K = \dfrac{[0.3 \times 10^2]\,[0.6 \times 10^2]^2}{[0.15]^2} = 1.6 \times 10^{-5}$; yes, we do.

On your AP exam, there is a chance you will be confronted with an equilibrium calculation that is more complex. Just remember to draw a table and follow the basic steps, which include:

1. Writing a balanced equation for the reaction.

2. Listing the initial concentrations.

3. Defining the change needed to reach equilibrium (this is x), and defining equilibrium concentrations by applying this change to reactants and/or products.

4. Substituting these concentrations, writing the equilibrium expression, and solving for x.

5. Checking your work by making sure that your calculated equilibrium concentrations give the correct value of K. Be sure to reread the question and to make sure that your answer is a proper response to what the question asks.

LE CHATELIER'S PRINCIPLE

For any system at equilibrium, we can predict the impact of changes in concentration, pressure, and temperature using **Le Chatelier's principle**, which states that if any change is applied to a system at equilibrium, the equilibrium will shift so as to reduce that change. This principle sounds abstract, but there are really only three specific cases you'll need to understand on the AP exam.

First, you'll need to be able to predict how an equilibrium will shift in response to a change in concentration of one of the components. For example, suppose we have an equilibrium defined by the following reaction equation:

$PCl_5\,(g) \rightleftarrows PCl_3\,(g) + Cl_2\,(g)$

Let's suppose the system is at equilibrium and we look at the following three cases:

1. Adding PCl_5: If we increase the concentration of PCl_5, the system is no longer at equilibrium ($Q < K$). Le Chatelier's principle states that the system will react so as to oppose the change, so we expect some of the PCl_5 we added will be consumed and the equilibrium will shift toward products. When equilibrium is re-established, the concentrations of PCl_5, PCl_3, and Cl_2 will all be higher than those we started with before the extra PCl_5 was added.

2. Adding PCl_3: If we increase the concentration of PCl_3, the system is no longer at equilibrium ($Q > K$). Le Chatelier's principle states that some of the PCl_3 we added must be consumed, so the equilibrium must shift toward reactants. By looking at the balanced chemical equation, we see that if some of the added PCl_3 is consumed, some Cl_2 must also be consumed in the process. At equilibrium, the concentrations of PCl_3 and PCl_5 will be higher than we started with, and the concentration of Cl_2 will be lower.

3. Adding Cl_2: If we increase the concentration of Cl_2, the system is no longer at equilibrium ($Q > K$). We must consume some of the added Cl_2 to reach equilibrium, so the equilibrium will shift toward reactants. At equilibrium, the concentrations of Cl_2 and PCl_5 will be higher than we started with, and the concentration of PCl_3 will be lower.

Notice in each case that if we add a component, the system shifts away from that component. In contrast, if we were to take away a component, the system would shift toward that component.

You'll also need to be able to predict how an equilibrium will shift in response to a change in pressure. The system responds to a change in pressure by shifting *only* if the pressure change is caused by a volume change. When we decrease the volume our system is allowed to occupy, Le Chatelier's principle predicts that *the system will respond by minimizing its own volume*. How would a system at equilibrium do this? The equilibrium position will shift to the side of the reaction involving the fewer number of moles of gaseous components. Let's have a look at an example:

$$N_2 \text{ (g)} + 3\,H_2 \text{ (g)} \rightleftarrows 2\,NH_3 \text{ (g)}$$

In this case, there are a greater number of moles of gas in the reactants than in the products. An increase in pressure (decrease in volume) would cause the equilibrium to shift toward the products because this is the side with fewer moles of gas. Conversely, a decrease in pressure (increase in volume) would create more "room" for gas molecules and cause a shift back to the starting materials.

Finally, you will need to predict how a system at equilibrium will react to a change in temperature. In this case, you will need to know whether the reaction is endothermic or exothermic as written, either in the forward direction or the reverse direction. What do we predict raising the temperature will do? In this case, we are adding thermal energy to the system, so according to Le Chatelier's principle the system must react so as to consume some of that energy. Therefore, the *equilibrium will shift in the direction of the endothermic process*, since energy is consumed in an endothermic reaction. Conversely, if we decrease the temperature of a system, we predict that the equilibrium will shift in the direction of the exothermic process.

SOLUBILITY EQUILIBRIA

Solubility equilibria relate closely to those equilibrium situations we have already discussed. These equilibrium expressions are often used to describe a situation in which an ionic compound is only minimally soluble in water. For example, let's consider the dissolution of BaF_2 (s):

$$BaF_2 \text{ (s)} \rightleftarrows Ba^{2+} \text{ (aq)} + 2F^- \text{ (aq)}$$

As before, we do not include the concentrations of pure solids or liquids in our equilibrium expression, so the **solubility product constant**, or K_{sp}, is given by the following:

$$K_{sp} = [Ba^{2+}][F^-]^2$$

It's important to remember the difference between solubility product constant and solubility. For a given solid, there is only one solubility product constant at a given temperature. However, solubility of an ionic compound may vary depending on temperature or the presence of a common ion in solution, as we will see later.

The measured solubility can be used to calculate K_{sp} for a given temperature. For example, suppose that the solubility of BaF_2 at 25°C is 0.0182 mol/L. This means that 0.0182 moles of solid BaF_2 will dissolve per 1.0 L of solution to come to equilibrium with the excess solid. We can make a table corresponding to this situation:

Initial Concentration (mol/L)	Change (mol/L)	Equilibrium Concentration (mol/L)
$[Ba^{2+}]_0 = 0$	+0.0182	$[Ba^{2+}] = 0.0182$
$[F^-]_0 = 0$	+2(0.0182)	$[F^-] = 0.0364$

Notice that according to the balanced chemical equation, there are two moles of F^- formed for every mole of BaF_2 dissolved in solution, so the F^- at equilibrium is double the molar solubility of BaF_2.

What is the K_{sp} for BaF_2 at this temperature? We just need to substitute our calculated equilibrium values into the expression for K_{sp}:

$$K_{sp} = [Ba^{2+}][F^-]^2 = [0.0182][0.0364]^2 = 2.41 \times 10^{-5}$$

REVIEW QUESTION

The K_{sp} of lead iodide is 1.4×10^{-8}. Which of the following is the solubility of PbI_2 in mol/L?

(A) 3.4×10^{-4} M

(B) 1.5×10^{-3} M

(C) 9.6×10^{-3} M

(D) 3.2×10^{-2} M

(E) 8.5×10^{-2} M

SOLUTION

The correct answer is choice (B). In this case, we start by writing a balanced equation for the dissolution of lead iodide (PbI_2) in water:

$$PbI_2\ (s) \rightleftarrows Pb^{2+}\ (aq) + 2I^-\ (aq)$$

If we set the solubility of PbI_2 equal to x, we can fill out a table as follows:

Initial Concentration (mol/L)	Change (mol/L)	Equilibrium Concentration (mol/L)
$[Pb^{2+}]_0 = 0$	$+x$	x
$[I^-]_0 = 0$	$+2x$	$2x$

We know that K_{sp} is given by the following:

$$K_{sp} = [Pb^{2+}][I^-]^2 = [x][2x]^2 = 1.4 \times 10^{-8}$$

$$4x^3 = 1.4 \times 10^{-8}$$

$$x = 1.5 \times 10^{-3}$$

Therefore, the solubility of PbI_2 is 1.5×10^{-3} mol/L.

How would the solubility of PbI_2 be affected by the presence of Pb^{2+} already in solution, a common ion? The solubility of PbI_2 would be markedly decreased. Let's calculate the solubility of PbI_2 in a 0.10 M solution of $Pb(NO_3)_2$. Since $Pb(NO_3)_2$ is a highly soluble compound, it will dissociate completely:

$$Pb(NO_3)_2\ (s) \rightarrow Pb^{2+}\ (aq) + NO_3^-\ (aq)$$

If we set up a table again to determine the solubility of PbI_2 in this solution, we must take into account that there is already an initial concentration of Pb^{2+} present in solution ($[Pb^{2+}]_0 = 0.10$ M):

Initial Concentration (mol/L)	Change (mol/L)	Equilibrium Concentration (mol/L)
$[Pb^{2+}]_0 = 0.10$	$+x$	$0.10 + x$
$[I^-]_0 = 0$	$+2x$	$2x$

We know that the K_{sp} of PbI_2 is 1.4×10^{-8}, so:

$$K_{sp} = [Pb^{2+}][I^-]^2 = [x + 0.10][2x]^2 = 1.4 \times 10^{-8}$$

Remember that this is not an algebra question, so there must be some way of simplifying this expression. Since K_{sp} is so small, we can assume that the value of x is small compared to 0.10 and that $x + 0.10 \approx 0.10$. In fact, we know this to be true, since the solubility in this solution will be less than the solubility in pure water (0.0015 M), so:

$$K_{sp} = [x + 0.10][2x]^2 = 1.4 \times 10^{-8}$$
$$[0.10][4x^2] = 1.4 \times 10^{-8}$$
$$x = 1.9 \times 10^{-4} \text{ mol/L}$$

Finally, although acid-base reactions will be covered in the next section, let's think qualitatively about how the solubility of a compound will be affected by the pH of a solution. In other words, how will the solubility of a compound be affected by whether or not the solution is acidic or basic?

First, consider the dissolution of calcium hydroxide, or $Ca(OH)_2$:

$$Ca(OH)_2 \text{ (s)} \rightleftharpoons Ca^{2+} \text{ (aq)} + 2OH^- \text{ (aq)}$$

The K_{sp} for this reaction is 1.3×10^{-6} in water at 25°C. How would the solubility of $Ca(OH)_2$ be affected by the pH of the solution? In basic solution, $[OH^-]$ is higher than that in a neutral solution. Therefore, by the common ion effect, we expect that $Ca(OH)_2$ will be less soluble in basic solution. If the solution is acidic, there will be more H^+ ions in solution, which can react by the following neutralization reaction:

$$OH^- \text{ (aq)} + H^+ \text{ (aq)} \rightarrow H_2O \text{ (l)}$$

Therefore, OH^- anions will be removed from solution, further driving the solubility equilibrium to the right and allowing more $Ca(OH)_2$ to dissolve.

Finally, the **common ion effect** can often be used to purify a single compound from a mix of compounds in solution. This is an example of **selective precipitation**. We can define an **ion product** Q that is similar to the Q we defined earlier for gaseous equilibrium. For example, for the dissolution of PbI_2 in solution:

$$PbI_2 \text{ (s)} \rightleftharpoons Pb^{2+} \text{ (aq)} + 2I^- \text{ (aq)}$$
$$Q = [I^-]_0^2[Pb^{2+}]_0$$

where $[I^-]_0$ and $[Pb^{2+}]_0$ are initial, rather than equilibrium, concentrations of these ions. Q is related to K_{sp} by the following:

1. If $Q > K_{sp}$, then **precipitation** occurs and will continue until equilibrium is established.

2. If $Q = K_{sp}$, then the solution is **saturated** and no more compound can be dissolved in solution.

3. If $Q < K_{sp}$, then no precipitation occurs, and if solid is present it will dissolve until equilibrium is established.

Selective precipitation is often performed by continuous addition of a common ion in solution, as you will see in the review questions.

REVIEW QUESTIONS

1. The following components are at equilibrium in a 1 L flask at 25°C:

$$P_4 \, (g) + 5 \, O_2 \, (g) \rightleftarrows P_4O_{10} \, (g)$$

If 0.25 mol of gaseous chlorine is added to the flask, which of the following is true?

I. $[P_4]$ will increase.

II. $[P_4O_{10}]$ will increase.

III. $[O_2]$ will decrease.

(A) I only

(B) II only

(C) I and II only

(D) II and III only

(E) I, II, and III

2. For the following reaction, the standard enthalpy change is 16.1 kJ:

$$2 \, NOBr \, (g) \rightleftarrows 2 \, NO \, (g) + Br_2 \, (g)$$

Which of the following will be true if the temperature of the system is decreased, with pressure held constant?

(A) K will increase.

(B) The concentration of NO will increase.

(C) The concentration of Br_2 will decrease.

(D) The concentration of NOBr will decrease.

(E) The reaction $2 \, NO \, (g) + Br_2 \, (g) \rightleftarrows 2 \, NOBr \, (g)$ is endothermic.

3. Which of the following is *NOT* true?

(A) The molar solubility of CaF_2 ($K_{sp} = 4.0 \times 10^{-11}$) is less than that of BaF_2 ($K_{sp} = 2.4 \times 10^{-5}$).

(B) The solubility of $MnCO_3$ will be higher in acidic solution than in basic solution.

(C) The solubility of AgCl will be higher in acidic solution than in basic solution.

(D) The solubility of Ag_3PO_4 will be lower in basic solution than in acidic solution.

(E) The solubility of $Ca(OH)_2$ will be lower in basic solution than in acidic solution.

4. A solution is prepared in which $[Sr^{2+}] = [Ba^{2+}] = 4.0 \times 10^{-4}$ M. NaF is slowly added to the solution at 25°C. The K_{sp} for BaF_2 is 2.4×10^{-5}, and the K_{sp} for SrF_2 is 7.9×10^{-10} at this temperature. Which of the following is true?

I.   The first compound that will precipitate is SrF_2.

II. There is a concentration of F^- at which SrF_2 will precipitate but not BaF_2.

III. There is no concentration of F^- at which SrF_2 and BaF_2 will both precipitate.

(A) I only

(B) II only

(C) I and II only

(D) II and III only

(E) I, II, and III

5. $2 H_2O_2 (g) \rightleftarrows 2 H_2O (g) + O_2 (g)$

   After the equilibrium represented above is established, some pure H_2O vapor is injected into the reaction vessel at constant temperature. When equilibrium is re-established, which of the following has a lower value compared to its value at the original equilibrium?

   (A) K

   (B) $[H_2O_2]$

   (C) $[H_2O]$

   (D) $[O_2]$

   (E) The total pressure in the reaction vessel

Questions 6–9 refer to the following equilibrium:

$2 SO_3 (g) \rightleftarrows 2 SO_2 (g) + O_2 (g) \; \Delta H° = 197 \text{ kJ}$

	$[SO_3]$	$[SO_2]$	$[O_2]$	K
(A)	increase	increase	decrease	–
(B)	decrease	increase	increase	increase
(C)	increase	decrease	decrease	decrease
(D)	increase	decrease	increase	–
(E)	increase	increase	increase	–

Assume that the reaction is initially at equilibrium. Determine how the following changes will affect the quantities once a new equilibrium is established.

6. Pure O_2 is added to the reaction vessel.

7. The temperature is increased from 25°C to 75°C.

8. The pressure of the reaction vessel is increased by compression to a smaller volume at constant temperature.

9. Pure SO_3 is added to the reaction vessel.

FREE-RESPONSE QUESTION

1. A solution is prepared by adding 3.0×10^{-4} moles of PbI_2 to distilled water to form 500 mL of solution at 25°C. The K_{sp} of PbI_2 at this temperature is 1.4×10^{-8}.

(a) Explain why this solution is not saturated using the solubility product constant and the ion product.

(b) NaI is slowly added to the solution. How many moles of NaI must we add to induce precipitation of PbI_2?

ANSWERS AND EXPLANATIONS

1. D

Remember that even though chlorine is highly reactive, it won't easily form molecules with P or O because they are all electron donors. The effect of adding chlorine will be to increase the pressure of the system. We predict that the equilibrium will shift so as to favor the direction in which fewer moles of gaseous products are produced, in this case to the right. Therefore, we expect that the concentrations of both P_4 and O_2 will decrease, and the concentration of P_4O_{10} will increase as the new equilibrium is reached.

2. C

As the temperature is decreased, the equilibrium will shift so as to produce thermal energy, i.e., in the direction of the exothermic reaction. Since the reaction is endothermic as written, the equilibrium will shift to the left. Therefore, when the new equilibrium is reached, [NO] will have decreased, [B_2] will have decreased, and [NOBr] will have increased. The equilibrium constant $K = \dfrac{[NO]^2}{[NOBr]^2}$ will also have decreased.

3. C

Since Cl^- is the conjugate base of a strong acid, it is not basic and therefore will not be removed from solution by the presence of H^+. The solubility of AgCl will not be significantly affected by the pH of the solution. Choice (A) is correct since a lower K_{sp} implies that the solubility is lower, as long as the ratios of dissolved ions are the same between the two species being compared. Choice (B) is true since the dissolution of $MnCO_3$ produces CO_3^{2-}, which

can undergo a neutralization reaction with H^+ and drive the equilibrium to the right. Choice (D) is true since Ag_3PO_4 will dissolve to form PO_4^{3-}, which is a basic anion. This compound should therefore be more soluble in acidic solution than in basic solution. Choice (E) is true since $Ca(OH)_2$ will form OH^- when dissolved, and will therefore be less soluble in basic solution as a result of the common ion effect.

4. C

We can tell by the relative K_{sp} that SrF_2 has a much smaller solubility than BaF_2, and therefore will precipitate first as the common ion F^- is added to the solution. So there will be a concentration of F^- at which we can precipitate SrF_2 but not BaF_2.

5. D

$$2\,H_2O_2\,(g) \rightleftarrows 2\,H_2O\,(g) + O_2\,(g)$$

When H_2O is added to the vessel, the equilibrium is shifted to the left, increasing the concentration of H_2O_2. In the process, O_2 is consumed, decreasing its concentration in the vessel. K will remain unchanged in the process.

6. D

Adding O_2 will cause the equilibrium to shift to the left, consuming SO_2 and some of the added O_2 and making more SO_3. As long as the temperature is constant, K does not change.

7. B

The reaction is endothermic, so increasing the temperature will cause a shift toward products and an increase in the value of K.

8. E

Increasing the pressure causes a shift toward the side with fewer moles of gas, but all species will have higher concentrations after the change (though there will be fewer moles of SO_2 and O_2 than before the change, their concentrations will be higher because of the smaller space).

9. E

Adding SO_3 causes a shift toward products and increases the concentrations of SO_2 and O_2 but does not completely use up the added SO_3, so all concentrations are higher after the change.

FREE-RESPONSE QUESTION

1. Begin by writing the balanced chemical equation:

$$PbI_2 \rightleftarrows Pb^{2+} + 2I^-$$

(a) We need to explain why this solution is not saturated by showing that $Q < K$. The solubility product constant, K_{sp} of PbI_2 at this temperature, is 1.4×10^{-8} and needs to be compared with the ion product, Q (calculated from initial concentrations of Pb^{2+} and I^-). According to the balanced chemical equation above, we see that dissolution of 3.0×10^{-4} moles of PbI_2 will produce 3.0×10^{-4} moles of Pb^{2+} and 6.0×10^{-4} moles of I^-. Since the volume of solution is 500 mL:

$$[I^-]_0 = \frac{6.0 \times 10^{-4} \text{ mol}}{0.500 \text{ L}} = 1.2 \times 10^{-3} \text{ M}$$

$$[Pb^{2+}]_0 = \frac{3.0 \times 10^{-4} \text{ mol}}{0.500 \text{ L}} = 6.0 \times 10^{-4} \text{ M}$$

$$Q = [I^-]_0^2[Pb^{2+}]_0 = [1.2 \times 10^{-3}]^2[6.0 \times 10^{-4}] = 8.6 \times 10^{-10}$$

Since $Q < K$, we have shown that the solution is NOT saturated. This is because the ion product is smaller than the solubility constant.

(b) We need to assess how much I^- needs to be added to the solution in order to begin precipitation of PbI_2. Precipitation will occur when $Q > K$, so we need to find the point at which $Q = K$ to find the concentration of I^- at which precipitation will begin. Remember here that the principle we are using is the common ion effect. So:

$$Q = [I^-]_0^2[Pb^{2+}]_0 = K_{sp} = 1.4 \times 10^{-8}$$

We know that the concentration of Pb^{2+} has not changed. So:

$$[I^-]_0^2[6 \times 10^{-4}] = 1.4 \times 10^{-8}$$

$$[I^-]_0 = 4.8 \times 10^{-3} \text{ M}$$

This is the total concentration of I^- required to begin concentration of PbI_2. The corresponding number of moles in our 500 mL solution is given by:

moles I^- = $(4.8 \times 10^{-3} \text{ moles/L})$ $(0.500 \text{ L}) = 2.4 \times 10^{-3}$ moles I^-

We already had 6.0×10^{-4} moles of I^- due to dissolution of PbI_2, so the number of moles of NaI that need to be added is:

$$(2.4 \times 10^{-3}) - (0.6 \times 10^{-3}) = 1.8 \times 10^{-3} \text{ moles of NaI}$$

CHAPTER 14: ACID-BASE REACTIONS

IF YOU LEARN ONLY 12 THINGS IN THIS CHAPTER . . .

1. Acid-base reactions can be understood using the Arrhenius, Lewis, or Brønsted-Lowry models. On the AP exam, most questions will relate to the Brønsted-Lowry definition, which states that acids are proton (H^+) donors and bases are proton (H^+) acceptors.

2. The strength of an acid refers to the extent of dissociation in aqueous solution.

3. For a weak acid in solution:

$$HA\ (aq) \rightleftarrows H^+\ (aq) + A^-\ (aq) \qquad K_a = \frac{[H^+][A^-]}{[HA]}$$

4. The relative strength of acids can be explained in part by the electronegativities of the atoms attached to the proton.

5. The autoionization of water occurs according to the following equilibrium:

$$2H_2O\ (l) \rightleftarrows H_3O^+\ (aq) + OH^-\ (aq) \qquad K_w = [H_3O^+][OH^-]$$

Or more simply, $K_w = [H^+][OH^-]$.

6. For any aqueous solution, $K_w = 1 \times 10^{-14}$ (at 25°C).

7. The pH and $[H^+]$ of an aqueous solution are related by the following:

$$pH = -\log[H^+]$$

8. For any titration, remember that there are two distinct steps you must consider: stoichiometric conversion that goes essentially to completion, and the resulting equilibrium, if any.

9. For a strong acid-strong base titration, the pH will be determined by the presence of OH⁻ (aq) or H⁺ (aq) in solution that has not been neutralized. The pH at the equivalence point is equal to 7.

10. For a weak acid-strong base titration, please review the chapter for a summary of how to calculate pH at different points in the pH curve. The pH at the equivalence point will be >7 due to the presence of conjugate base in solution.

11. For a weak base-strong acid titration, the pH at the equivalence point will be <7 due to the presence of conjugate acid.

12. Buffers represent a solution containing a conjugate acid/conjugate base pair. The buffering capacity will be determined by the absolute concentrations of these components in solution, with dilute buffers possessing a lower buffering capacity.

Acid-base reactions are essential in the normal functioning of the human body and will be equally important on your AP Chemistry exam. It's important to notice that some of the more complicated calculations you will be asked to perform on test day are found in this section. We will practice these as well as provide you with a qualitative understanding of the most difficult topics, including titrations and buffering.

In the last section, we discussed the meaning of chemical equilibrium, which defines the relationship between reactants and products in a reversible process. It is almost impossible to discuss equilibrium without mentioning acid-base chemistry. Since our bodies do not tolerate extreme variations in pH, the majority of proton exchanges in the body will take place between weak acids and bases under equilibrium conditions.

If we were to take a tiny sample of blood from the human bloodstream and determine the pH, it would most likely be 7.40 for a young, healthy person. In fact, a pH of less than 7.35 or greater than 7.45 would be extremely unusual. The pH of the bloodstream is under tight control by the lungs and kidneys, and significant deviation from a pH of 7.40 (even a pH of 7.10) can be life-threatening.

Under what circumstances would the pH of the bloodstream change? Production of an acid, for example lactic acid ($HC_3H_5O_3$), formed during athletic activity or severe infection, can produce protons (H⁺) in solution.

Fortunately, the body is equipped with defenses against such toxic insults. The bloodstream is **buffered** by the presence of bicarbonate ion (HCO_3^-). The acid can be neutralized by the following reaction:

$$HCO_3^- \text{ (aq)} + H^+ \text{ (aq)} \rightleftarrows H_2CO_3 \text{ (aq)}$$

In this section, we will start by defining what we mean when we talk about acids and bases. We will then discuss the meaning of pH and define the acid dissociation constant K_a. Finally, we will look at the changes in pH that occur during acid-base titrations and in buffered solutions. By the end of this section, you should be able to perform the calculations associated with acids and bases at equilibrium, as well as have a qualitative understanding of why acid-base titrations proceed the way they do.

ARRHENIUS, LEWIS, AND BRØNSTED-LOWRY ACIDS

Let's start with some basic definitions. There are three basic ways of understanding the nature of acids and bases in solution:

According to the **Arrhenius definition**, an acid is a substance that produces hydrogen ions (H^+) in aqueous solution, while a base is a substance that produces hydroxide ions (OH^-) in solution. You won't find yourself applying this definition often on the AP exam, but it is important to remember that dissolving an acidic substance (such as citric acid, the principal acid in lemon juice) results in a solution in which $[H^+] > [OH^-]$, whereas dissolving a basic substance (think drain cleaner) results in a solution in which $[OH^-] > [H^+]$.

According to the **Lewis model**, an acid (also known as a Lewis acid) is a molecule that can accept an electron lone pair, whereas a base is a molecule that is an electron-pair donor. For example, the reaction of NH_3 with H^+ to form the ammonium cation can be considered a Lewis acid-base reaction:

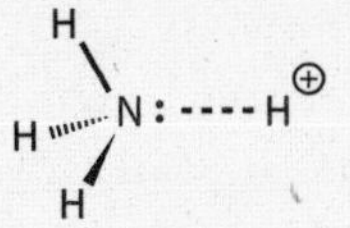

In this case, we notice that we have an interaction between an electron-deficient atom (H^+ has no electrons) and an electron-rich atom (N has an electron lone pair, as indicated by its correctly drawn Lewis structure). In this case, H^+ is the Lewis acid (electron-pair acceptor) and NH_3 is the Lewis base (electron-pair donor).

The only tricky thing about this definition is that Lewis acids and bases do not always involve H^+. The reaction between BF_3 and NH_3 is an example of a Lewis acid and base. Remember that BF_3 represents an exception to the octet rule; in this molecule, B is electron-deficient and has an empty p orbital instead of an electron lone pair in its correctly drawn Lewis structure. NH_3 is electron-rich, and can donate its lone pair to BF_3 to form a stable bonding interaction. In this case, BF_3 is the Lewis acid and NH_3 is the Lewis base:

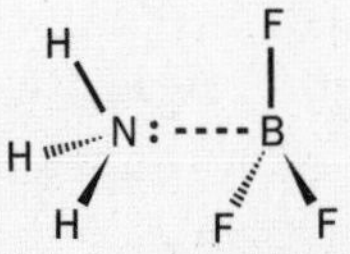

Remember that metal ions commonly behave as Lewis acids in aqueous solution. If you see a transition metal cation in solution on the AP exam, chances are it's behaving as a Lewis acid.

The **Brønsted-Lowry model** states that an acid is a proton donor, whereas a base is the proton acceptor in an acid-base reaction. This is the most useful and commonly applied definition on your AP exam. For example, consider the reaction of HBr (hydrobromic acid) and water:

$$HBr\ (aq) \rightarrow H^+\ (aq) + Br^-\ (aq)$$

In this reaction, HBr dissociates completely (more or less) in water to form Br^- and H^+ ions. The H^+ atom can then react with the lone pair of H_2O to form the hydronium ion, or H_3O^+. According to the Brønsted-Lowry model, H^+ is acting as the proton donor while H_2O is acting as the proton acceptor or base in this reaction. Remember that since this reaction always goes to completion, $[H^+] = [H_3O^+]$.

$$H^+\ (aq) + H_2O\ (l) \rightarrow H_3O^+\ (aq)$$

Alternatively, H_2O can act as an acid in an aqueous solution. Consider the dissociation of NaOH (sodium hydroxide) in aqueous solution.

$$NaOH\ (s) \rightarrow Na^+\ (aq) + OH^-\ (aq)$$

NaOH completely dissociates into sodium and hydroxide ions as it becomes surrounded by water molecules. Water will then act as the proton donor, or acid, in the reaction shown below.

$$H_2O\ (l) \rightarrow OH^-\ (aq) + H^+\ (aq)$$

ACID-BASE EQUILIBRIUM AND K_a

Some acids do not dissociate completely in water. In this case, an equilibrium is reached. For example, consider the dissolution of acetic acid ($HC_2H_3O_2$) in water:

$$HC_2H_3O_2\ (aq) + H_2O\ (l) \rightleftarrows H_3O^+\ (aq) + C_2H_3O_2^-\ (aq)$$

We define the acid as the proton donor, and its corresponding **conjugate base** as everything that remains of the proton donor once the proton has been lost. A **conjugate acid** refers to a base once a proton has been transferred to it. For any reaction, you must be able to identify the **conjugate acid-base pair**. In the expression above, $HC_2H_3O_2$ and $C_2H_3O_2^-$ represent a conjugate acid-base pair; $HC_2H_3O_2$ is the conjugate acid with $C_2H_3O_2^-$ the corresponding conjugate base. Similarly, H_3O^+ and H_2O represent a conjugate acid-base pair in the reaction above.

And we can define an equilibrium constant K_a. Again, since the concentration of liquid water is essentially constant when an acid is dissolved, we can exclude it from our equilibrium expression:

$$K_a = \frac{[H_3O^+][C_2H_3O_2^-]}{[HC_2H_3O_2]}$$

Since we exclude water from the equilibrium constant, we can simplify our expression for the dissolution of acetic acid as follows:

$$HC_2H_3O_2 \ (aq) \rightleftarrows H^+ \ (aq) + C_2H_3O_2^- \ (aq)$$

$$K_a = \frac{[H^+][C_2H_3O_2^-]}{[HC_2H_3O_2]}$$

Since $[H^+] = [H_3O^+]$, both expressions for K_a are identical. On your AP exam, it's easier to use the simplified dissociation expression, so in general:

$$HA \ (aq) \rightleftarrows H^+ \ (aq) + A^- \ (aq) \qquad K_a = \frac{[H^+][A^-]}{[HA]}$$

ACID STRENGTH

What does the strength of an acid refer to? This term refers to the extent to which an acid dissociates in acidic solution. As we previously mentioned, HBr will dissolve more or less completely in water to form H^+ and Br^-. If we were to express the dissolution of HBr as an equilibrium:

$$HBr \ (aq) \rightleftarrows H^+ \ (aq) + Br^- \ (aq) \qquad K_a = \frac{[H^+][Br^-]}{[HBr]}$$

the value of K_a would be exceedingly large—for practical purposes, infinite—and the equilibrium lies far to the right.

Therefore, HBr is a **strong acid** and can be considered to dissociate completely. In contrast, remember our expression for the dissolution of acetic acid:

$$HC_2H_3O_2 \ (aq) \rightleftarrows H^+ \ (aq) + C_2H_3O_2^- \ (aq)$$

The K_a for this reaction is 1.8×10^{-5}, implying that the equilibrium lies far to the left. This implies that the majority of $HC_2H_3O_2$ in solution is not dissociated. $HC_2H_3O_2$ is therefore a **weak acid**.

It is very important to understand how the strength of an acid corresponds to the strength of its corresponding conjugate base. The stronger the acid, the weaker its corresponding conjugate base. In contrast, for a weak acid, the conjugate base will be relatively strong. If we compare the base strength in each case to water, we note the following:

- The conjugate base of a strong acid is less basic than H_2O.
- The conjugate base of a weak acid is more basic than H_2O.

- Sometimes you can predict the strength of an acid based on molecular structure. Some acids are **oxyacids**, in which the acidic proton is attached to an oxygen atom. Low molar mass oxyacids (HNO_3) and the conjugate acids of halides (HBr, HI, etc.) are more likely to be strong, whereas **carboxylic acids** (citric acid, etc.) are usually weak.

There are some pretty tough concepts here, but if the AP exam asks you to relate molecular structure to acid strength, you should use **electronegativity** arguments to make your case. Essentially, it is easier to pull off a proton when there are more electronegative atoms in the acid "hogging" all the electrons. We would expect the acid to be stronger when this is the case.

WATER AND K_w

As always, we must give special attention to water, as your AP exam almost certainly will. One special feature of H_2O in acid-base reactions is that it is **amphoteric**, which means that it can behave as an acid or as a base. This can be seen in the autoionization of water:

$$2H_2O \ (l) \rightleftarrows H_3O^+ \ (aq) + OH^- \ (aq)$$

The equilibrium constant for this expression is K_w:

$$K_w = [H_3O^+][OH^-] \text{ or more simply, } K_w = [H^+][OH^-]$$

It turns out that in a solution at pH = 7, $[H_3O^+] = [OH^-] = 1 \times 10^{-7}$ M. The corresponding K_w is 1×10^{-14}. You will need to remember the following:

- For any aqueous solution, $K_w = [H_3O^+][OH^-] = 1 \times 10^{-14}$.

- In a neutral solution, $[OH^-] = [H_3O^+]$.

- In an acidic solution, $[H_3O^+] > [OH^-]$.

- In a basic solution, $[OH^-] > [H_3O^+]$.

THE pH SCALE

The most common way to represent solution acidity is using the pH scale, and you will be expected to relate $[H^+]$ to pH frequently on the AP exam. Neutral solutions have a pH of exactly 7, while acidic solutions have a pH < 7 and basic solutions have a pH > 7. Your calculator will come in handy when using the following formula:

$$pH = -\log[H^+]$$

or $[H^+] = 10^{-pH}$

In addition to being able to use the above formula, it's useful to consider what the above logarithmic relationship means. A logarithmic relationship means that a single unit of pH corresponds to a tenfold difference in $[H^+]$. In other words, compared to a solution with pH = 7, a solution with pH = 6 will have 10 times the H^+ concentration, and a solution with pH = 5 will have 100 times the H^+ concentration.

CALCULATING THE pH OF AN ACIDIC SOLUTION

How do we calculate the pH of a strong acid solution? Since we predict complete dissociation of the acid in solution, this calculation is straightforward. What is the pH of a 0.040 M solution of HCl? In this case, we have complete dissociation of HCl according to the following expression:

$$HCl\ (aq) \rightarrow H^+\ (aq) + Cl^-\ (aq)$$

We can ignore the contribution to $[H^+]$ of the autoionization of water, so we predict that $[H^+] =$ 0.040 M. The corresponding pH is given by:

$$pH = -\log[H^+] = -\log[0.040] = 1.40$$

The rule for significant figures here is that the number of *decimal places* in the log is equal to the number of *significant figures* in the original number.

How do we go about calculating the pH for a weak acid in solution? The answer is that we can use similar techniques to those we applied previously to equilibrium situations. For example, suppose we want to calculate the pH of a solution formed by dissolving 0.25 moles of propanoic acid, CH_3CH_2COOH, to form a 1.0 L solution at 25°C. If the K_a for propanoic acid is 1.0×10^{-5}, what is the pH of the resulting solution? To solve this problem, we start by writing a balanced chemical equation corresponding to the dissociation:

$$CH_3CH_2COOH\ (aq) \rightleftarrows H^+\ (aq) + CH_3CH_2COO^-\ (aq)$$

We then use a table to summarize the changes that occur as equilibrium is reached, setting the change in concentration of the undissociated propanoic acid equal to x. According to the balanced chemical equation, for every mole of propanoic acid that dissociates, one mole of H^+ and one mole of $CH_3CH_2COOO^-$ will be formed. Therefore, the equilibrium concentrations of both of these species will be equal to x.

Initial Concentration (mol/L)	Change (mol/L)	Equilibrium Concentration (mol/L)
$[CH_3CH_2COOH]_0 = 0.25$	$-x$	$0.25 - x$
$[H^+]_0 = 0$	$+x$	x
$[CH_3CH_2COOO^-]_0 = 0$	$+x$	x

We can then use the expression for K_a to calculate the value of x when equilibrium is reached:

$$K_a = \frac{[H^+][CH_3CH_2COOO^-]}{[CH_3CH_2COOH]} = \frac{[x][x]}{[0.25 - x]}$$

Since you won't be doing a lot of complicated algebra on your AP Chemistry exam, there must be a way to simplify this expression. We know that since propanoic acid is a weak acid, the majority of propanoic acid remains undissociated in solution. In other words, the equilibrium lies far to the left. Therefore, we know that x will be very small relative to the initial concentration of CH_3CH_2COOH. We can make the following approximation:

$$\frac{[x][x]}{[0.25 - x]} \cong \frac{[x][x]}{[0.25]} = 1 \times 10^{-5}$$
$$x^2 = 2.5 \times 10^{-6}$$
$$x = [H^+] = 0.0016 \text{ M}$$

Notice that x did turn out to be pretty small compared to 0.25, the initial concentration of CH_3CH_2COOH. What is the corresponding pH?

$$pH = -\log [H^+] = -\log [0.0016] = 2.80$$

How would we define the **percent dissociation** for this case? This is another way of expressing how much the weak acid has dissociated in achieving equilibrium:

$$\text{Percent dissociation} = \frac{\text{Amount dissociated (M)}}{\text{Initial concentration (M)}} \times 100\%$$

In the above case, we figured out that the amount dissociated, or x, was equal to 0.0016 M. Since the initial concentration of CH_3CH_2COOH was 0.25 M, the percent dissociation is given by:

$$\text{Percent dissociation} = \frac{0.0016}{0.25} \times 100 = 0.64\%$$

One interesting feature of percent dissociation is that it tends to increase as the solution becomes more dilute. So for any weak acid HA, as $[HA]_0$ decreases, the percent dissociation increases, even though $[H^+]$ at equilibrium is decreasing.

POLYPROTIC ACIDS

One important type of calculation you may be asked to perform on the AP exam involves **polyprotic acids**, which can furnish more than one proton and dissociate in a stepwise manner. For example, H_2CO_3 (carbonic acid) is formed in the human body by the reaction of CO_2 (carbon dioxide) with H_2O. Carbonic acid can then dissociate twice:

$$H_2CO_3 \text{ (aq)} \rightleftharpoons H^+ \text{ (aq)} + HCO_3^- \text{ (aq)} \qquad K_{a1} = \frac{[H^+][HCO_3^-]}{[H_2CO_3]} = 4.3 \times 10^{-7}$$

$$HCO_3^- \text{ (aq)} \rightleftharpoons H^+ \text{ (aq)} + CO_3^{2-} \text{ (aq)} \qquad K_{a2} = \frac{[H^+][CO_3^{2-}]}{[HCO_3^-]} = 5.6 \times 10^{-11}$$

Notice that $K_{a1} > K_{a2}$. This is a general trend for polyprotic acids; the acid in each step gets successively weaker as protons are lost. As the negative charge of an acid increases, it becomes weaker since electrostatic forces make it more difficult to remove a proton.

What is the pH of a 0.15 M solution of carbonic acid? Remember that we must calculate the total concentration of H^+, and therefore need to take into account both equilibria. For the first dissociation step:

Initial Concentration (mol/L)	Change (mol/L)	Equilibrium Concentration (mol/L)
$[H_2CO_3]_0 = 0.15$	$-x$	$0.15 - x$
$[H^+]_0 = 0$	$+x$	x
$[HCO_3^-]_0 = 0$	$+x$	x

Using our expression for K_{a1} and again using the approximation that x is small relative to the initial concentration of carbonic acid:

$$\frac{[H^+][HCO_3^-]}{[H_2CO_3]} \cong \frac{[x][x]}{[0.15]} = 4.3 \times 10^{-7}$$
$$[x] = 0.00025 \text{ M} = [HCO_3^-]^- = [H]^+$$

We can then set up a second table corresponding to the second dissociation step, in order to determine how much of the HCO_3^- will further dissociate:

Initial Concentration (mol/L)	Change (mol/L)	Equilibrium Concentration (mol/L)
$[HCO_3^-]_0 = 0.00025$	$-x$	$0.00025 - x$
$[H^+]_0 = 0.00025$	$+x$	$0.00025 + x$
$[CO_3^{2-}]_0 = 0$	$+x$	x

Using the expression for K_{a2}:

$$\frac{[H^+][CO_3^{2-}]}{[HCO_3^-]} = \frac{[0.00025 + x][x]}{[0.00025 - x]} = 5.6 \times 10^{-11}$$

Since the equilibrium constant is so small, x will be incredibly small compared to 0.00025, so we can simplify the expression:

$$\frac{[0.00025 + \cancel{x}][x]}{[0.00025 - \cancel{x}]} \cong 5.6 \times 10^{-11}$$

$$[x] = 5.6 \times 10^{-11} M = [CO_3^{2-}]$$

Since x is so small, this result implies that for a polyprotic acid, only the first dissociation step makes an important contribution to $[H^+]$. Therefore, if you are only asked to calculate the pH, you only need to calculate $[H^+]$ for the first dissociation step. In this case, $[H^+] = 0.00025$ M, so in order to calculate the pH:

$$pH = -\log[H^+] = -\log[0.00025] = 3.60$$

CALCULATING THE pH OF A BASIC SOLUTION

What about bases? The calculations involved are similar to those for acids, although it's easy to get confused if you don't practice a few questions prior to test day.

Let's start by calculating the pH of a strong base in solution. What is the pH of a 0.020 M solution of KOH? Since KOH is a strong base, the dissociation occurs essentially to completion:

$$KOH\ (aq) \rightarrow K^+\ (aq) + OH^-\ (aq)$$

Therefore, $[OH^-]$ should be 0.020 M once complete dissociation occurs. In order to calculate the pH, we need to determine the $[H^+]$ in this solution. Remember that for any aqueous solution:

$$K_w = 1 \times 10^{-14} = [H^+][OH^-]$$

So in our case:

$$[H^+][0.020] = 1 \times 10^{-14}$$

$$[H^+] = 5 \times 10^{-13} M$$

$$pH = -\log[5 \times 10^{-13}] = 12.30$$

If we want to calculate the pH of a weak base in solution, the situation is slightly more complicated and we must define a few new terms. For the reaction of a weak base with water, we have the following general expression:

$$A^-\ (aq) + H_2O\ (l) \rightleftarrows HA\ (aq) + OH^-\ (aq) \qquad K_b = \frac{[HA][OH^-]}{[A^-]}$$

where K_b is the base equilibrium constant. If we are given K_b for a weak base, it's pretty straightforward to calculate $[OH^-]$ at equilibrium to determine the pH. However, often we must determine K_b using the K_a associated with the corresponding conjugate acid. These two constants are related as follows:

$$K_w = K_a \times K_b$$

On your AP exam, this is most likely to come up when you are asked to calculate the pH of a solution formed by dissolving a **salt** containing a weak base. In this case, you will need to know the K_a of the corresponding conjugate acid to solve the problem.

REVIEW QUESTION

Calculate the pH of a 0.50 M solution of sodium acetate ($NaC_2H_3O_2$). The K_a of acetic acid is 1.8×10^{-5}.

SOLUTION

Since sodium acetate is an ionic compound, it will dissociate completely in water as follows:

$$NaC_2H_3O_2 \text{ (s)} \rightarrow Na^+ \text{ (aq)} + C_2H_3O_2^- \text{ (aq)}$$

We know that $C_2H_3O_2^-$ is the conjugate base of a weak acid (acetic acid), so it should be basic in aqueous solution. We start by writing the balanced chemical equation for the reaction of the acetate anion with water:

$$C_2H_3O_2^- \text{ (aq)} + H_2O \text{ (l)} \rightleftarrows H C_2H_3O_2 \text{ (aq)} + OH^- \text{ (aq)}$$

The equilibrium constant for this reaction is K_b, given by:

$$K_b = \frac{K_w}{K_a} = \frac{(1 \times 10^{-14})}{(1.8 \times 10^{-5})}$$

$$K_b = 5.6 \times 10^{-10} = \frac{[HC_2H_3O_2]\,[OH^-]}{[C_2H_3O_2^-]}$$

We can then set up a table corresponding to the situation:

Initial Concentration (mol/L)	Change (mol/L)	Equilibrium Concentration (mol/L)
$[C_3H_3O_2^-]_0 = 0.50$	$-x$	$0.50 - x$
$[OH^-]_0 = 0$	$+x$	x
$[HC_3H_3O_2]_0 = 0$	$+x$	x

Using our expression for K_b and again simplifying since $x <<< [C_3H_3O_2^-]$:

$$K_b = 5.6 \times 10^{-10} = \frac{[HC_2H_3O_2][OH^-]}{[C_2H_3O_2^-]} \cong \frac{[x][x]}{0.50}$$

$$x^2 = 2.8 \times 10^{-10}$$

$$x = 1.7 \times 10^{-5} \, M = [OH^-]$$

Since $[OH^-][H^+] = 1 \times 10^{-14}$ for any aqueous solution:

$$[H^+] = \frac{(1 \times 10^{-14})}{(1.7 \times 10^{-5})} = 5.9 \times 10^{-10} \, M$$

$$pH = -\log[5.9 \times 10^{-10}] = 9.23$$

ACID-BASE TITRATIONS

During your AP Chemistry class, you most likely spent some time in the laboratory performing a **titration**. In acid-base chemistry, a titration is often performed to determine the amount of acid or base in solution. On your exam, you may or may not be asked to do a lot of titration calculations. However, it will always be high yield to understand what is going on qualitatively for the following three cases:

- Strong acid–strong base
- Weak acid–strong base
- Weak base–strong acid

For each case, you must be familiar with how the pH curve looks, including the position of the **equivalence point** and the pH at which it is reached. The equivalence point refers to the point in the titration where we have added exactly enough acid/base to react completely with the acid/base present initially in solution. Remember that the equivalence point in any acid-base titration is defined by the stoichiometry, not by the pH.

First, let's consider the strong acid–strong base titration. An example would be the titration of 50.0 mL of a 0.250 M HCl solution with 0.100 M NaOH. What reactants are present? We know that H^+, Cl^-, OH^-, and Na^+ will all be formed in solution by complete dissociation of the strong acid and strong base. In this case, Cl^- and Na^+ will be innocent bystanders during the titration, which involves the neutralization reaction between H^+ and OH^- as follows:

$$H^+ \, (aq) + OH^- \, (aq) \rightarrow H_2O \, (l)$$

Notice that we write this reaction as going to completion. For any of the titrations that may appear on your AP exam, this will be the case. For each scenario, we must consider the initial acid-base reaction stoichiometrically, then consider any equilibria that may exist once this initial reaction has taken place.

What will be the equivalence point for this titration? The equivalence point will occur when we have added the number of moles of OH^- needed to neutralize the H^+ in solution. Therefore, we need to calculate how many moles of H^+ we started with:

moles H^+ = (0.250 mol/L)(0.050 L) = 0.0125 mol H^+

This implies that the equivalence point will occur when we have added 0.0125 moles of OH^-. Since the concentration of NaOH solution is 0.100 M, we can calculate the volume required by the following:

0.0125 mol OH^- = (0.100 mol/L)(x)

x = 0.125 L = 125 mL

Since there are no acids or bases remaining in solution at the equivalence point, the pH = 7.

What would the pH of the solution be if we were only to add 100.0 mL of NaOH? In this case, the number of moles of NaOH we have added will be given by:

moles OH^- = (0.100 mol/L)(0.1000 L) = 0.0100 mol OH^-

Remember that since we had 0.0125 moles of H^+ initially in solution, there are (0.0125 – 0.0100) or 2.5×10^{-3} moles of H^+ left over once the neutralization reaction has taken place. The total volume of the solution is now 0.0500 L + 0.100 L = 0.150 L, so the concentration of H^+ is:

2.5×10^{-3} mol/0.150 L = 0.017 M

The pH is given by –log [0.017] = 1.78.

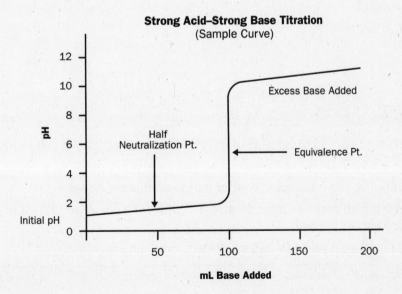

The second type of titration we must consider is the titration of a weak acid using a strong base. To determine the pH at any given point in the titration, follow these two simple steps:

1. Consider the reaction of a weak acid (for example, HCN) with a strong base (for example, NaOH) as going to completion. Remember that you must treat any titration first as a **stoichiometric** problem.

2. Calculate the concentrations of all acids and bases in solution.

Before the equivalence point is reached, both the weak acid and its conjugate base will be present in solution. You may use the Henderson-Hasselbalch equation to calculate the pH of the solution. This is also known as the **buffer region** of the titration curve.

At the equivalence point, the pH will be determined by the conjugate base of the acid you are titrating.

After the equivalence point, the pH of the solution will be determined by the excess concentration of strong base added, taking into account the new total solution volume.

The Henderson-Hasselbalch equation is extremely useful in calculating the pH of a solution in which both an acid and its conjugate base are present. For the general equilibrium situation:

$$HA\ (aq) \rightleftarrows H^+\ (aq) + A^-\ (aq)$$

$$pH = pK_a + \log \frac{[A^-]}{[HA]}$$

where $pK_a = -\log (K_a)$

In a buffer solution, the conjugate acid and conjugate base are both in the same volume of solution. Therefore, the ratio of their concentrations is the same as the ratio of moles in each:

$$\frac{[A^-]}{[HA]} = \frac{mol\ A^-}{mol\ HA}$$

$$pH = pK_a + \log \left(\frac{mol\ A^-}{mol\ HA} \right)$$

REVIEW QUESTION

400.0 mL of a 0.150 M solution of HF is titrated with a 0.300 M solution of NaOH. The K_a of HF is 7.2×10^{-4}.

(a) Calculate the pH when 100 mL of NaOH solution are added.

(b) Will the pH of the solution be acidic, neutral, or basic when 200 mL of NaOH are added? Explain.

(c) Calculate the pH when 250 mL of NaOH solution are added.

SOLUTION

To start, we need to determine how many moles of HF are present in the solution. Since dissociation of this weak acid is negligible:

moles of HF = (0.150 mol/L)(0.400 L) = 0.0600 mol

For (a), we have added 100.0 mL of NaOH solution. The number of moles we have added is given by (0.300 mol/L)(0.100 L) = 0.0300 mol. Remember that the reaction of a weak acid and strong base is considered to go to completion:

$$HF\,(aq) + OH^-\,(aq) \rightarrow F^-\,(aq) + H_2O\,(l)$$

Because we started with 0.0600 moles of HF and added only 0.0300 moles of OH^-, according to the balanced chemical equation above we will form 0.0300 moles of F^-, with 0.0300 moles of HF left over. Since this is a conjugate acid/conjugate base pair, we can calculate the pH using the Henderson-Hasselbalch equation:

$$pH = pK_a + \log \frac{(mol\ F^-)}{(mol\ HF)}$$

For HF: $pK_a = -\log(K_a) = -\log(7.2 \times 10^{-4}) = 3.14$

The pH of the solution is therefore given by:

$$pH = 3.14 + \log \frac{[0.0300]}{[0.0300]} = 3.14$$

For (b), we have added 200 mL of NaOH solution. The number of moles we have added is given by (0.300 mol/L)(0.200 L) = 0.0600 mol. This is equal to the number of moles of HF, so we are at the equivalence point. Be sure that you don't assume that the pH is 7.0. It's easy to make that mistake, but look again at the balanced chemical equation:

$$HF\,(aq) + OH^-\,(aq) \rightarrow F^-\,(aq) + H_2O\,(l)$$

At the equivalence point, all 0.0600 moles of HF have reacted to form 0.0600 moles of F^-. Since F^- is the conjugate base of a weak acid, it will be basic in solution. Therefore, the pH at the equivalence point will be basic. This is true for all weak acid-strong base titrations.

For (c), we are now beyond the equivalence point. The number of moles of NaOH we have added is given by (0.300 mol/L)(0.250 L) = 0.0750 mol. According to the balanced chemical equation, this means that we have added 0.0150 moles more of

NaOH than was required to react with all 0.0600 moles of HF present in solution. The pH in this situation is determined only by this excess concentration of OH^-, so:

$$[OH^-] = \frac{0.0150 \text{ mol}}{0.650 \text{ L}} = 0.0231 \text{ M}$$

$$K_w = [OH^-][H^+] = 1 \times 10^{-14}$$

$$[H^+] = \frac{K_w}{[OH^-]} = \frac{(1 \times 10^{-14})}{0.0231} = 4.33 \times 10^{-13} \text{ M}$$

$$pH = -\log[H^+] = 12.363$$

The overall pH curve for this type of titration is shown here.

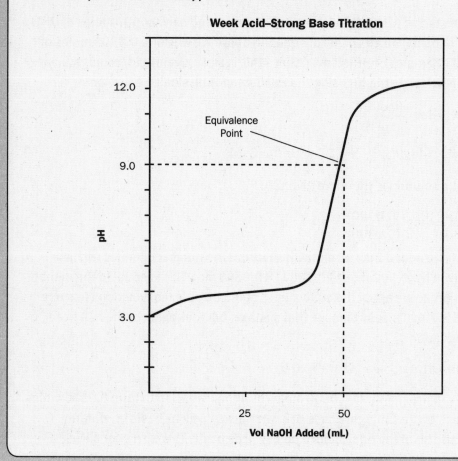

Week Acid–Strong Base Titration

The third case we will consider is the titration of a weak base using a strong acid. For example, consider the titration of a solution of NH_3 using HCl. The reaction will be considered to go to completion as follows:

$$NH_3 \text{ (aq)} + H^+ \text{ (aq)} \rightarrow NH_4^+ \text{ (aq)}$$

Again, there will be three important cases to consider:

1. Prior to the equivalence point: Again, in this case, NH_3 and NH_4^+ will both be present in solution. And again, since an acid and its conjugate base are both present, the Henderson-Hasselbalch equation may be used to calculate the pH. This is also known as the buffer region of the titration.

2. At the equivalence point: All NH_3 will be consumed and converted to NH_4^+. Since NH_4^+ is a weak acid, the pH at the equivalence point will be less than 7.

3. Beyond the equivalence point, the pH will be determined by the $[H^+]$ due to excess HCl added and must account for the new total solution volume.

The overall pH curve for this type of titration is shown below.

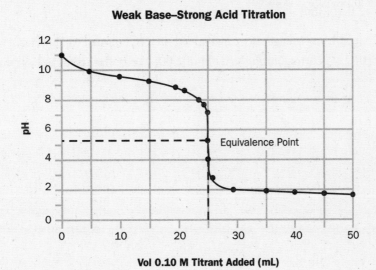

BUFFERED SOLUTIONS

The final concept we need to consider in this section is **buffers**. A buffered system is one that resists changes in pH and consists of a conjugate acid/base pair in solution. We can construct a buffer by combining a weak acid and the salt of its conjugate base or by combining a weak base and the salt of its conjugate acid. For example, suppose we generated a buffer by dissolving 0.350 moles of acetic acid ($HC_2H_3O_2$) and 0.200 moles of sodium acetate ($NaC_2H_3O_2$) in water to form 1 L of solution. Given that the K_a of acetic acid is 1.8×10^{-5}, what is the pH of the resulting solution? In this case, the pH can be calculated using the Henderson-Hasselbalch equation:

$$pH = pK_a + \log \frac{[C_2H_3O_2^-]}{[HC_2H_3O_2]}$$

where $pK_a = -\log (K_a) = -\log (1.8 \times 10^{-5}) = 4.74$

thus, $pH = 4.74 + \log \dfrac{[0.200]}{[0.350]} = 4.50$

Let's suppose we added 0.010 moles of HCl to the solution without changing the volume of the solution significantly. What would happen to the pH? This is the equivalent situation to a titration of a weak base with a strong acid. In other words, the following reaction will take place to completion:

$$C_2H_3O_2^- \text{ (aq)} + H^+ \text{ (aq)} \rightarrow HC_2H_3O_2 \text{ (aq)}$$

The result is that the concentration of $C_2H_3O_2^-$ will decrease by 0.010 M to 0.190 M, while the concentration of $HC_2H_3O_2$ will increase by 0.010 M to 0.360 M. The new pH is given by:

$$pH = 4.74 + \log \dfrac{[0.190]}{[0.360]} = 4.46$$

This is not a very significant change. If we had added the HCl instead to 1L of pure water (pH = 7), $[H^+]$ would be 0.010 M, corresponding to a pH of 2.00. The buffer seems to have done its job.

REVIEW QUESTIONS

1. HI is a strong acid whereas HF is a weak acid. Which of the following is *NOT* true?

   (A) If a 1 M solution of HI is prepared, $[I^-]$ is approximately 1 M.

   (B) For an HF solution at equilibrium, $[HF] > [H^+]$.

   (C) For an HF solution at equilibrium, $[H^+] > [OH^-]$.

   (D) A solution of NaF is expected to be basic.

   (E) A solution of NaI is expected to be basic.

2. A 0.25 M solution of hydrocyanic acid (HCN) is prepared. Which of the following is/are true at equilibrium?

   I. $[H^+] > 1 \times 10^{-7}$ M

   II. $[OH^-] < 1 \times 10^{-7}$ M

   III. $[H^+][OH^-] = 1 \times 10^{-14}$

   (A) I only

   (B) II only

   (C) I and II only

   (D) II and III only

   (E) I, II, and III

3. The dissociation of sulfurous acid (H_2SO_3) in aqueous solution occurs as follows:

$$H_2SO_3 \text{ (aq)} \rightleftarrows H^+ \text{ (aq)} + HSO_3^- \text{ (aq)}$$

$$K_{a1} = \frac{[H^+][HSO_3^-]}{[H_2SO_3]} = 1.5 \times 10^{-2}$$

$$HSO_3^- \text{ (aq)} \rightleftarrows H^+ \text{ (aq)} + SO_3^{2-} \text{ (aq)}$$

$$K_{a2} = \frac{[H^+][SO_3^{2-}]}{[HSO_3^-]} = 1.0 \times 10^{-7}$$

If 0.50 moles of sulfurous acid are dissolved to form a 1 L solution, which of the following concentrations will be LEAST at equilibrium?

   (A) $[H_2SO_3]$

   (B) $[H^+]$

   (C) $[H_3O^+]$

   (D) $[HSO_3^-]$

   (E) $[SO_3^{2-}]$

4. You wish to construct a buffer of pH = 7.0. Which of the following weak acids (with corresponding conjugate base) would you select?

   (A) $HClO_2$ (chlorous acid) $K_a = 1.2 \times 10^{-2}$

   (B) HF (hydrofluoric acid) $K_a = 7.2 \times 10^{-4}$

   (C) HOCl (hypochlorous acid) $K_a = 3.5 \times 10^{-8}$

   (D) HCN (hydrocyanic acid) $K_a = 4.0 \times 10^{-10}$

   (E) C_2H_5OH (ethanol) $K_a = 1.0 \times 10^{-25}$

5. A buffer is generated using NH_3 and NH_4Cl. Which of the following is/are true?

    I. Any buffer generated for which $\dfrac{[NH_3]}{[NH_4Cl]}$ is the same will have an identical pH.

    II. If $[NH_3] = [NH_4Cl]$, the pH will equal the pK_a for NH_4^+.

    III. Any buffer generated for which $\dfrac{[NH_3]}{[NH_4Cl]}$ is the same will have an identical buffering capacity.

    (A) I only

    (B) II only

    (C) I and II only

    (D) II and III only

    (E) I, II, and III

6. When you exercise, the burning sensation that sometimes occurs in your muscles represents the buildup of lactic acid ($HC_3H_5O_3$). In a 0.20 M aqueous solution, lactic acid is 2.6% dissociated. What is the value of K_a for this acid?

    (A) 4.3×10^{-6}

    (B) 8.3×10^{-5}

    (C) 1.4×10^{-4}

    (D) 5.2×10^{-3}

    (E) 9.8×10^{-3}

7. At 25°C, an aqueous solution has a proton concentration, $[H^+]$, of 1.0×10^{-10} M. The hydroxide ion concentration, $[OH^-]$, is which of the following?

    (A) 0 M

    (B) 1.0×10^{-2} M

    (C) 1.0×10^{-4} M

    (D) 5.0×10^{-6} M

    (E) 1.0×10^{-10} M

Questions 8–11 refer to aqueous solutions containing 1:1 mole ratios of the following pairs of substances. Assume all concentrations are 1 M.

    (A) NaCl and NaOH

    (B) NaOAc and HOAc

    (C) HNO_3 and $NaNO_3$

    (D) NH_3 and NH_4Cl

    (E) NaCl and K_2SO_4

8. The solution with pH = 7

9. A buffer at pH > 8

10. A buffer at pH < 6

11. The solution with the lowest pH

12. A molecule or ion is classified as a Lewis base if it

    (A) accepts a proton from water.

    (B) donates a proton to water.

    (C) accepts a pair of electrons.

    (D) donates a pair of electrons.

    (E) results in a basic solution when dissolved in water.

FREE-RESPONSE QUESTION

1. Consider the weak acids HIO and HIO_3.

   (a) Draw the Lewis structures for both acids.

   (b) HIO has a pK_a of 10.50, whereas HIO_3 has a pK_a of 0.78.

      i. Calculate the K_a of each acid.

      ii. Use structural arguments to explain the difference in acid strength between these two compounds.

      iii. Which is a stronger base in aqueous solution, IO^- or IO_3^-? Explain.

   (c) Calculate the pH of a 0.12 M solution of HIO.

ANSWERS AND EXPLANATIONS

1. E

This statement is not true, since I^- is the conjugate base of a strong acid and therefore will not be appreciably basic in solution. Choice (A) is true since HI is a strong acid and therefore will be expected to dissociate completely in aqueous solution to H^+ and I^-. Choice (B) is true since HF is a weak acid and therefore dissociates only minimally in solution. Choice (C) is true since according to the Arrhenius definition of acids and bases, for any acid in solution $[H^+] > [OH^-]$. The solution will therefore have a pH < 7. Choice (D) is true because F^- is the conjugate base of a weak acid and will therefore be a stronger base than H_2O.

2. E

All three statements are correct regarding the concentrations of ions at equilibrium.

3. E

On the AP exam, you'll need to recognize questions that can be answered correctly without performing long and tedious calculations. In this case, if you actually calculated all of the above concentrations, you would be wasting a lot of time. We saw for carbonic acid that the second dissociation is very small (and its contribution to the total $[H^+]$ is not significant). Therefore, we expect that $[SO_3^{2-}]$, the conjugate base of the second dissociation of H_2SO_3, will be much lower than the other concentrations listed.

4. C

It is always best to use a buffer system for which the conjugate acid has a pK_a close to the desired pH. In this case, the pK_a of hypochlorous acid is $-\log [K_a] = 7.46$.

5. C

According to the Henderson-Hasselbalch equation, both I and II will be true. Buffering capacity refers to the amount of protons or hydroxide ions the buffer can absorb without a significant change in pH. We expect that the buffering capacity will be higher when the buffer is more concentrated and lower when the buffer system is more dilute. The ratio $\frac{[NH_3]}{[NH_4Cl]}$ will determine the pH, but the buffering capacity will be determined by the magnitudes of $[NH_3]$ and $[NH_4Cl]$.

6. C

As in any equilibrium problem, we start by writing the balanced chemical equation for the dissociation described:

$$HC_3H_5O_3 \text{ (aq)} \rightleftarrows H^+ \text{ (aq)} + C_3H_5O_3^- \text{ (aq)}$$

We can then construct a table for the dissociation:

Initial Concentration (mol/L)	Change (mol/L)	Equilibrium Concentration (mol/L)
$[HC_3H_5O_3]_0 = 0.20$	$-x$	$0.20 - x$
$[H^+]_0 = 0$	$+x$	x
$[C_3H_5O_3^-]_0 = 0$	$+x$	x

We know that in order to calculate the equilibrium constant for this reaction, we will need to know the concentrations of all of the above at equilibrium, but we are only given the percent dissociation. We can calculate the amount of acid dissociated as follows:

$$\text{Percent dissociation} = \frac{x}{[HC_3H_5O_3]_0} \times 100$$

$$2.6\% = \frac{x}{(0.20 \text{ M})} \times 100$$

$$x = 0.0052 \text{ M}$$

Now that we know x, we can fill out the table above:

Initial Concentration (mol/L)	Change (mol/L)	Equilibrium Concentration (mol/L)
$[HC_3H_5O_3]_0 = 0.20$	-0.0052	0.20
$[H^+]_0 = 0$	$+0.0052$	0.0052
$[C_3H_5O_3^-]_0 = 0$	$+0.0052$	0.0052

We can then substitute the above equilibrium concentrations into our expression for K_a:

$$\frac{[H^+][C_3H_5O_3^-]}{[HC_3H_5O_3]} = \frac{[0.0052]^2}{0.20} = 1.4 \times 10^{-4}$$

7. C

For any aqueous solution, $K_w = 1.0 \times 10^{-14} = [H^+][OH^-]$. Therefore, the concentration of OH^- in this solution is given by:

$$[OH^-] = \frac{K_w}{[H^+]} = \frac{[1.0 \times 10^{-14}]}{[1.0 \times 10^{-10}]} = 1.0 \times 10^{-4} \text{ M}$$

8. E

A solution of NaCl and K_2SO_4 will have a neutral pH, since none of the ions present in solution are appreciably acidic or basic. Cl^- and SO_4^{2-} are the conjugate bases of strong acids, and therefore not appreciably basic in aqueous solution.

9. D

A basic buffer can be identified by comparing the K_a of the weak acid to the K_b of the corresponding conjugate base. If K_a is less than 1×10^{-7} (implying that K_b is greater than 1×10^{-7}), the buffer will be basic. Because the K_a of NH_4^+ is 5.6×10^{-10}, we expect that the buffer solution in question will be basic. We could confirm this using the Henderson-Hasselbalch equation.

10. B

The HOAc (with a K_a of 1.75×10^{-5}) may be used to generate a buffer with acidic pH when combined with its conjugate base.

11. C

HNO_3 is a strong acid while $NaNO_3$ has no acid/base properties. Therefore, this represents an unbuffered acidic solution with a low pH.

12. D

A Lewis base donates a lone pair of electrons to an electron-deficient atom to form a bond.

FREE-RESPONSE QUESTION

1. **(a)** The correct Lewis structures of these two oxyacids are shown below. Notice that in order to draw a correct Lewis structure for HIO_3, we needed to draw I with an expanded octet.

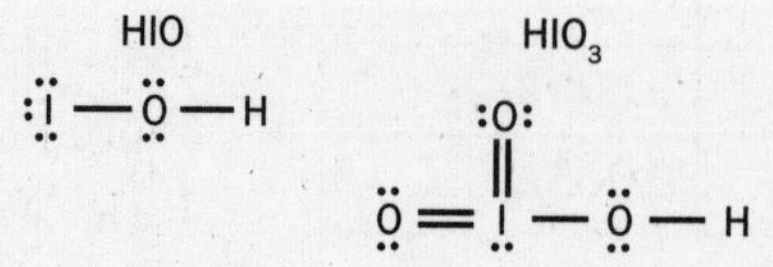

(b) **i.** The K_a is related to the pK_a by the following formula:

$pK_a = -\log[K_a]$
Therefore, $K_a = 10^{-pK_a}$
The K_a for $HIO = 3.2 \times 10^{-11}$
The K_a for $HIO_3 = 1.7 \times 10^{-1}$

ii. Looking at the conjugate bases, in IO^-, the negative charge is localized on the O atom, making it interact more strongly with the positive H^+. In IO_3^-, the negative charge is delocalized through resonance to give a net $-\frac{1}{3}$ charge on each O atom, so the O atoms are much less basic. Therefore, HIO_3 is a stronger acid.

iii. The stronger the acid, the weaker the conjugate base in solution. Therefore, IO_3^- will be a much weaker base than IO^-.

(c) The dissociation of HIO will occur as follows:

$HIO \text{ (aq)} \rightleftarrows H^+ \text{ (aq)} + IO^- \text{ (aq)}$

If we set the change in concentration equal to x and assume that x is significantly smaller than the initial concentration, then at equilibrium:

$K_a = 3.2 \times 10^{-11} = \dfrac{[x][x]}{[0.12 - x]} \approx \dfrac{x^2}{0.12}$

$x = 2.0 \times 10^{-6} \text{ M} = [H^+]$

Therefore, the pH of the solution is $-\log[2.0 \times 10^{-6}] = 5.71$.

CHAPTER 15: REDOX REACTIONS

IF YOU LEARN ONLY NINE THINGS IN THIS CHAPTER . . .

1. Redox reactions can be summarized using half-reactions. The reduction half-reaction describes the compound being reduced, also known as the oxidizing agent. The oxidation half-reaction describes the compound being oxidized, also known as the reducing agent.

2. In an electrochemical cell, reduction occurs at the cathode while oxidation occurs at the anode. The most important feature of any electrochemical cell is the choice of materials used.

3. For an electrochemical cell:

$$E^{\circ}_{cell} = E^{\circ}_{red} \text{ (reduction process)} + E^{\circ}_{ox} \text{ (oxidation process)}$$

4. If E°_{cell} is positive for an oxidation/reduction reaction, the process is spontaneous and will occur in the forward direction.

5. The relationship between E°_{cell} and Gibbs free energy is as follows:

$$\Delta G^{\circ} = -nFE^{\circ}_{cell}$$

6. The course of any redox reaction can be determined by looking at standard reduction potentials. A process with a more positive E° will always occur more readily than a process with a less positive E°. This basic rule can be used to predict which compound will be oxidized and which compound will be reduced in any redox reaction.

7. The Nernst equation can look at the voltage of an electrochemical cell when the components of the cell are *not* at standard conditions:

$$E_{cell} = E^{\circ}_{cell} - \ln Q$$

8. Electrolysis is a process that forces reactant-favored reactions to occur by applying a current. Remember that in electrolysis, you must always determine whether or not a given oxidation/reduction reaction is favored with respect to water (under aqueous conditions).

9. Corrosion is a process by which metals are oxidized. This process can be limited by anodic or cathodic protection. Cathodic protection involves use of a sacrificial anode that is more readily oxidized than the metal in question.

In this section, we will tackle the intimidating topic of redox chemistry. Don't panic when you encounter a redox chemistry question on the AP exam. It's likely that the redox questions you will see on test day will be relatively straightforward. The most common question posed will be, "Predict which redox reaction will occur." Read this section carefully, and you will be able to answer this basic question. This will help you to collect as many points as possible on the day of the exam.

Oxidation and reduction reactions are extremely important in the human body, but in this section we will focus on some redox chemistry relevant to industrial processes. How is a metal extracted from the earth? Ionic materials are reduced to form pure metals. This process might seem pretty obscure, but remember that *ages* were defined by the technology available to extract a given metal (e.g., the Bronze Age). You can read more about the extraction of metals during these periods by doing an Internet search, thereby taking advantage of our recent entry into the "Silicon Age."

In this chapter, we will start with some basic definitions that will help us to describe what is going on during an oxidation-reduction reaction. We will then consider the representation of redox reactions using half-reactions and the use of standard reduction potentials to determine the most favorable path for electrons to flow. Finally, we will look at some important applications of redox chemistry, including electrochemical cells, electrolysis, and corrosion technologies.

To answer a redox question correctly on your AP Chemistry exam, it helps to know when the reaction in question involves oxidation and reduction. You will know that a redox reaction is taking place by noticing one or all of the following:

- A change in oxidation number. You will need to determine the oxidation number of each element as it appears in a reactant or product.

- The presence of an uncombined element as a reactant or product

- The presence of a strong oxidizing or reducing agent as a reactant or product (for example, $KMnO_4$)

For example, the reaction

$$Cl^- \, (aq) + Cr_2O_7^{2-} \, (aq) \rightarrow Cl_2 \, (g) + Cr^{3+} \, (aq)$$

clearly satisfies all these conditions.

DETERMINING OXIDATION STATES

How do we go about determining the oxidation states for all elements in a reaction? The following rules must be obeyed:

- *The oxidation number of an atom of a pure element is zero.* For example, the oxidation number is zero for O_2, S_8, Cu, and Br_2.

- *The oxidation number of a monoatomic ion equals its charge.* The oxidation number of Fe^{3+} (aq) is +3 while the oxidation number of Cl^- (aq) is −1.

- *Some common elements almost always have the same oxidation number.* Fluorine always has an oxidation number of −1, as do the rest of the halogens unless they are attached to a more electronegative element. In compounds, atoms of the first three groups have positive oxidation numbers +1, +2, and +3 respectively. The oxidation number of O is −2 except in peroxides (H_2O_2, Na_2O_2), for which the oxidation number of O is −1 (notice that H and Na can't be +2, so O can't be −2). When combined with fluorine, the oxidation number for O becomes +2. The oxidation number of H is +1 except in hydrides (NaH) for which the oxidation number of H is −1 (again, notice that Na won't be −1, so H can't be +1 in this case).

- *In a binary compound, an atom of the more electronegative element has an oxidation number equal to the charge it would have if it were an ion.* Fortunately, in all ionic binary compounds, the more electronegative element is the second element in the formula, so it is easy to tell which element is more electronegative. In ionic compounds, this means that the cation is written first and the anion is second. Consider NaH from the previous example. The H in NaH is the more electronegative element and acts as the anion, which is why H has an oxidation number of −1 in this compound.

- *The sum of the oxidation numbers in a neutral compound is zero. In a polyatomic ion, the sum of the oxidation numbers is equal to the charge of the ion.* For example, consider the ion $Cr_2O_7^{2-}$ above. What is the oxidation number of Cr in this compound? We know that the sum of all the oxidation numbers must be −2, so: $2(x) + 7(-2) = -2$. Solving, $x = 6$, so the oxidation number of chromium in this compound is 6.

In any redox reaction, it's important to be able to identify which species are being oxidized and which are being reduced. Let's have a look at a sample redox reaction and identify which elements are being oxidized and which are being reduced:

$$Zn \, (s) + 2 \, HCl \, (aq) \rightarrow ZnCl_2 \, (aq) + H_2 \, (g)$$

- On the reactant side, the oxidation numbers of Zn, H, and Cl are 0, +1, and −1 respectively. On the product side, the oxidation numbers of Zn, H, and Cl are +2, 0, and −1 respectively. Since the oxidation number of Zn is increasing, we say that Zn is **oxidized** as the reaction occurs. In contrast, the oxidation number of H is decreasing, so we say that H is **reduced** as the reaction proceeds. The compound that gets oxidized during the course of a reaction, in this case Zn, is known as the **reducing agent**. In contrast, the compound that gets reduced during the course of the reaction, in this case H^+, is known as the **oxidizing agent**.

- What does all this mean about the flow of electrons in the sample redox reaction? Oxidation corresponds to loss of electrons, whereas reduction corresponds to gain of electrons. Therefore, in the above redox reaction, electrons are flowing from Zn to HCl as the reaction proceeds.

You may also come across reactions in which an element in a compound is both oxidized and reduced in the course of a reaction. This type of reaction is called a **disproportionation reaction**. The element begins in one oxidation state and reacts to form separate compounds, with the element in new oxidation states in each compound. Let's take a look at an example.

$$3FeO \rightarrow Fe + Fe_2O_3$$

Iron (II) oxide reacts to produce iron and iron (III) oxide. Iron starts in an oxidation state of +2 and is reduced to form iron with an oxidation state of 0 and oxidized to form iron (III) oxide with an oxidation state of +3.

HALF-REACTIONS

One way of summarizing this electron flow is by representing the above reaction using **half-reactions**. Because the reaction is occurring in aqueous solution (in which HCl and $ZnCl_2$ dissociate completely), we can simplify the reaction by canceling Cl^- on both sides of the equation:

$$Zn \ (s) + 2 \ H^+ \ (aq) + \cancel{2 \ Cl^- \ (aq)} \rightarrow Zn^{2+} \ (aq) + \cancel{2 \ Cl^- \ (aq)} + H_2 \ (g)$$

We can then represent the oxidation and reduction reactions separately:

$Zn \ (s) \rightarrow Zn^{2+} \ (aq) + 2 \ e^-$	(oxidation half-reaction)
$2 \ H^+ \ (aq) + 2 \ e^- \rightarrow H_2 \ (g)$	(reduction half-reaction)
$Zn \ (s) + 2 \ H^+ \ (aq) \rightarrow Zn^{2+} \ (aq) + H_2 \ (g)$	(net reaction)

BALANCING REDOX REACTIONS

One important role that half-reactions play is in balancing redox equations, a process that can be a source of frustration for the AP Chemistry student. Let's use the following redox equation (unbalanced) as an example and see if we can use a systematic process to balance it:

$$Al\ (s) + MnO_4^-\ (aq) \rightarrow Al^{3+}\ (aq) + Mn^{2+}\ (aq) \qquad \text{(acidic conditions)}$$

To balance this redox equation, we use the following steps:

- *Determine what is oxidized and what is reduced.* In this case, the oxidation states of Al and Mn are 0 and +7, respectively, on the reactant side and +3 and +2, respectively, on the product side. Since the oxidation number of Al is increasing, it is oxidized as the reaction proceeds. Since the oxidation number of Mn is decreasing, it is reduced as the reaction proceeds.

- *Break the overall unbalanced reaction into half-reactions.*

$$Al\ (s) \rightarrow Al^{3+}\ (aq) \qquad \text{(oxidation half-reaction)}$$
$$MnO_4^-\ (aq) \rightarrow Mn^{2+}\ (aq) \qquad \text{(reduction half-reaction)}$$

- *Balance the atoms in each half-reaction.* Balance all atoms except O and H first, and then add H^+ and H_2O as needed (even for basic conditions, which we will correct later). In our case:

$$Al\ (s) \rightarrow Al^{3+}\ (aq)$$
$$MnO_4^-\ (aq) + 8\ H^+\ (aq) \rightarrow Mn^{2+}\ (aq) + 4\ H_2O\ (l)$$

- *Balance the half-reactions for charge using electrons:*

$$Al\ (s) \rightarrow Al^{3+}\ (aq) + 3\ e^-$$
$$MnO_4^-\ (aq) + 8\ H^+\ (aq) + 5e^- \rightarrow Mn^{2+}\ (aq) + 4\ H_2O\ (l)$$

- *Multiply the half-reactions by appropriate factors so that the oxidizing agent accepts as many electrons as the reducing agent produces.* In this case, we must multiply the oxidation reaction by 5 and the reduction reaction by 3:

$$5Al\ (s) \rightarrow 5Al^{3+}\ (aq) + 15\ e^-$$
$$3\ MnO_4^-\ (aq) + 24\ H^+\ (aq) + 15e^- \rightarrow 3\ Mn^{2+}\ (aq) + 12\ H_2O\ (l)$$

- *For basic solutions, you need to add an appropriate amount of OH^- on both sides of the equation to neutralize H^+, which does not exist in appreciable concentrations in basic solutions.* The H^+ and OH^- will combine to make an equal number of moles of water. This correction does not apply to this problem.

- *Add the half-reactions together to give the overall reaction:*

$$5\ Al\ (s) \rightarrow 5\ Al^{3+}\ (aq) + 15\ e^-$$
$$3\ MnO_4^-\ (aq) + 24\ H^+\ (aq) + 15e^- \rightarrow 3\ Mn^{2+}\ (aq) + 12\ H_2O\ (l)$$
$$5\ Al\ (s) + 3\ MnO_4^-\ (aq) + 24\ H^+ \rightarrow 5\ Al^{3+}\ (aq) + 3\ Mn^{2+}\ (aq) + 12\ H_2O\ (l)$$

- *Check your balanced equation to make sure that atoms and charge are in fact balanced.* Reactant charge $= 3(-1) + 24(+1) = 5(+3) + 3(+2) = +21 =$ product charge. All atoms are balanced.

THE ELECTROCHEMICAL CELL

Now that we understand the basics of redox reactions, let's turn to some applications of redox chemistry that are sure to turn up on your AP exam. The first application of redox chemistry we will look at is the **electrochemical cell** (also called a voltaic cell), an arrangement of an oxidizing and reducing agent that can only react if electrons flow through an outside conductor.

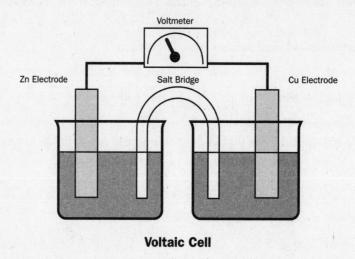

Voltaic Cell

An example of how this could be done for Zn and Cu is shown here. A **salt bridge** or other porous barrier is necessary to allow ions to flow between the electrodes. The overall reaction that occurs is as follows:

$$Zn \ (s) + Cu^{2+} \ (aq) \rightarrow Zn^{2+} \ (aq) + Cu \ (s)$$

In an electrochemical cell, two half-reactions are allowed to occur in separate beakers. The individual half-reactions that occur are as follows:

$$Zn \ (s) \rightarrow Zn^{2+} \ (aq) + 2e^- \quad \text{(oxidation half-reaction)}$$
$$Cu^{2+} \ (aq) + 2e^- \rightarrow Cu \ (s) \quad \text{(reduction half-reaction)}$$

As Zn atoms are oxidized, the electrons produced pass through the wire and are able to reduce Cu^{2+} in the other beaker. The resulting electric current produced can be detected using a lamp or other current-sensitive device. The zinc and copper metal strips in this cell are also known as **electrodes**, the general term for a conductor that conducts electrical current in or out of something. The site of reduction (in this case, the Cu strip) is also known as the **cathode**, whereas the site of oxidation (in this case the Zn strip) is also known as the **anode**. This implies that electrons in the wire are flowing away from the anode and toward the cathode.

What is the voltage of an electrochemical cell? Since electrons are flowing from the anode toward the cathode in an electrochemical cell, they can be thought of as driven by an **electromotive force** or **emf**. The emf of an electrochemical cell is also called the **cell voltage** or **cell potential** and shows how much work (in joules) can be produced for each coulomb of charge that the chemical reaction produces.

Understanding the relationship between voltage, current, charge, and work can be confusing and is best addressed in an AP Physics course. However, we will need to discuss these relationships when we cover electroplating later in this chapter. For now, the most important idea is that the *voltage of an electrochemical cell depends on the substances that make up the cell*. The cell voltage is symbolized by $E°$, which is defined under **standard conditions**. These conditions are the same as those defined for $\Delta H°$: all reactants and products are pure, all gases are at 1 atm, and all solutes are at 1 M concentrations. For any electrochemical cell, the cell voltage is given by the sum of the oxidation and reduction half-reactions:

$$E°_{cell} = E°_{red} \text{ (reduction process)} + E°_{ox} \text{ (oxidation process)}$$

If $E°_{cell}$ is positive, the reaction is product-favored (spontaneous in the forward direction). If $E°_{cell}$ is negative, the reaction is reactant-favored (spontaneous in the reverse direction). In fact, $E°_{cell}$ and $\Delta G°$ are related by the following expression:

$$\Delta G° = -nFE°_{cell}$$

where n is the number of moles of electrons transferred in the overall balanced chemical reaction and F is the Faraday constant (96,500 C/mol of electrons). Notice that this equation at least makes some sense. If $E°_{cell}$ is positive, $\Delta G°$ will be negative; both indicate a spontaneous redox reaction in the forward direction. We can also substitute the expression $\Delta G° = -RT\ln K$ to give the following expression:

$$E°_{cell} = \frac{RT\ln K}{nF}$$

USING STANDARD CELL POTENTIALS

As we mentioned before, the substances chosen to construct an electrochemical cell are by far the most important determinant of cell voltage. How do we predict in which direction (and with what electromotive force) electrons will flow in a voltaic cell? We can use *standard cell potentials*, which define the reduction potential of substances against the hydrogen electrode (chosen to be zero) under standard conditions. You can interpret standard reduction potentials according to the following rules:

- All reactions are written as reduction half-reactions, even though all half-reactions listed in a typical table can occur in either direction.

- The more positive the reduction potential, the more likely reduction of the substance in question will occur. It is less likely that an oxidation of this substance will take place.

- The more negative the reduction potential, the more likely oxidation of the substance will occur. In other words, the reaction is more likely to take place in the opposite direction.

- When presented with two redox-active substances, you can predict in which direction electron flow will occur by looking at the relative values for $E°_{red}$. The species reduced will be the one with the more positive value for $E°_{red}$. Since

$$E°_{cell} = E°_{red} + E°_{ox}$$

the value for $E°_{cell}$ will always be positive (and therefore the reaction spontaneous) when this is the case.

- Multiplication by stoichiometric coefficients does not change $E°_{red}$. For example:

$$Hg^{2+} (aq) + 2e^- \rightarrow Hg (l) \qquad E°_{red} = +0.855$$

$$2 Hg^{2+} (aq) + 4e^- \rightarrow 2 Hg (l) \qquad E°_{red} = +0.855$$

Therefore, we can predict whether or not a reaction will occur by breaking it down into its component half-reactions and calculating $E°_{cell}$. Remember that since you will typically be given standard reduction potentials only, you will need to reverse the sign of $E°$ for the oxidation half-reaction. For example:

REVIEW QUESTION

Will a 1 M solution of gold (III) spontaneously react with silver metal (Ag)?

SOLUTION

The question has already identified which substance will be reduced and which may be oxidized, if any. In this case, we have the following half-reactions:

$$Au^{3+} (aq) + 3e^- \rightarrow Au (s) \qquad E°_{red} = +1.50$$

$$3 Ag (s) \rightarrow 3 Ag^+ (aq) + 3 e^- \qquad E°_{ox} = -0.80$$

$$Au^{3+} (aq) + 3 Ag (s) \rightarrow Au (s) + 3 Ag^+ \qquad E°_{cell} = +0.70$$

Therefore, since $E°_{cell}$ is positive, we expect that the reaction will indeed occur. Notice that when we balanced the number of electrons being transferred by multiplying the equation corresponding to the oxidation of Ag (s) by three, we did not change the corresponding value of $E°_{ox}$. Also remember that we had to reverse the sign of the standard reduction potential of Ag (aq) since it is the oxidation and not the reduction of this species that is occurring in this reaction.

CONCENTRATION AND CELL POTENTIAL

Since a battery consists of a single (or multiple) electrovoltaic cell, it makes sense that the voltage produced by the cell will eventually drop to zero. This occurs when the reactants and products are at equilibrium and means you will be headed to the store to buy new batteries. In other words, a cell potential will only exist when there is an electrochemical gradient at work, and when equilibrium is reached the cell potential will be zero.

The Nernst equation relates the cell potential to actual, as opposed to standard, conditions. To calculate this potential, you will use the quantity Q, which we discussed earlier. Recall that Q represents a relationship between reactants and products identical to that defined by K, but not at equilibrium. To calculate the potential of a cell for any given concentrations of oxidizing and reducing agents:

$$E_{cell} = E°_{cell} - \frac{RT}{nF} \ln Q$$

There are two important things to notice about this equation:

1. When $Q = K$, then the second term in the equation above is equal to $E°_{cell}$, and therefore $E_{cell} = 0$. This means that your battery has run out.

2. When reactants and products are at standard conditions, the second term is equal to zero and therefore $E_{cell} = E°_{cell}$.

ELECTROLYSIS

When electrons are forced into an electrochemical system, we can force reactions that are non-spontaneous under standard conditions to occur. Redox reactions that are reactant-favored can occur in a process called **electrolysis**. An electrolysis cell is similar to a voltaic cell, but the electrodes themselves are usually not involved in any redox chemistry. Instead, these **inert** electrodes are connected to a source of electrical current that can be used to force oxidation and reduction half-reactions to occur in aqueous solution. Remember the following key points:

- You always need to take into account the medium in which electrolysis is occurring. For example, electrolysis of **molten salts** will lead to a straightforward oxidation and reduction of the corresponding ions, while electrolysis of **aqueous solutions** may be complicated by the fact that it's sometimes easier to reduce or oxidize water than the metal in question.

- When two or more electrochemical reactions are possible at an electrode, the reaction with the more positive (or less negative) $E°$ will occur.

- Water has a **reduction potential** of -0.83 V. In other words, $E°_{red}(H_2O) = -0.83$ V. Therefore, in order for the reduction of a material to occur in aqueous solution, that material must have a reduction potential more positive than -0.83 V. The half-reaction corresponding to the reduction of water is:

$$2 H_2O \text{ (l)} + 2 e^- \rightarrow H_2 \text{ (g)} + 2 OH^- \text{ (aq)} \qquad E° = -0.83 \text{ V}$$

- Water has an **oxidation potential** of -1.23 V. Therefore, $E°_{ox}(H_2O) = -1.23$ V. In order for a material to be oxidized in aqueous solution, the material must have an oxidation potential more positive than -1.23 V. The half-reaction corresponding to the oxidation of water is:

$$6 H_2O \text{ (l)} \rightarrow O_2 \text{ (g)} + 4 H_3O^+ \text{ (aq)} + 4 e^- \qquad E° = -1.23 \text{ V}$$

Let's use these basic rules to solve a typical electrolysis problem.

REVIEW QUESTION

Which of the following statements is true regarding the electrolysis of a 1 M CuCl$_2$ (aq) solution?

(A) At the cathode, Cl$_2$ (g) will be formed.

(B) At the anode, Cl$_2$ (g) will be formed.

(C) At the cathode, O$_2$ (g) will be formed.

(D) At the anode, O$_2$ (g) will be formed.

(E) At the anode, Cu (s) will be formed.

SOLUTION

The correct answer is choice (D). The first step is to identify which ions are present in solution and the half-reactions that might occur in solution. We know that CuCl$_2$ dissociates to form Cu^{2+} (aq) and Cl$^-$ (aq) ions in solution. The oxidation and reduction reactions that might occur are:

$$Cu^{2+} \text{ (aq)} + 2 e^- \rightarrow Cu \text{ (s)} \qquad E° = +0.34 \text{ V}$$

$$2 Cl^- \text{ (aq)} \rightarrow 2 e^- + Cl_2 \text{ (g)} \qquad E° = -1.36 \text{ V}$$

In an aqueous solution, you must always determine whether or not these reactions are favored relative to the corresponding oxidation or reduction of water itself. Remember that the reduction potential of water is -0.83 V, and therefore the reduction potential of Cu^{2+} is more positive ($+0.34$). Therefore, reduction of Cu^{2+} to form Cu will occur at the cathode, which is the site of reduction.

For the oxidation of Cl$^-$, we must compare the above oxidation potential to that of water. In this case, the oxidation potential of water (-1.23 V) is more positive than that of Cl$^-$ (-1.36). Therefore, oxidation of water at the anode will occur instead of oxidation of Cl$^-$ according to the following half-reaction:

$$6 H_2O \text{ (l)} \rightarrow O_2 \text{ (g)} + 4 H_3O^+ \text{ (aq)} + 4 e^- \qquad E° = -1.23 \text{ V}$$

ELECTROPLATING

In the above example, notice that Cu (s) is produced. This is a potentially useful reaction in manufacturing copper metal. If we were to run an external current through the electrolysis chamber for a given amount of time, we could actually calculate the mass of Cu (s) produced. Don't be fazed by this type of problem on the AP Chemistry exam. If you are confronted with an electroplating question, use units analysis and the following basic steps:

- If you are given the current applied for a given amount of time, remember the following relationship:

    (Current) × (Time) = Charge

    where current is typically given in amps (A), time in seconds (s), and charge in coulombs (C).

- Now you know the total charge applied, but how many electrons does this correspond to? One mole of electrons corresponds to 96,500 C (this is the Faraday constant). Therefore:

    $$\text{\# of moles of electrons} = \frac{(\text{Charge})}{96,500}$$

- We know the number of moles of electrons produced and can relate this to the number of moles of metal formed in the balanced redox reaction.

- Finally, we can calculate the mass of the metal produced by simply multiplying the number of moles by the molar mass.

Let's try an example:

REVIEW QUESTION

A current of 0.30 A is passed through a solution of $CuCl_2$ for 30 minutes (1,800 seconds). What mass of Cu (s) will be deposited at the cathode?

SOLUTION

Although this question looks frightening, we can use the steps explained previously in order to solve it. Remember that if you get stuck on the actual AP exam, you can use unit analysis to determine the correct relationships between current, charge, etc. We can use the balanced half-reactions from the previous question to determine the correct stoichiometric relationships.

- Charge = (Current) × (Time) = (0.30) × (1,800) = 540 C

- Number of moles of $e^- = \dfrac{540}{96,500} = 5.596 \times 10^{-3}$ mol e^-

- 5.596×10^{-3} mol $e^- \times \left(\dfrac{1 \text{ mol Cu}}{2 \text{ mol } e^-} \right) = 2.798 \times 10^{-3}$ mol Cu

- 2.798×10^{-3} mol Cu $\times \left(\dfrac{63.546 \text{ g Cu}}{1 \text{ mol Cu}} \right) = 0.18$ g Cu

Therefore, 0.18 g of Cu can be plated using the above technique.

CORROSION AND CORROSION PROTECTION

Most of us are familiar with **corrosion**, which is the process by which metals are oxidized to form their corresponding metal oxides. What does the *Titanic* look like these days? If you went diving in search of it, you might find that the old ocean liner is a rusty heap. In fact, salt water is an excellent catalyst for the formation of rust or $Fe_2O_3 \cdot nH_2O$. This compound is otherwise known as hydrated iron (III) oxide, where n varies from 2 to 4.

It's important to remember that this anodic process must be coupled with a corresponding reduction or cathodic process in order to occur. In corrosion, usually it is oxygen that is reduced:

$$2 H_2O \text{ (l)} + 2 e^- \rightarrow H_2 \text{ (g)} + 2 OH^- \text{ (aq)} \qquad E° = -0.8 \text{ V}$$

Dissolved O_2 in the water or air is the oxidizing agent. Water is only reduced directly by structural metals in anaerobic conditions.

How might we prevent corrosion from occurring? In order to prevent corrosion, we must inhibit the anodic process, the cathodic process, or both. The two options are described as follows:

1. In **anodic inhibition**, we can directly prevent the oxidation half-reaction by coating the material with paint, grease, etc. This is the basic approach to preventing the corrosion of planes, trains, and automobiles.

2. In **cathodic inhibition**, we can use a metal more readily oxidized than the one being protected. The net result is that the metal being protected becomes the cathode, with the "sacrificial" material getting oxidized instead. The key feature here is that the $E°$ for this oxidation-reduction reaction must be more favorable than that occurring between the protected material and water.

REVIEW QUESTIONS

1. For the reaction below, indicate which elements are reduced and which ones are oxidized:

   $$2Cu(NO_3)_2 \rightarrow 2CuO + 4NO_2 + O_2$$

oxidized	reduced
(A) nitrogen	oxygen
(B) copper	oxygen
(C) copper	nitrogen
(D) nitrogen	copper
(E) oxygen	nitrogen

2. When the following redox equation is balanced and all coefficients reduced to lowest whole-number terms, the coefficient for H^+ is:

   $$H_3O^+ (aq) + MnO_4^- (aq) + Fe^{2+} (aq)$$
   $$\rightarrow Fe^{3+} (aq) + Mn^{2+} (aq) + H_2O (l)$$

   (A) 1

   (B) 3

   (C) 5

   (D) 7

   (E) 9

3. An electrochemical cell is constructed using Cd and Ag. Which of the following is true?

   (A) Ag must be the anode.

   (B) Cd must be the cathode.

   (C) The $E°_{cell}$ is 1.20 V.

   (D) The $E°_{cell}$ is 0.40 V.

   (E) Cd is reduced in the cell according to the following half-reaction: $Cd^{2+} + 2e^- \rightarrow Cd (s)$.

4. The Nernst equation is given by the following:

   $$E_{cell} = E°_{cell} - \ln Q$$

   Which of the following is/are true?

   I. E_{cell} is always greater than $E°_{cell}$.

   II. When $Q = K$, $E_{cell} = 0$.

   III. When all concentrations are equal to 1 M (standard state), $E_{cell} = 0$.

   (A) I only

   (B) II only

   (C) I and II only

   (D) II and III only

   (E) I, II, and III

5. Which of the following metals could be used for cathodic protection of iron (Fe)?

   (A) Sn

   (B) Ni

   (C) Ag

   (D) Zn

   (E) Hg

6. Consider the following reaction:

   $$Cl_2 (g) + 2OH^- (aq) \rightarrow Cl^- (aq) + OCl^- (aq) + H_2O$$

   Which of the following is/are true regarding this reaction?

   I. Cl_2 undergoes both oxidation and reduction.

   II. The oxidation state of O in OCl^- is −1.

   III. OH^- is the reducing agent.

   (A) I only

   (B) II only

   (C) II and III only

   (D) I and III only

   (E) I, II, and III

7. In which of the following compounds does manganese have the same oxidation number as it does in $KMnO_4$?

   (A) $MnCl_2$

   (B) MnO_2

   (C) Mn_2O_3

   (D) Mn_2O_7

   (E) Mn

8. In the electroplating of tin, 0.500 faraday of electrical charge is passed through a solution of $SnSO_4$ at 25°C. What is the mass of tin deposited?

   (A) 14.8 g

   (B) 29.7 g

   (C) 59.4 g

   (D) 119 g

   (E) 237 g

9. Which of the following metals would you expect to corrode more readily than Al?

   (A) Ag

   (B) Hg

   (C) Sn

   (D) Cd

   (E) Mg

FREE-RESPONSE QUESTION

1. $2\,H_2O\,(l) + 2\,Br_2\,(l) \rightarrow O_2\,(g) + 4\,H^+\,(aq) + 4\,Br^-\,(aq)$

    Provide the following information regarding the reaction shown above:

    (a) Identify the oxidizing and reducing agents.

    (b) Write down the corresponding oxidation and reduction half-reactions.

    (c) Calculate the value of $E°$ for the reaction. Is the reaction spontaneous under standard conditions?

    (d) Calculate the pH values at which the reaction is spontaneous, assuming the concentrations of all reactants/products other than $[H^+]$ are in their standard states and the temperature is 25°C.

ANSWERS AND EXPLANATIONS

1. E

The oxidation number of O is -2 in $Cu(NO_3)_2$, but 0 in its elemental form O_2. Therefore, since the oxidation number of O is becoming more positive, it is getting oxidized. The oxidation number of N is $+5$ in $Cu(NO_3)_2$ since the polyatomic anion NO_3^- must have a net negative charge. On the right side of the equation, the oxidation number of N is $+4$ in the neutral compound NO_2. Therefore, nitrogen is being reduced in the reaction.

2. D

Begin this question by determining what is oxidized and what is reduced. Since the oxidization number of Fe increases from $+2$ to $+3$, it is oxidized, and since the oxidization number of Mn decreases from $+7$ to $+2$, it is reduced.

Next, break the unbalanced reaction into two half-reactions:

$$Fe^{2+} (aq) \rightarrow Fe^{3+} (aq) \qquad \text{(oxidation half-reaction)}$$

$$H_3O^+ (aq) + MnO_4^- (aq) \rightarrow Mn^{2+} (aq) + H_2O (l)$$
(reduction half-reaction)

Balance all atoms except O and H. Then, add H^+ and H_2O as needed:

$$Fe^{2+} (aq) \rightarrow Fe^{3+} (aq)$$

$$H_3O^+ (aq) + MnO_4^- (aq) + 7H^+ \rightarrow Mn^{2+} (aq) + 5 H_2O (l)$$

Balance the half-reactions for charge using electrons:

$$Fe^{2+} (aq) \rightarrow Fe^{3+} (aq) + e^-$$

$$H_3O^+ (aq) + MnO_4^- (aq) + 7H^+ + 5 e^- \rightarrow Mn^{2+} (aq) + 5 H_2O (l)$$

Multiply the half-reactions so that the oxidizing agent accepts as many electrons as the reducing agent produces. Here, that means multiplying the first equation by 5 and leaving the second alone:

$$5 Fe^{2+} (aq) \rightarrow 5 Fe^{3+} (aq) + 5 e^-$$

$$H_3O^+ (aq) + MnO_4^- (aq) + 7H^+ + 5 e^- \rightarrow Mn^{2+} (aq) + 5 H_2O (l)$$

Add the half-reactions together to give the overall reaction:

$$H_3O^+ (aq) + MnO_4^- (aq) + 5 Fe^{2+} (aq) + 7H^+ \rightarrow Mn^{2+} (aq) + 5 H_2O (l) + 5 Fe^{3+} (aq)$$

Check to make sure charge and the amount of atoms are balanced. They are, so the correct answer is choice (D).

3. C

The standard reduction potentials for the two metals are given in the table on page 282:

$$Ag^+ + e^- \rightarrow Ag (s) \qquad E° = 0.80 \text{ V}$$

$$Cd^{2+} + 2e^- \rightarrow Cd (s) \qquad E° = -0.40 \text{ V}$$

Since the standard reduction potential for Ag^+ is more positive, this reduction will occur at the cathode. Cd will be the anode (site of oxidation). The $E°_{cell}$ is obtained by reversing the second potential above, then adding the two potentials.

4. B

When $Q = K$, the second term in the Nernst equation is equal to $E°_{cell}$, and therefore $E_{cell} = 0$. Remember that batteries don't work at equilibrium. They require a gradient and will cease to function once equilibrium is reached. The other statements are false.

5. D

The table on page 282 indicates two different possible oxidations of Fe (to Fe^{2+} and Fe^{3+}). The oxidation to Fe^{2+} is more spontaneous (has a more positive oxidation potential). The oxidation half-reaction for Fe is as follows:

$$Fe\ (s) \rightarrow Fe^{2+} + 2e^- \qquad E° = 0.44\ V$$

In cathodic protection, another metal is used as a sacrificial anode. In other words, we must choose a metal that is more readily oxidized than Fe. Of the metals listed, only Zn meets this criteria, since $E° = 0.76\ V$ for the oxidation of Zn.

6. A

The oxidation state of Cl in Cl^- and OCl^- is −1 and +1, respectively. Therefore, Cl_2 (which has an oxidation state of 0) is both the oxidizing and reducing agent in this reaction. Hydroxide ion, OH^-, is consumed in the reaction, and the oxidation state of O in OCl^- is −2, not −1.

7. D

In $KMnO_4$, the oxidation state of Mn is +7. Mn_2O_7 is the only compound listed containing Mn in the same oxidation state. The oxidation states of Mn in the other compounds are: choice (A) +2, choice (B) +4, choice (C) +3, choice (E), 0.

8. B

To solve this question, we need to remember what a faraday is. A faraday is the charge on one mole of electrons. Therefore, half a faraday represents half a mole of electrons. The reduction of Sn^{2+} takes place according to the following half-reaction:

$$Sn^{2+} + 2e^- \rightarrow Sn$$

According to the stoichiometry of the above expression, two moles of electrons are required to reduce one mole of Sn^{2+} to metallic tin. Therefore, we can determine the mass of tin deposited by the following:

$$0.500\ mol\ e^- \times \left(\frac{1\ mol\ Sn}{2\ mol\ e^-}\right) \times \left(\frac{118.7\ g\ Sn}{1\ mol\ Sn}\right) = 29.7\ g\ Sn\ deposited$$

9. E

We expect that metals that corrode more readily than Al will have more negative reduction potentials. For the reaction:

$$Al\ (s) \rightarrow Al^{3+}\ (aq) + 3\ e^- \qquad E° = 1.66\ V$$

Since this oxidation has a high (positive) value for $E°$, this implies that oxidation occurs readily for this metal. Only Mg has an oxidation half-reaction with a more positive potential:

$$Mg\ (s) \rightarrow Mg^{2+}\ (aq) + 2\ e^- \qquad E° = 2.37\ V$$

FREE-RESPONSE QUESTION

1. (a) In the reaction shown, the oxidation state of Br is going from 0 (in Br_2) to -1 (in Br^-). Therefore, Br is being reduced, making Br_2 the oxidizing agent. The oxidation state of oxygen is going from -2 (in H_2O) to 0 (in O_2). Therefore, oxygen is being oxidized, making H_2O the reducing agent.

    (b) The oxidation and reduction half-reactions are as follows:

    Oxidation: $2\,H_2O \rightarrow O_2 + 4\,H^+ + 4\,e^-$ $E° = -1.23\,V$

    Reduction: $Br_2 + 2\,e^- \rightarrow 2\,Br^-$ $E° = 1.07\,V$

    (c) The value of $E°$ for the reaction is therefore $(-1.23 + 1.07) = -0.16\,V$. Since $E°$ is negative, the reaction is not spontaneous under standard conditions.

    (d) In order to determine the pH values at which the reaction is spontaneous, we need to use the Nernst equation, which has the following form at 25°C:

    $$E = E° - \left(\frac{0.0592}{n}\right)(\ln Q) \qquad \text{where } Q = [H^+]^4[Br^-]^4$$

    In our case, $n = 4$ since there are four electrons being transferred. Also, $[Br^-] = 1$ since this component is in the standard state. We want to find the value of Q for which E is just positive, in other words, spontaneous. Thus, we start by setting E equal to zero:

    $$0 = -0.16 - \left(\frac{0.0592}{4}\right)(\ln Q)$$
    $$-10.81 = \log(Q) = \log([H^+]^4) = 4\log[H^+]$$
    $$[H^+] = 2.0 \times 10^{-3}$$

    The corresponding pH is 2.70. In other words, although the reaction is not spontaneous at $[H^+] = 1\,M$ (standard state, where pH $= 0$), the reaction will be spontaneous at any pH above 2.70.

CHAPTER 16: ORGANIC CHEMISTRY— NOMENCLATURE, ISOMERISM, ORGANIC REACTIONS

IF YOU LEARN ONLY FOUR THINGS IN THIS CHAPTER . . .

1. The four classes of hydrocarbons are alkanes, alkenes, alkynes, and aromatic hydrocarbons.

2. The chemistry and physical properties of organic compounds are highly dependent on the functional groups present in these compounds.

3. Compounds with identical chemical formulas but different geometries and/or connectivities are known as isomers. These isomers typically have different physical properties.

4. Important reactions of organic compounds include substitution, condensation, and hydrolysis.

Organic chemistry does not appear much on the AP Chemistry exam. You will have to wait for college to start your love affair with "orgo," a rite of passage for students going into the health-care professions or other fields. For now, we will look at only the most basic things you will need to know. Read the following section carefully, but remember that if you are low on time your efforts may be better spent elsewhere.

When we speak of **organic chemistry**, we are referring to the branch of chemistry that deals with compounds for which the principal element is carbon.

In this section, we will define some of the basic terminology used when describing organic compounds. Most importantly, you may need to identify an alcohol or an acid on the

AP Chemistry exam. Once we've defined the basic terms, we will look at structural isomerism, which describes compounds with the same chemical formulas that are connected in different ways. Finally, we will look at a few important organic reactions.

NOMENCLATURE

Hydrocarbons are chemical compounds that contain only two elements, carbon and hydrogen. (Contrast this with **carbohydrates**, which are composed of carbon, hydrogen, and oxygen, with the hydrogen and oxygen in a 2:1 ratio as in water.) Gasoline is composed entirely of hydrocarbons of varying molecular weights and is utilized for energy by the process of **combustion**. Hydrocarbons are classified by the presence of single, double, or triple bonds. These compounds are called **alkanes** (general formula: C_nH_{2n+2}), **alkenes** (general formula: C_nH_{2n}), and **alkynes** (general formula: C_nH_{2n-2}), respectively. The chemical structures of some representative hydrocarbons are shown here. In each case, R stands for a chain of any number of carbon atoms bonded to an appropriate number of hydrogen atoms.

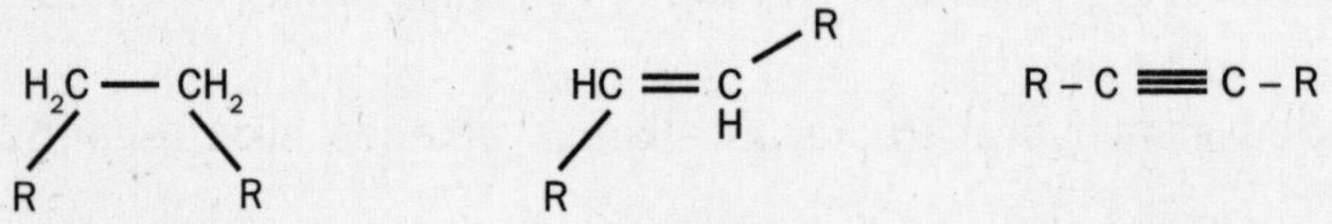

Alkane   Alkene   Alkyne

Aromatic hydrocarbons are defined by a ring structure, usually a six-membered ring with alternating double and single bonds. These structures have a higher stability than their corresponding straight-chain counterparts. The simplest aromatic hydrocarbon is benzene, the structure of which is shown below.

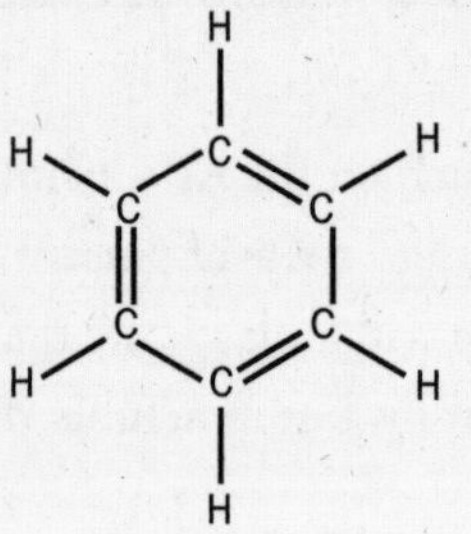

Benzene

In general, hydrocarbons are **volatile**, meaning that they have weak intermolecular forces and low boiling and melting points. The attractive forces between hydrocarbon molecules are nonpolar interactions or London dispersion forces. Many hydrocarbons will exist as gases at room temperature. In general, remember that there is a direct relationship between molecular weight and melting and boiling points for hydrocarbons.

FUNCTIONAL GROUPS

The chemical properties of organic molecules are determined principally by the **functional groups** present in those molecules. The most important of these are alcohols, carboxylic acids, ketones, aldehydes, esters, ethers, amides, and amino groups. In general, the presence of functional groups on any organic molecule will contribute significantly to its polarity.

Alcohols have an oxygen atom with a single bond to hydrogen. This basic structure is found in methanol, ethanol, and glucose.

Glucose   Ethanol   Methanol   General Form

Carboxylic acids are important since most weak acids (for example lactic and citric acid) are carboxylic acids. Carboxylic acids contain a carbon atom double-bonded to an oxygen atom and single-bonded to an alcohol group.

General Form

In lactic acid below, the carboxyl functional group is to the far right of the molecule. When present in aqueous solution at neutral pH, these typically exist as a mixture of the acid and its corresponding conjugate base.

Lactic Acid   Lactate

$CH_3\ CHOH\ COOH$ $CH_3\ CHOH\ COO^-$

Ketones and **aldehydes** are important in the chemistry of the human body and are represented by the following structures.

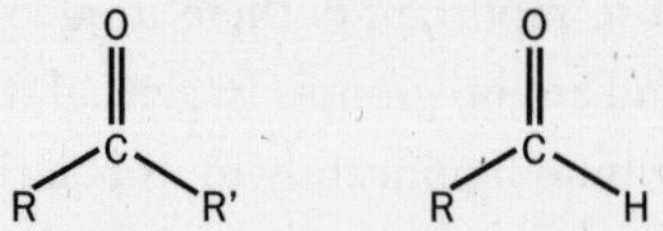

Ketone   Aldehyde

Esters are fruity-smelling compounds that consist of a carbon double-bonded to one oxygen and single-bonded to another, which is in turn bonded to another carbon chain. Esters are similar to carboxylic acids in which the acidic proton has been replaced with a carbon group. **Ethers** are organic molecules in which an oxygen atom is single-bonded to each of two carbon atoms. **Amides** are functional groups that form the basic backbone of proteins; they are similar to esters, but with the single-bonded oxygen replaced with a nitrogen atom.

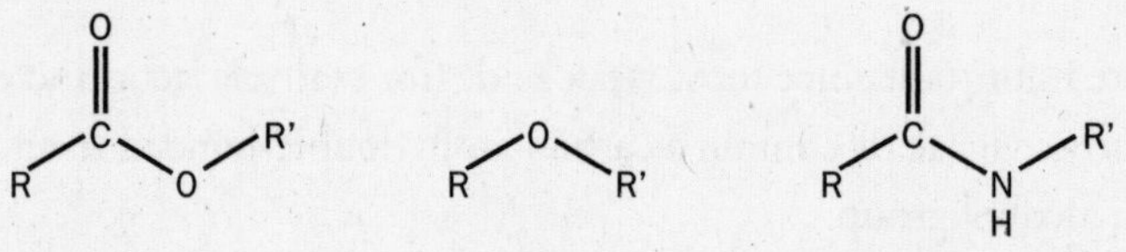

Ester   Ether   Amide

Finally, **amines** consist of nitrogen atoms with single bonds to one or more carbon atoms. A primary amine is bonded to one carbon atom, while secondary and tertiary amines are bonded to two and three carbon atoms, respectively. When present in aqueous solution, amines typically exist in equilibrium with their corresponding conjugate acids, just as ammonia does.

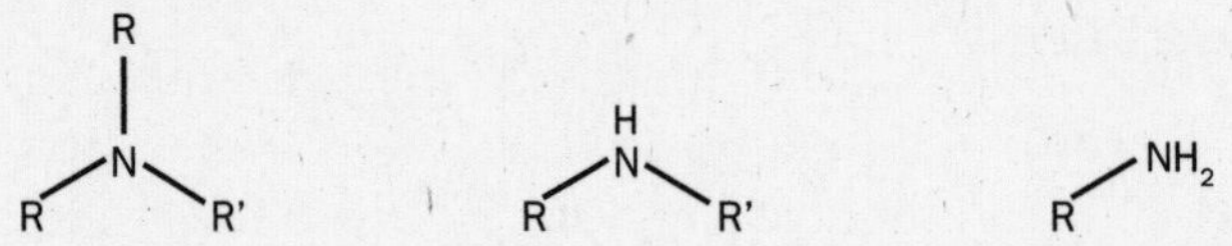

Tertiary   Secondary   Primary

REVIEW QUESTION

The structure of glutamic acid (an important amino acid) is shown. Which of the following is NOT true?

(A) Glutamic acid contains two carboxylic acid groups and one amino group.

(B) In neutral solution, the amine group will be positively charged.

(C) In neutral solution, the net charge of glutamic acid will be slightly negative.

(D) The amino group of glutamic acid is a basic group.

(E) Glutamic acid is also known as glutamine and designated by the single letter G.

Glutamic Acid

SOLUTION

The correct answer is choice (E). All other statements about this amino acid are true. There is another amino acid by the name of glutamine that is different from glutamic acid in many ways.

ISOMERISM

Sometimes, you can draw several different Lewis structures that have the same chemical formula. This phenomenon is known as **structural isomerism**. You may be asked on the AP exam to draw the structural isomers for a given chemical formula. For example, let's consider the structural isomers of pentane, C_5H_{12}. You will find that there are three different ways to connect the carbon atoms. Therefore, there are three structural isomers of pentane. Each will have *different physical properties*. For hydrocarbons of similar molar mass, a greater degree of branching means a lower boiling point:

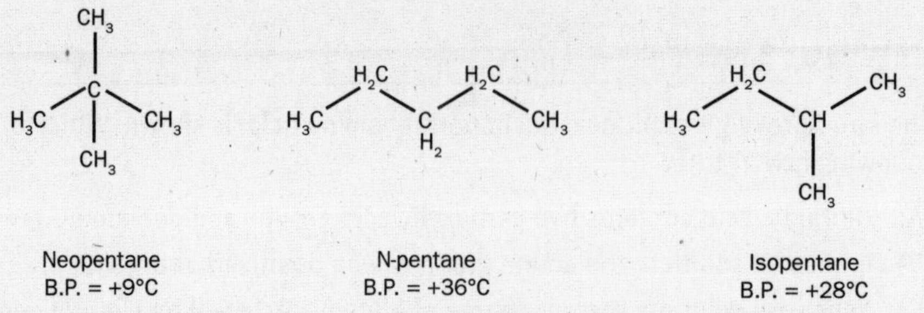

Neopentane
B.P. = +9°C

N-pentane
B.P. = +36°C

Isopentane
B.P. = +28°C

Another type of isomerism is **geometric isomerism**. It's unlikely that you will be asked anything about this on the AP exam. However, you should be aware that geometric isomerism refers to two molecules with identical connectivity but different geometries. These isomers are distinct because rotation of the two double-bonded carbon atoms cannot occur without breaking the double bond; the two atoms are locked into place. For example, consider the following two alkenes, which have different melting points, boiling points, densities, and chemical reactivities:

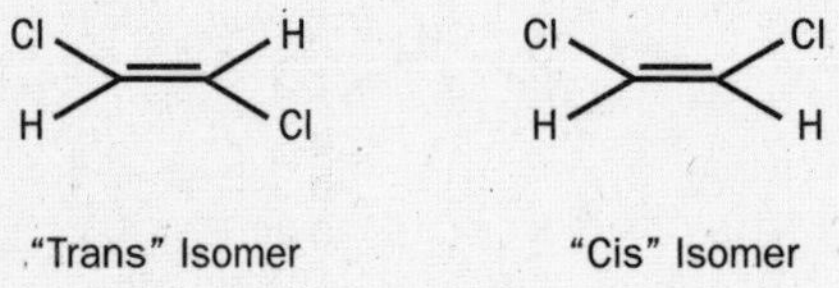

"Trans" Isomer

"Cis" Isomer

ORGANIC REACTIONS

There are thousand-page books on organic reactions, but here we will address four types of chemical reactions you may see on the AP Chemistry exam: **substitution, condensation, hydrolysis,** and **radical reactions**.

A **substitution** reaction occurs when one or more hydrogens of a hydrocarbon are replaced by another atom or functional group. Consider the following substitution reaction involving an aromatic hydrocarbon:

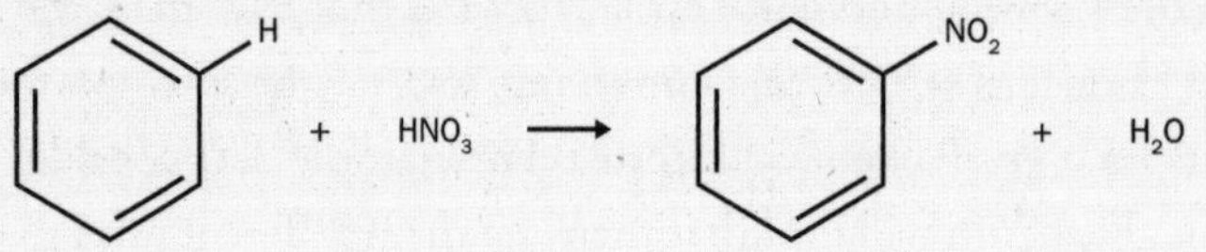

A **condensation** reaction occurs when two functional groups come together, often releasing water. For example, during the formation of a **peptide** bond between two amino acids, a molecule of water is released, resulting in an **amide** bond:

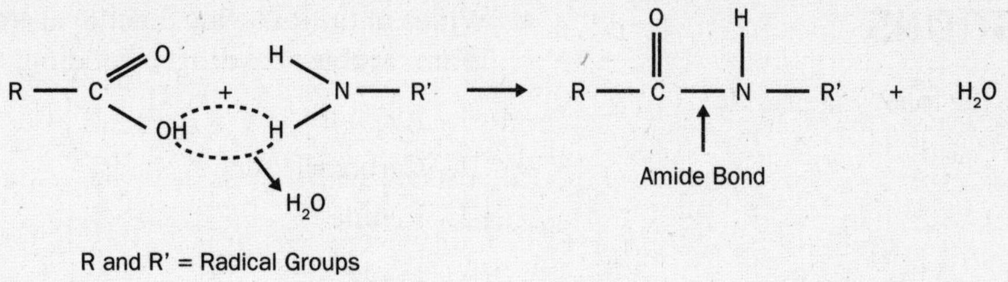

R and R' = Radical Groups

Third, a **hydrolysis** reaction is the opposite reaction and consists of a water molecule breaking apart a molecule. For example, the hydrolysis of fat follows this general form:

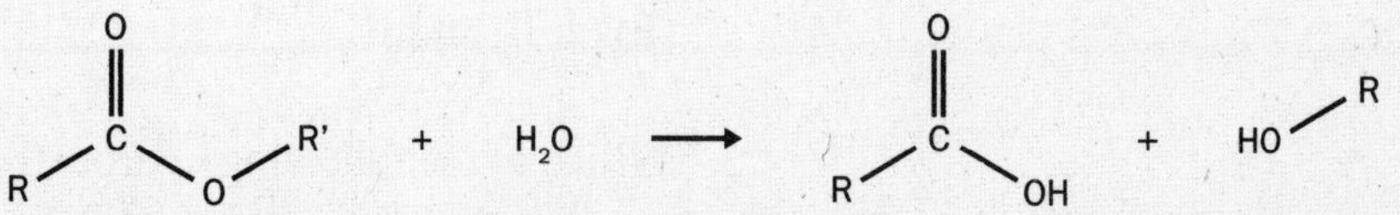

Finally, a quick note about **radical reaction** chemistry. This is not likely to be covered in depth on the AP Chemistry exam, but it is useful to know the mechanism of radical reactions. Radicals are highly reactive atoms, molecules, or ions with unpaired electrons in an open shell. These radicals cause compounds to be highly unstable and can create three-step reactions.

Initiation: When the number of radicals increases, the step is termed initiation and is typically the first step of the reaction.

Propagation: When the number of radicals remains unchanged, the step is termed propagation and may occur over and over again as an intermediate of the overall reaction.

Termination: When the number of radicals decreases, the step is termed termination and is typically the final step of the reaction.

For example, when chlorine gas (Cl_2) breaks into two radicals, it can chlorinate methane to produce chloromethane and hydrogen chloride in the following three steps:

1. Initiation:

   $Cl - Cl \rightarrow Cl \cdot + Cl \cdot$

2. Propagation:

   $Cl \cdot + CH_4 \rightarrow HCl + CH_3$

   $CH_3 + Cl - Cl \rightarrow Cl - CH_3 + Cl \cdot$

3. Termination:

   $Cl \cdot + CH_3 \rightarrow Cl - CH_3$

   $CH_3 + CH_3 \rightarrow H_3C - CH_3$

REVIEW QUESTIONS

Questions 1–2 refer to the following organic structures:

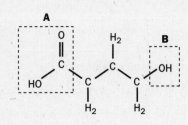

Which of the outlined functional groups represents:

1. An alcohol

2. An amide

3. How many structural isomers are possible for $C_2H_3Cl_3$?

(A) One

(B) Two

(C) Three

(D) Four

(E) Five

4. Which of the following functional groups is/are capable of hydrogen bonding?

   I. Alcohol

  II. Carboxylic Acid

 III. Ketone

(A) I only

(B) II only

(C) I and II only

(D) II and III only

(E) I, II, and III

5. Which of the following hydrocarbons could be an alkyne?

(A) C_2H_4

(B) C_3H_4

(C) C_3H_6

(D) C_4H_8

(E) Benzene, C_6H_6

ANSWERS AND EXPLANATIONS

1. B

Choice (B) represents an alcohol functional group; it shows an oxygen atom with a single bond to hydrogen.

2. C

Choice (C) represents an amide; it shows a carbon atom double-bonded to one oxygen and single-bonded to nitrogen and a carbon chain.

3. B

The first step is to draw a correct Lewis dot structure for the molecule. You will then notice that there are only two structural isomers possible: one in which all three Cl atoms are bonded to a single carbon and one in which two Cl atoms are bonded to one carbon and the third is bound to the other carbon. Because there is free rotation around single bonds, any other structures (Cl atoms pointing down or up or out) are equivalent to one of these two.

4. C

Since alcohols and carboxylic acids contain an electronegative atom (oxygen) bonding directly to hydrogen, these groups are capable of hydrogen bonding.

5. B

This compound is the only hydrocarbon that could contain a C-C triple bond. However, it is worth noting that C_3H_4 could also have two double bonds instead of one triple bond.

CHAPTER 17: DESCRIPTIVE CHEMISTRY

IF YOU LEARN ONLY TWO THINGS IN THIS CHAPTER . . .

1. When presented with a question regarding an element's reactivity, remember the trends in the periodic table and also extrapolate from more familiar elements in the same group.

2. Know that many of the multiple-choice questions can be tackled by using the process of elimination. Be sure to remember the periodic table groups within which there are the most similarities: 1, 2, 17, and 18.

Descriptive chemistry appears throughout the AP Chemistry exam, both in multiple-choice questions and as the basis for questions addressing other topics. In Section II, Part B, there is a series of questions asking you to write the formulas and products of some descriptions of chemical reactions. Generally, the information you will use to answer all these questions comes from many different experiences: some you may have learned in the lab or learned from demonstrations while other facts were collected through reading and from anecdotes. Some common conceptual threads do run through most of these questions (e.g., periodic trends or precipitation), and these can be mastered quickly. Others can be found in descriptions of experiments that are common to most textbooks. Please review chapter 5 (The Elements and the Periodic Table) before continuing, because the elements of the periodic table form the basis of this chapter.

It is often difficult to see the chemical changes and effects that occur at the atomic level. For example, two clear solutions can be mixed to form another clear solution, offering no changes to the naked eye. Other changes are much more obvious, such as precipitation, evolution of a gas, heat release, or color change. It is these macroscopic changes that form the basis for descriptive chemistry. With some knowledge about these changes and the starting reactants or products, we can understand what is happening at the atomic level. The AP Chemistry exam will require you to possess this type of specific knowledge. We'll break this large topic down into a few smaller areas and give you that knowledge in this chapter.

PRECIPITATION

Precipitation questions are a common part of the AP exam. Normally, in these questions, you are told that two solutions are being mixed, and you are asked whether or not they give a precipitate; if they give a precipitate, you are asked to identify it. There can also be a series of steps in which a precipitate is formed and then redissolved into another solution. Some examples of this are shown below. Generally, the solutions in these questions are aqueous and made up of metallic cations and nonmetallic anions. Other solutions can be solutions of acids (H_2SO_4, HCl) or bases (NH_4OH, NaOH). Precipitations occur when the cation of one solution forms an insoluble precipitate with the anion of the other. So how do you know what is soluble and what is not? You must remember six solubility rules, summarized below:

1. All compounds containing alkali metal cations or the ammonium ion are soluble (e.g., Na_2SO_4, CsCl, and $NH_4C_2H_3O_2$ are soluble).

2. All compounds containing NO_3^-, ClO_4^-, ClO_3^-, and $C_2H_3O_2^-$ anions are soluble (e.g., $Fe(NO_3)_3$, $Ba(ClO_3)_2$, and $Pb(C_2H_3O_2)_2$ are soluble).

3. All compounds containing bromide, chloride, and iodide anions are soluble except those containing Ag^+, Pb^{2+}, or Hg_2^{2+} (e.g., $CaCl_2$, $FeBr_3$, and ZnI_2 are soluble; AgCl, PbI_2, and Hg_2Br_2 are insoluble).

4. All sulfates are soluble except those containing Ag^+, Hg_2^{2+}, Pb^{2+}, Sr^{2+}, Ca^{2+}, or Ba^{2+} (e.g., $PbSO_4$ and $CaSO_4$ are insoluble).

5. All hydroxides are insoluble except compounds of the alkali metals, Ca^{2+}, Sr^{2+}, and Ba^{2+} (e.g., $Fe(OH)_3$ and $Al(OH)_3$ are insoluble; KOH and $Ba(OH)_2$ are soluble). However, note that aluminum hydroxide, $Al(OH)_3$, will redissolve in concentrated base to give $Al(OH)_4^-$.

6. All compounds containing PO_4^{3-}, S^{2-}, CO_3^{2-}, and SO_3^{2-} ions are insoluble except those that contain alkali metals or NH_4^+ (e.g., $AlPO_4$ and $BaCO_3$ are insoluble; $(NH_4)_2S$ and K_3PO_4 are soluble).

A good way to remember these rules is to put them into two sets: those that say members of a class are mostly soluble (rules 1–4) and those that say most of a class are insoluble (rules 5 and 6). If you see a phosphate in a problem, think first that it might be part of a precipitate. If you see a nitrate or perchlorate, then it will probably be a spectator ion and not a precipitate. Questions involving a precipitate followed by adding another reagent to redissolve the precipitate will be covered under the gas evolution heading.

COMBUSTION REACTIONS

Combustion normally refers to combining a substance such as an element or compound with oxygen. It is actually a subset of redox reactions but comes up often enough to warrant a separate section. Combustions can be complete, involving an excess of oxygen, or incomplete, using a limited supply of oxygen. One example is the combustion of carbon or hydrocarbons. With excess oxygen, the carbon component is converted to carbon dioxide (CO_2), but with a limited amount of oxygen, it forms the deadly carbon monoxide (CO).

Oxygen is a reactive molecule with which many elements easily combine. These reactions can be slow (the case of the rusting of iron) or explosive (the case of hydrogen). Oxygen is electronegative and generally combines with elements to the left and below it on the periodic table (i.e., most of the periodic table). From its position in the periodic table, we can see that oxygen is usually in the -2 oxidation state. Some questions may ask you for molecular formulas from the combustion of elements or to come up with coefficients from balanced combustion reactions. From chapter 12, you already know how to balance reactions; the only additional information is the identity of the products. So long as you remember that oxygen normally has an oxidation state of -2, you can master most of these problems.

REDOX REACTIONS

Many of the questions regarding descriptive chemistry in Section II, Part B, are going to be redox-based. The standard reduction potential table given on page 282 at the beginning of Section II serves as an important reference that is available to you. The answers to some of the questions can be directly read from this table, so you must be comfortable using it. Review chapter 15 (Redox Reactions), which will give you the basics on how to use that table. The most obvious type of question is one in which a more active *free metal* (lower down on the reduction table) is added to a solution of a less active *metal cation* (higher up on the table).

REVIEW QUESTION

Write the formulas to show the reactants and the products for the situation described on the next page. Assume that solutions are aqueous unless stated otherwise.

Represent substances in solution as ions if the substances are extensively ionized.

Omit formulas for any ions or molecules that are unchanged by the reaction. You need not balance the equations.

Do you expect this reaction to proceed? Why or why not?

A strip of magnesium metal is added to a solution of lead (II) acetate.

SOLUTION

$$Mg + Pb^{2+} \rightarrow Mg^{2+} + Pb$$

Looking at the standard reduction table on page 282, you can see that magnesium is lower (−2.37 V) than lead (−0.13 V), and thus magnesium is more likely to give up its electrons than lead, and the reaction proceeds to the right.

For this example, note that the acetate anion is a spectator and shouldn't be written in the reaction because it doesn't change; also, don't write $Mg(C_2H_3O_2)_2$ as the product because it is completely dissociated in solution.

Another situation dealing with redox reactions of active metals is the addition of alkali metals and alkaline earth metals to water. These metals are very reactive and generate hydrogen and hydroxide as they hydrolyze.

REVIEW QUESTION

Write the formulas to show the reactants and the products for the situation described below. Assume that solutions are aqueous unless stated otherwise.

Represent substances in solution as ions if the substances are extensively ionized.

Omit formulas for any ions or molecules that are unchanged by the reaction. You need not balance the equations.

Do you expect this reaction to proceed? Why or why not?

Calcium metal is added to warm water.

SOLUTION

$$Ca + H_2O \rightarrow Ca^{2+} + OH^- + H_2$$

Calcium reacts with water to form calcium hydroxide and hydrogen gas. Calcium loses two electrons and goes from an oxidation state of 0 to +2, so it is oxidized. Water splits into hydroxide ions and hydrogen gas. The H in OH^- retains the same oxidation state as the H in H_2O, but the H that forms the H_2 gas has been reduced from an oxidation state of +1 to 0. This reaction will proceed spontaneously because the reduction potential of H_2O to OH^- and H_2 is −0.83, while the reduction potential of Ca to Ca^{2+} is + 2.87.

REVIEW QUESTION

Write the formulas to show the reactants and the products for the situation described below. Assume that solutions are aqueous unless stated otherwise.

Represent substances in solution as ions if the substances are extensively ionized.

Omit formulas for any ions or molecules that are unchanged by the reaction. You need not balance the equations.

Do you expect this reaction to proceed? Why or why not?

Magnesium is burned in an atmosphere of chlorine.

SOLUTION

$$Mg + Cl_2 \rightarrow MgCl_2$$

As in the combustion reactions outlined previously, questions may involve "burning" or reacting an element with a halogen such as chlorine. The halogen takes the place of oxygen in the combustion reaction and will be reduced from an oxidation state of 0 to -1. The reaction proceeds to the right as this is a highly exothermic reaction.

Finally, remind yourself of common oxidizing and reducing agents and the normal products of their reactions:

OXIDIZING

Permanganate ion	$MnO_4^- + 8H^+ + 5e^- \rightarrow Mn^{2+} + 4H_2O$ (acidic conditions)
Halogens (rarely I_2)	$Cl_2 + 2e^- \rightarrow 2Cl^-$
Dichromate	$Cr_2O_7^{2-} + 14H^+ + 6e^- \rightarrow 2Cr^{3+} + 7H_2O$ (acidic)

REDUCING

Iron (II)	$Fe^{2+} \rightarrow Fe^{3+} + e^-$
Tin (II)	$Sn^{2+} \rightarrow Sn^{4+} + 2e^-$
Copper (I)	$Cu^+ \rightarrow Cu^{2+} + e^-$

REVIEW QUESTION

Write the formulas to show the reactants and the products for the situations described below. Assume that solutions are aqueous unless stated otherwise.

Represent substances in solution as ions if the substances are extensively ionized.

Omit formulas for any ions or molecules that are unchanged by the reaction. You need not balance the equations.

(a) Solutions of sodium dichromate and iron (II) sulfate are mixed.

(b) Aluminum metal is added to a solution of hydrogen chloride.

SOLUTION

The correct answers are:

(a) $Cr_2O_7{}^{2-} + Fe^{2+} + H^+ \rightarrow Cr^{3+} + Fe^{3+} + H_2O$

(b) $Al + H^+ \rightarrow Al^{3+} + H_2$

Recognizing that dichromate is an oxidant and iron (II) is a reductant will allow you to solve the first part of the question. Remembering the products of their reactions gives the answer. The second problem can be solved by looking at the standard reduction potential table and noticing that aluminum is an active metal. It is so active that it should react with neutral water, but a hard coating of aluminum oxide prevents it. The dilute acid dissolves this protective layer, and the redox reaction occurs with the proton being reduced to free hydrogen.

GAS-FORMING REACTIONS

Another common reaction type for descriptive chemistry questions is one that generates a gas. Release of a gas is obvious to an observer and also serves to drive reactions forward to completion. Two common ways this is presented is generating unstable acids and strongly heating some substances.

Some acids that are unstable in the free form (but fine as anions) are listed below, some with their gaseous products:

$$H_2CO_3 \text{ (s)} \rightarrow CO_2 \text{ (g)} + H_2O \text{ (g)}$$
$$H_2SO_3 \text{ (s)} \rightarrow SO_2 \text{ (g)} + H_2O \text{ (g)}$$
$$H_2S \text{ (g)}$$

All of these are weak acids and are formed from their salts when they are treated with a strong acid. Some examples follow (unbalanced):

$$CaCO_3 \text{ (s)} + HCl \text{ (l)} \rightarrow Ca^{2+} \text{ (aq)} + Cl^- \text{ (aq)} + H_2O \text{ (l)} + CO_2 \text{ (g)}$$

$$NaHSO_3 \text{ (s)} + H_2SO_4 \text{ (l)} \rightarrow Na^+ \text{ (aq)} + H_2O \text{ (l)} + SO_4^{2-} \text{ (aq)} + SO_2 \text{ (g)}$$

$$Al_2S_3 \text{ (s)} + HClO_4 \text{ (l)} \rightarrow Al^{3+} \text{ (aq)} + ClO_4^- \text{ (aq)} + H_2S \text{ (g)}$$

$$NaH \text{ (s)} + HCl \text{ (l)} \rightarrow Na^+ \text{ (aq)} + Cl^- \text{ (aq)} + H_2 \text{ (g)}$$

Notice that all of these compounds are insoluble, except for $NaHSO_3$, but dissolve when treated with acid. These reactions can be part of a multistep question in which a precipitate is formed then re-dissolved via treatment with acid, and a gas is formed.

Some acids and bases are gases when there is no water around to dissolve them, including HCl and NH_3:

$$NaCl \text{ (s)} + H_2SO_4 \text{ (s)} \rightarrow NaHSO_4 \text{ (s)} + HCl \text{ (g)}$$

$$NaOH \text{ (s)} + NH_4NO_3 \text{ (s)} \rightarrow NaNO_3 \text{ (s)} + NH_3 \text{ (g)}$$

Another way to form gases is by strongly heating some substances to decomposition. Some examples follow:

Carbonates and bicarbonates:	$CaCO_3$ (or HCO_3^-) $\rightarrow CO_2$ (g) + CaO (and H_2O)
Oxygenated ions and molecules:	ClO_3^- (s) $\rightarrow Cl^-$ (s) + O_2 (g)
	NO_3^- (s) $\rightarrow NO_2^-$ (s) + O_2 (g)
	H_2O_2 (aq) $\rightarrow H_2O$ (l) + O_2 (g)

COLORS OF ELEMENTS AND COMPOUNDS

Color is another observable characteristic of systems that is important in descriptive chemistry. Colors are the result of low-energy electronic transitions; colors are seen in a few elements and gases and, more frequently, in transition metals. The energies required for a transition to absorb visible light are in a very small range. The result is that many substances are not colored and, for compounds that are colored, a slight change in the chemical environment can have a very drastic effect on the perceived color.

Only a few elements are colored, including the halogens: F_2 – pale yellow; Cl_2 – green; Br_2 – brown; I_2 – violet; sulfur – yellow; phosphorus – white (P_4) and red (P_n). The rest of the elemental substances are either colorless or metallic in appearance.

Some common **gaseous compounds** are colored. Other than the halogens, the most likely one you'll come across is NO_2 – brown/red (smog-colored).

Most important, however, are the **transition metal cations**. Unfortunately, they are also very diverse and their color depends not only on their oxidation state, but also on their ligand environment and on the solvent. On the other hand, their complexity helps you on the AP exam because there are only a few well-known examples that are likely to be tested. A review of chapter 5 will help refresh your memory of the electronic states of these metals. Remember that the color of these compounds is due to electronic excitement of the *d*-electrons. This can also be termed ***d*-to-*d* transition** or *d*-to-*d* excitement. Consequently, only transition metals with partially filled *d* orbitals can take part in it; the zinc group normally forms colorless compounds.

Ni^{2+} – green

Mn^{2+} – pink; Mn^{7+} (permanganate – MnO_4^-) – dark purple

Co^{2+} – pink; Co^{3+} – variable from green with $[Co(NH_3)_4Cl_2]Cl$ to red with $[Co(NH_3)_5Cl]Cl_2$

Cu^{2+} – generally blue

TRENDS

Remembering trends is very useful in questions that ask you to order a set of substances. For example, you are never going to remember the ionization energies of all the elements, but you should know that it is easier to remove valence electrons from sodium than oxygen and that potassium fluoride is more ionic than carbon dioxide.

PERIODIC TRENDS AND GROUP CHARACTERISTICS

The most important periodic trend is electron affinity or **electronegativity**. The most electronegative element is fluorine and the least is cesium (francium is not stable and decays by nuclear reactions, so its chemistry can be ignored). As you move down or left in the periodic table, elements generally have less electron affinity and are less electronegative. Once again, this has been covered in chapter 5. Beyond the direct questions of electron affinity, this concept can also be tested by asking about the type of bonding in molecules. For instance, a question might present a series of molecules and ask you to order their bonding from covalent to ionic. The way to answer this type of question is to compare the relative electronegativities of the two elements. If they are close in electronegativity (close in the periodic table), then the bonding is likely to be covalent; this will certainly be the case if both elements are nonmetals. If they are far apart, then they are likely to have ionic-type bonds.

The various groups, or columns, in the periodic table have some characteristics which most or all members of that group possess. Electronegativity changes more as we go across a period than it does as we move down a group; electronegativity changes between adjacent elements are about twice as large moving across a period as moving down a group. So members of groups share a lot of common characteristics, which can help you more effectively answer questions about them. You may remember the reactivity of one member of a group and be able to extrapolate from this information facts about other members of the group. Let's start at the left and move to the right of the periodic table.

Group 1, **alkali metals**, such as sodium, are very reactive as free metals and very unreactive as ions. They always have an oxidation state of +1 in their compounds. Hydrogen heads the group, but as a nonmetal it is nothing like the other members, so it won't be discussed here. The reactivity of the free metals increases as we go from lithium to cesium, and their melting points decrease (Li: 181°C to Cs: 29°C). They react with most elements, including nitrogen, and their compounds are mostly ionic in character. Of all the groups, this one is the most homogeneous.

Group 2, **alkaline earth metals**, can be thought of as a toned down version of their alkali metal neighbors. They are less electropositive than the alkalis but still react with water (with the exception of magnesium, which will react with steam) and form ionic compounds with halogens. They always have an oxidation state of +2 in their compounds. Their melting points are much higher than the alkalis and do not show a trend.

Groups 3–12, **transition metals**, are a large and diverse group of metals in which the *d*-shell is being filled. Being metals, they almost always give up their electrons to form compounds although they have anywhere from +1 to +7 oxidation states. They are less reactive than the alkaline earth metals, and some even are found in the free form in nature (gold, silver, etc.).

Lanthanides and **Actinides** are in a subgroup of the transition metals and have very similar properties. The *f* orbitals are filled in this series, but filling the *f* orbitals doesn't change the reactivity of the members to any great extent, so it is sometimes very difficult to separate one lanthanide from another. Each element can have a number of oxidation states, and their reactivity is similar to that of many of the transition metals.

In groups 13–16, the assumption that members of a group show similar characteristics starts to fail. As one progresses down the group, the metallic character of the elements increases, and the difference between the head of the group and the tail can be like night and day. For example, group 15 is headed by nitrogen, a gaseous element, but ends with bismuth, a metallic solid. We'll look at a few examples from each of these groups.

Group 13 starts with a nonmetal, boron, yet the rest of the group is made up of metals. Aluminum is a very reactive metal, although a thin, strong oxide coating makes it appear unreactive in the metallic state. Gallium is interesting in that it has a very low melting point, 30°C.

Group 14 is headed by carbon, which is integral to life because it can form chains of essentially infinite length and strong bonds to many different elements. Silicon is below it and is considered a metalloid. Its bonding is quite different from that of carbon. The remaining elements in group 14 are metals.

Group 15 starts with the very unreactive nitrogen gas, N_2. In other compounds, nitrogen has common oxidation states of −3 (ammonia), +3 (nitrite), and +5 (nitrate). Phosphorus is also a nonmetal and has three main allotropes (red P_n; white P_4, which is reactive and toxic; and black, which is unreactive and rare). Antimony (Sb) has two allotropes, a metallic and an amorphous grey form. Finally, bismuth is metallic.

Group 16 is also known as the **chalcogens** (from the Greek for "ore former"), as many ores contain elements from this group. The most common oxidation state for this group is −2. Oxygen is very electronegative although the electronegativity decreases as you go down the group. Oxygen has two allotropes—the familiar O_2 and ozone (O_3), which forms a protective barrier in our upper atmosphere and absorbs harmful UV rays. Oxygen is the only element in this group to form double bonds, although frequently sulfur is drawn in compounds with a double bond. Sulfur and selenium are nonmetallic solids, with sulfur having important allotropes. Sulfur can form both orthorhombic and monoclinic crystals whose molecular formula is S_8, as well as a polymeric liquid when heated.

Elements in group 17, known as the **halogens**, have similar reactivities. They are electronegative and, in keeping with the usual trend, electronegativity decreases as we move down the group. Fluorine and chlorine are gases, bromine is a volatile liquid, and iodine is a volatile solid. They are all colored as discussed above and tend to form compounds in which they have an oxidation state of −1. Also, chlorine, bromine, and iodine can have oxidation states of up to +7 (ClO_4^-, BrO_4^-, and IO_4^-). The elements range in reactivity from extreme (F_2) to mild (I_2), although none are ever found free in nature. Compounds of metals and the halogens are generally ionic. Compounds can also be formed within the group and are called interhalogens. Their reactivity is similar to that of the component halogens.

Group 18 contains the **noble gases**, and as their name implies, they do not generally interact with the other elements. As we learned with the octet rule, this makes sense. Krypton and xenon have been forced to form compounds with fluorine and oxygen, and recently even argon has been shown to react to form a hydrogen fluoride (HArF).

ACID/BASE REACTIONS AND TRENDS

For the AP exam, the most useful way to think about acids and bases is with the Brønsted-Lowry definition first covered in chapter 14. This definition deals only with proton donors (acids) and proton acceptors (bases). This is most useful for an aqueous solution, the most common situation in this exam. By now you should be familiar with strong and weak acids and bases. Strong acids and bases fully ionize in aqueous solution, whereas weak acids and bases only partially ionize. Let's review some common ones here and some trends that will help you order them in terms of strength:

Strong acids: Sulfuric (H_2SO_4), hydrochloric (HCl), hydrobromic (HBr), hydroiodic (HI), perchloric ($HClO_4$), or nitric (HNO_3) acid. Only the first proton in polyprotic acids (such as sulfuric) is considered strong. Note the large number of electronegative oxygens in these acids.

Weak acids: Acetic ($HC_2H_3O_2$)—or any organic acid, carbonic (H_2CO_3), sulfurous (H_2SO_3), chloric ($HClO_3$), nitrous (HNO_2), hydrocyanic (HCN), or boric (H_3BO_3) acid. Note that some of these are less oxygenated versions of strong acids.

Acid strength: Acid strength is a measure of hydrogen ion dissociation. Weak acids actually have strongly bound hydrogen ions and release only a few in solution, whereas strong acids with weak hydrogen bonds almost totally dissociate, basically releasing all their hydrogen ions. Do not confuse the terms acid strength and pH, however. The strength of an acid has to do with initial proportion of acid molecules that are ionized, not with the total number. If a high percentage of acid molecules are in the ionized form, the acid will be stronger. Conversely, the weaker the acid, the lower the percentage of hydrogen ions. Additionally, weak acids can exhibit low pH readings just like strong acids.

Strong bases: The strongest bases in water are metallic hydroxides of the alkaline and alkaline earth series: NaOH, KOH, $Ca(OH)^2$, etc. Notice that LiOH, $Be(OH)_2$, and $Mg(OH)_2$ are not considered strong bases because they have limited solubility in water.

Weak bases: Alkali metal salts of weak acids are weak bases: NaCN, K_2CO_3, NaOCl. Most amines are weak bases, such as ammonia (NH_3 or NH_4OH) or methylamine (CH_3NH_2).

Base strength: A good way to get a handle on the relative strength of bases is to look at their corresponding conjugate acids. This concept was covered in chapter 14. The weaker the conjugate acid, the stronger the conjugate base will be and vice versa. Let's look at some examples:

H_2O—very weak acid, NaOH—strong base

H_2CO_3—weak acid, $NaHCO_3$—weak base, HCO_3^-—weaker acid, Na_2CO_3—stronger base

H_2—very weak acid, NaH—very strong base (reacts with water)

$HClO_4$—strong acid, $NaClO_4$—very weak base/neutral

pH of salts: The pH of a solution of a salt can be estimated if we know the acidity and basicity of both the anion and cation. We need to know the strength of the conjugate base of the cation as well as the strength of the conjugate acid of the anion. From this, we can quickly say whether a solution of ammonium chloride, potassium nitrate, or sodium cyanide is acidic or basic. Let's look at a simple 2×2 matrix:

	Conjugate Base—Strong	Conjugate Base—Weak
Conjugate Acid—Strong	Neutral	Acidic
Conjugate Acid—Weak	Basic	Undetermined (requires K_a & K_b values)

Let's reexamine the previous examples using this table. Ammonium chloride (NH_4Cl) is a salt formed from a weak base (NH_4OH) and a strong acid (HCl). Looking at this table, we see that a solution of NH_4Cl is acidic. Potassium nitrate (KNO_3) is a salt formed from a strong base (KOH) and a strong acid (HNO_3) and therefore should be neutral. Finally, sodium cyanide is a salt of a strong base ($NaOH$) and a weak acid (HCN) and would produce a basic solution. A simple way to remember this table is that the strong component wins out. When both the acidic and basic components of the salt are equal, the result is either a tie (two strong components yield a neutral solution), or it depends on their relative strength when both are weak.

FLAME TESTS

Flame tests are performed by exposing a solution of a salt to the high temperatures of a flame. Alkali and alkaline earth metals impart a characteristic color to the flame, and their identity can be inferred from this color. The color is due to the ions becoming electronically excited in the hot flame and, upon returning to the ground state, giving off a characteristic set of wavelengths of light. The energies of these emissions (and thus their color) do not follow any pattern, so the following list must be committed to memory. Two everyday instances of this phenomenon are the yellow color of many sodium street lights and the red color of highway flares (strontium).

Li—Red	Na—Yellow	K—Light purple	Cs—Blue
Ca—Red/Orange	Sr—Red	Ba—Green/Yellow	

Beryllium and magnesium do not impart any color to a flame test.

REVIEW QUESTIONS

1. The pairs of solutions shown below are mixed. Which pair(s) show(s) a precipitate?

   I. Na_3PO_4 and Cs_2S
   II. $AgClO_3$ and NH_4OH
   III. $CsC_2H_3O_2$ and $Hg(NO_3)_2$

   (A) I only
   (B) II only
   (C) III only
   (D) II and III only
   (E) I and III only

2. Allene (C_3H_4) is burned in a torch with excess oxygen. For each mole of allene used, how many moles of oxygen are consumed?

   (A) 8
   (B) 7
   (C) 5
   (D) 4
   (E) 2.5

3. Order the following molecules from covalent to ionic with respect to their bonding.

   I. H_2O
   II. RbBr
   III. N_2

	Covalent				Ionic
(A)	I	>	II	>	III
(B)	I	>	III	>	II
(C)	II	>	III	>	I
(D)	III	>	II	>	I
(E)	III	>	I	>	II

4. A solution of a pure white crystalline compound shows a red flame test and a precipitate when treated with a solution of Na_2SO_4. This compound could be

   (A) $Ba(NO_3)_2$.
   (B) $Mg(NO_3)_2$.
   (C) $Sr(NO_3)_2$.
   (D) $CsNO_3$.
   (E) $LiNO_3$.

ANSWERS AND EXPLANATIONS

1. B

Choice I will not show a precipitate because both Na_2S and Cs_3PO_4 are soluble. Also, choice III will not show a precipitate because both $Hg(C_2H_3O_2)_2$ and $CsNO_3$ are soluble. If you are having trouble remembering the rules, look at all of the solutions given in the question. All of these must be soluble (the AP exam is not that tricky), and it may help jog your memory about the rules governing solubility.

2. D

Because the question states that oxygen is in excess, we know that the carbon-containing product is CO_2 (not CO). We know that hydrogen will be converted to water (H_2O), so we can now write an unbalanced equation: $C_3H_4 + O_2 \rightarrow H_2O + CO_2$. After balancing it using the procedure outlined in chapter 12, we have the full equation: $C_3H_4 + 4O_2 \rightarrow 2H_2O + 3CO_2$ and the answer: each mole of allene requires four moles of oxygen.

3. E

The most covalent molecule is nitrogen, III, because its constituent elements are the same and the bonding electrons are shared equally between the two atoms. Then comes water, I, whose bonding is best described as polar covalent. Finally comes rubidium bromide, II, whose constituent elements have a wide difference in their electronegativity. Its bonding, and that of all compounds formed between alkali metals and halogens, is ionic in nature.

4. C

Of the possibilities, only strontium and barium have insoluble sulfates, but only strontium would show a red flame test. Barium exhibits a pale green emission.

CHAPTER 18: LABORATORY SKILLS

> ## IF YOU LEARN ONLY THREE THINGS IN THIS CHAPTER . . .
>
> 1. Remember the basic laboratory techniques you learned in chemistry class and follow common safety guidelines.
>
> 2. When working in the lab, it is important to be accurate in your measurements. Be sure to follow the proper measurement techniques.
>
> 3. You should be familiar with the general instrument-based problems that may come up in Section II.

There are not many questions regarding good laboratory practice in the multiple-choice section, though you should expect a few. Laboratory-type questions are guaranteed to appear on either Part A or Part B of Section II. The laboratory component of your AP Chemistry course will have prepared you for these topics. We'll hit the highlights in this chapter and go over some specific lab-based questions at the end.

GOOD LABORATORY PRACTICE

SAFETY

From day one in lab, you've been educated about what to do and what not to do there. Let's just go over some of the major items:

- Never wear contact lenses in the lab, because chemicals can be trapped between your eye and the lens. Instead, wear corrective eye glasses and goggles.

- Always wear goggles, closed-toed shoes, and a lab coat or apron when in the lab.

- Never eat or drink in the lab and never taste anything, even if you're sure it is harmless.

- Always listen to your lab instructor and inform that person if there is a problem.

MEASURING LIQUIDS AND SOLIDS

Another important consideration in lab is to be accurate in your measurements. By following a few simple rules, your results can improve and you can become a better scientist. Actual lab work can't be tested on the AP Chemistry exam, but techniques can and will be tested. So let's go over some of these techniques.

When determining the mass of solids, the substance you are weighing should be static-free and its temperature equilibrated with the ambient air. Weighing something that is too hot or too cold will create air currents that can give you the wrong reading.

Preparing a solution from a solid is another place where improper technique can give the wrong result. Let's say that we are to make 1.0 L of a 1.0 M NaCl solution. We first calculate the molecular weight of NaCl (58.5 g/mol), which is also the mass of NaCl required to prepare this solution. The following are the correct steps to follow when making this solution:

1. Add 500.0 mL of room temperature, deionized (D.I.) water into a clean, dry, room-temperature 1 L volumetric flask.

2. Add the 58.5 g of NaCl and swirl to dissolve.

3. Fill the volumetric flask with more D.I. water so that the bottom of the meniscus is at the 1 L line.

Dissolving solutes can change the volume of a solvent, so you need to dissolve the NaCl first in less than 1 L of water before filling to the 1 L line. The volume of solutions is also temperature-dependent, so the solution should be made with room-temperature water. Most volumetric flasks have an operating temperature written on them. The meniscus is the lowest part of the top of the solution and should always be read at eye level. These considerations also apply whenever you are preparing a solution, whether it is diluting a stock solution or dissolving a solid. Remember to keep track of significant figures; making up a solution to a certain precision requires making *all* measurements to at least that precision. You don't want to lose points on something as trivial as significant figures.

The glassware required for precision measurements must also be treated with care. Never leave volumetric flasks in a drying oven. The glass can distort in the hot oven and be ruined. Also, beware of corrosive solutions in this glassware. Concentrated alkali hydroxide (such as NaOH) solutions and hydrogen fluoride can etch the glass and also give wrong readings.

When preparing solutions of compounds that release a lot of heat when diluted with water, careful precautions must be taken. Always slowly add the substance to a large volume of water. This is especially important when diluting strong mineral acids such as sulfuric acid and also when making solutions of strong bases such as sodium or potassium hydroxides (NaOH or KOH). Adding the acid or base slowly to water is done so that the heat generated is dispersed by a large amount of water, thus reducing the chance that it will boil or splatter. The hot solution must then be cooled completely to ambient temperature before the final dilution to the mark. The density of the solution and the volume of the glassware both change with temperature; for an accurate dilution, the solution must be at the temperature specified on the glassware (usually 25°C).

TITRATION

So you've done a good job of making standard solutions, say, for a pH titration. Without the correct technique, all of this good work will be for naught. Now we'll go over the correct procedure for calculating the molecular weight of a monoprotic acid using titration with a base:

1. Measure out a weighed amount of the unknown and make a solution according to the procedures outlined above.

2. Rinse a burette with the standardized base and then fill and record the level of the meniscus.

3. Add a very small amount of an indicator that covers the required pH range of the equivalence point of the unknown solution.

4. Add the base slowly until the indicator shows that the equivalence point is close and then add the base drop by drop, stirring until the color persists.

5. Record the level of the base in the burette, and calculate the molecular weight of the unknown.

USE OF A SPECTROPHOTOMETER

A spectrophotometer is an instrument used for measuring the absorbance of light by a solution. When used correctly, it can give very accurate measurements and is used to monitor reactions in which there is a change in absorbance. A question involving the use of a spectrophotometer can appear in Section II, normally in conjunction with measuring the rate of a reaction.

One important equation that we must know when using a spectrophotometer is the **Beer-Lambert law** or:

$$A = \varepsilon bc$$

where A is the absorbance at a particular wavelength, ε is the molar absorptivity of the substance in question, b is the path length the light has to travel through the solution that you are measuring (usually 1 cm), and c is concentration of the substance. (When solving an equation using the Beer-Lambert law, be sure to check the units in the molar absorptivity to make sure your answers are in the correct units.)

Let's go over a procedure for using a spectrophotometer in order to measure the molar absorptivity of permanganate anion.

Finding λ_{max} and ε for MnO_4^-:

1. Measure a blank solution (one without permanganate) to subtract out the absorbances of the glass, water, etc. In this step and all other steps, be very careful to keep fingerprints, dust, etc., off the glass.

2. Make a series of dilute known solutions of MnO_4^- ions, and find the wavelength maximum and their absorbances.

3. If absorbances are determined at several different concentrations, the best value for molar absorptivity is obtained from the slope of a calibration curve. Calculating the absorptivity from each solution will not receive full credit in this sort of case.

Note: Make sure that the absorbance is below two for all these data points. An absorbance of over two indicates that less than 1% of the light is making it through the solution, which is not enough for an accurate reading.

GRAVIMETRIC EVALUATION OF A HYDRATE

This experiment type involves analysis of a hydrated salt, for example, $CuSO_4 \cdot H_2O$. A sample is weighed before and after heating. The heating should be sufficient to drive off all the water of crystallization. A sample procedure follows:

1. Clean and dry a crucible.

2. Weigh out a sample of the blue hydrate and record the value.

3. Heat the hydrate over a medium-low Bunsen burner until all of the blue color is gone. Make sure all of the water is off the sides of the crucible by heating the entire crucible.

4. Cover the anhydrous white salt and allow to cool to room temperature.

5. Once the salt is cool, make sure that no blue hydrate has reformed. This procedure should include at least two heat-cool-weigh cycles to make sure it has been heated to constant weight. Color alone is not a sufficient indicator that all water has been driven off, even in a case like this one where there is a color change; many compounds don't have a color change between the hydrate and the anhydrous salt.

6. Weigh the anhydride and calculate the mass of water driven off and the formula for the hydrate.

MOLAR MASS FROM VAPOR PRESSURE

This type of experiment relies on the ideal gas law: $PV = nRT$ where P is pressure, V is volume, n is the number of moles, R is the gas constant, and T is the temperature. Looking at the units of the gas constant is critical; be sure to convert all of your data to the same units as your gas constant. Let's look at determining the molar mass of acetonitrile. The general idea in the experiment is to fill an Erlenmeyer flask of known volume with only the vapor of warm (100°C) acetonitrile and then cool it down to measure the mass of that vapor. From this experiment, we will have the volume of the gas, the temperature, the mass, and the pressure and can calculate the number of moles and hence the molar mass. So here's the procedure:

1. Weigh a clean, dry 1 L volumetric graduated flask with a tiny piece of aluminum foil covering the mouth.

2. Prepare a 2 L beaker with boiling water and float the volumetric flask so that it is mostly immersed in the water without water getting in the flask. Inject an excess (5 mL) of acetonitrile into the flask through the foil cap. The acetonitrile will turn to a gas and fill the flask. Excess acetonitrile will exit through the hole in the cap. After there is no more liquid in the flask, we have only 1L of acetonitrile vapor at the ambient pressure in the lab, which would be given in the problem, and 100°C, the temperature of the boiling water.

3. Remove the flask and cool rapidly to prevent acetonitrile vapor from escaping the flask. The acetonitrile vapor will condense in the flask, and the remaining volume will be replaced with air. Dry the outside and weigh the flask.

REVIEW QUESTIONS

1. To make 1.0 L of a 1.0 M solution of H_2SO_4 from the concentrate, which of the following steps are required and in what order?

   I.   Add 1.0 L of distilled water to the volumetric flask.

   II. Add concentrated acid.

   III. Add 0.50 L of distilled water to the volumetric flask.

   IV. Fill flask to 1.0 L total volume with distilled water.

   (A) III, II, then IV

   (B) IV, I, then II

   (C) III then I

   (D) I, IV, then II

   (E) I then II

2. Calculate the molecular weight of an unknown monoprotic acid with a mass of 0.343 g. The volume of 0.1 M NaOH required for equivalence is 28.1 mL.

3. A 1×10^{-4} M solution is analyzed in a 1 cm cell and found to have a λ_{max} at 525 nm and an absorbance of 0.22. Calculate the molar absorptivity of the solution using the Beer-Lambert equation.

4. Calculate the weight and number of moles of water driven off of hydrated $CuSO_4$ to form $CuSO_4$, an anhydrous salt, using gravimetric evaluation of the following values: mass of initial hydrate 4.0 g, mass of anhydrous salt 2.5 g.

5. Calculate the molar mass of acetonitrile from its vapor pressure at 100°C and 1 atm using the ideal gas law and the following values: weight of the initial volumetric flask: 1,231.132 g, weight of the final flask + acetonitrile: 1,232.470 g.

ANSWERS AND EXPLANATIONS

1. A

Always add the concentrated acid slowly to water, fully dissolve it, let it cool to room temperature, and then fill to the top of the volumetric flask.

2.

Because we know that our acid is monoprotic, we know that for each mole of base used, we have one mole of acid. Solving for the moles of $NaOH(n)$: $0.1\text{ M} = \dfrac{n}{0.0281\text{ L}}$ or $n = 0.00281$ mol. To find the molecular weight of the acid, we divide the weight by the number of moles or $\dfrac{0.343\text{ g}}{0.00281\text{ mol}} = 122$ g/mol.

3.

Plugging in to the Beer-Lambert equation, $A = \varepsilon bc$, gives $0.22 = \varepsilon \times 1\text{ cm} \times 1 \times 10^{-4}$ M. Solving for ε gives a value of 2.2×10^3 cm^{-1}M^{-1}.

4.

We can calculate the weight of water driven off using subtraction: $(4.0\text{ g} - 2.5\text{ g}) = 1.5$g. We can then calculate the number of moles of water driven off by dividing the weight by the molar mass: $\dfrac{1.6\text{ g}}{16\text{ g/mol}} = 0.09375$mol.

5.

The weight of acetonitrile is $1{,}232.470 - 1{,}231.132 = 1.330$ g. The number of moles of ideal gas in the flask at 100ºC and 1 atm is found from the ideal gas law: $1\text{ atm} \times 1\text{ L} = 0.0821$ (L · atm · mol^{-1} · K^{-1}) $\times n \times 373$ K, or 0.0326 moles. Dividing the weight by the number of moles gives the molar mass, or $\dfrac{1.33\text{ g}}{0.0326\text{ mol}} = 40.8$ g/mol.

Part Four

PRACTICE TESTS

Information in this table may be useful in answering the questions in the Practice Tests.

Periodic Table of the Elements

Atomic Number —— 6
Symbol —— C
Atomic Mass —— 12.01

* Numbers in parentheses are the *mass numbers* of the most stable isotope of the element.

	1	2	3	4	5	6	7	8	9	10	11	12	13	14	15	16	17	18
1	1 H 1.008																	2 He 4.003
2	3 Li 6.941	4 Be 9.012											5 B 10.81	6 C 12.01	7 N 14.01	8 O 16.00	9 F 19.00	10 Ne 20.18
3	11 Na 22.99	12 Mg 24.31											13 Al 26.98	14 Si 28.09	15 P 30.97	16 S 32.07	17 Cl 35.45	18 Ar 39.95
4	19 K 39.10	20 Ca 40.08	21 Sc 44.96	22 Ti 47.88	23 V 50.94	24 Cr 52.00	25 Mn 54.94	26 Fe 55.85	27 Co 58.93	28 Ni 58.69	29 Cu 63.55	30 Zn 65.39	31 Ga 69.72	32 Ge 72.61	33 As 74.92	34 Se 78.96	35 Br 79.90	36 Kr 83.80
5	37 Rb 85.47	38 Sr 87.62	39 Y 88.91	40 Zr 91.22	41 Nb 92.91	42 Mo 95.94	43 Tc (98)	44 Ru 101.1	45 Rh 102.9	46 Pd 106.4	47 Ag 107.9	48 Cd 112.4	49 In 114.8	50 Sn 118.7	51 Sb 121.8	52 Te 127.6	53 I 126.9	54 Xe 131.3
6	55 Cs 132.9	56 Ba 137.3	57–70	72 Hf 178.5	73 Ta 181.0	74 W 183.8	75 Re 186.2	76 Os 190.2	77 Ir 192.2	78 Pt 195.1	79 Au 197.0	80 Hg 200.6	81 Tl 204.4	82 Pb 207.2	83 Bi 209.0	84 Po (209)	85 At (210)	86 Rn (222)
7	87 Fr (223)	88 Ra 226.0	89–102	104 Rf (261)	105 Db (262)	106 Sg (263)	107 Bh (262)	108 Hs (265)	109 Mt (268)	110 Uun (269)	111 Uuu (272)	112 Uub (277)	113 Uut (284)	114 Uuq (289)	115 Uup (288)	116 Uuh (292)	117 Uus (294)	118 Uuo (294)

57 La 138.9	58 Ce 140.1	59 Or 140.9	60 Nd 144.2	61 Pm (145)	62 Sm 150.4	63 Eu 152.0	64 Gd 157.3	65 Tb 158.9	66 Dy 162.5	67 Ho 164.9	68 Er 167.3	69 Tm 168.9	70 Yb 173.0
89 Ac 227.0	90 Th 232.0	91 Pa 231.0	92 U 238.0	93 Np 237.0	94 Pu (244)	95 Am (243)	96 Cm (247)	97 Bk (247)	98 Cf (251)	99 Es (252)	100 Fm (257)	101 Md (258)	102 No (259)

Information in this table may be useful in answering the questions in Section II of the Practice Tests.

STANDARD REDUCTION POTENTIALS IN AQUEOUS SOLUTION AT 25°C

Half-reaction			$E°(V)$
$F_2(g) + 2\,e^-$	$\rightarrow$	$2F^-$	2.87
$Co^{3+} + e^-$	$\rightarrow$	Co^{2+}	1.82
$Au^{3+} + 3\,e^-$	$\rightarrow$	$Au(s)$	1.50
$Cl_2(g) + 2\,e^-$	$\rightarrow$	$2Cl^-$	1.36
$O_2(g) + 4\,H^+ + 4\,e^-$	$\rightarrow$	$2H_2O(\ell)$	1.23
$Br_2(\ell) + 2\,e^-$	$\rightarrow$	$2Br^-$	1.07
$2\,Hg^{2+} + 2\,e^-$	$\rightarrow$	$Hg_2^{\,2+}$	0.92
$Hg^{2+} + 2\,e^-$	$\rightarrow$	$Hg(\ell)$	0.85
$Ag^+ + e^-$	$\rightarrow$	$Ag(s)$	0.80
$Hg_2^{\,2+} + 2\,e^-$	$\rightarrow$	$2Hg(\ell)^-$	0.79
$Fe^{3+} + e^-$	$\rightarrow$	Fe^{2+}	0.77
$I_2(s) + 2\,e^-$	$\rightarrow$	$2I^-$	0.53
$Cu^+ + e^-$	$\rightarrow$	$Cu(s)$	0.52
$Cu^{2+} + 2\,e^-$	$\rightarrow$	$Cu(s)$	0.34
$Cu^{2+} + e^-$	$\rightarrow$	Cu^+	0.15
$Sn^{4+} + 2\,e^-$	$\rightarrow$	Sn^{2+}	0.15
$S(s) + 2\,H^+ + 2\,e^-$	$\rightarrow$	$H_2S(g)$	0.14
$2\,H^+ + 2\,e^-$	$\rightarrow$	$H_2(g)$	0.00
$Pb^{2+} + 2\,e^-$	$\rightarrow$	$Pb(s)$	−0.13
$Sn^{2+} + 2\,e^-$	$\rightarrow$	$Sn(s)$	−0.14
$Ni^{2+} + 2\,e^-$	$\rightarrow$	$Ni(s)$	−0.25
$Co^{2+} + 2\,e^-$	$\rightarrow$	$Co(s)$	−0.28
$Cd^{2+} + 2\,e^-$	$\rightarrow$	$Cd(s)$	−0.40
$Cr^{3+} + e^-$	$\rightarrow$	Cr^{2+}	−0.41
$Fe^2 + 2\,e^-$	$\rightarrow$	$Fe(s)$	−0.44
$Cr^{3+} + 3\,e^-$	$\rightarrow$	$Cr(s)$	−0.74
$Zn^{2+} + 2\,e^-$	$\rightarrow$	$Zn(s)$	−0.76
$2H_2O(\ell) + 2\,e^-$	$\rightarrow$	$H_2(g)+2OH^-$	−0.83
$Mn^{2+} + 2\,e^-$	$\rightarrow$	$Mn(s)$	−1.18
$Al^{3+} + 3\,e^-$	$\rightarrow$	$Al(s)$	−1.66
$Be^{2+} + 2\,e^-$	$\rightarrow$	$Be(s)$	−1.70
$Mg^{2+} + 2\,e^-$	$\rightarrow$	$Mg(s)$	−2.37
$Na^+ + e^-$	$\rightarrow$	$Na(s)$	−2.71
$Ca^{2+} + 2\,e^-$	$\rightarrow$	$Ca(s)$	−2.87
$Sr^{2+} + 2\,e^-$	$\rightarrow$	$Sr(s)$	−2.89
$Ba^{2+} + 2\,e^-$	$\rightarrow$	$Ba(s)$	−2.90
$Rb^+ + e^-$	$\rightarrow$	$Rb(s)$	−2.92
$K^+ + e^-$	$\rightarrow$	$K(s)$	−2.92
$Cs^+ + e^-$	$\rightarrow$	$Cs(s)$	−2.92
$Li^+ + e^-$	$\rightarrow$	$Li(s)$	−3.05

Information in this table may be useful in answering the questions in Section II of the Practice Tests.

AP CHEMISTRY EQUATIONS AND CONSTANTS

ATOMIC STRUCTURE

$$E = h\nu$$

$$c = \lambda\nu$$

$$\lambda = \frac{h}{mv}$$

$$p = mv$$

$$E_n = \frac{-2.178 \times 10^{-18}}{n^2} \text{ joule}$$

EQUILIBRIUM

$$K_a = \frac{[H^+][A^-]}{[HA]}$$

$$K_b = \frac{[OH^-][HB^+]}{[B]}$$

$$K_w = [OH^-][H^+] = 1.0 \times 10^{-14} \text{ @ } 25°C$$

$$= K_a \times K_b$$

$$pH = -\log [H^+]$$

$$pOH = -\log [OH^-]$$

$$14 = pH + pOH$$

$$pH = pK_a + \log \frac{[A^-]}{[HA]}$$

$$pOH = pK_b + \log \frac{[HB^+]}{[B]}$$

$$pK_a = -\log K_a$$

$$pK_b = -\log K_b$$

$$K_p = K_c(RT)^{\Delta n}, \text{ where}$$

$$\Delta n = \text{moles product gas} - \text{moles reactant gas}$$

EQUILIBRIUM CONSTANTS

K_a (weak acid)

K_b (weak base)

K_c (molar concentrations)

K_p (gas pressure)

K_w (water)

EQUILIBRIUM CONSTANTS (CONT.)

$S°$ = standard entropy

$H°$ = standard enthalpy

$G°$ = standard free energy

$E°$ = standard reduction potential

n = moles

m = mass

T = temperature

q = heat

c = specific heat capacity

C_p = molar heat capacity at constant pressure

Faraday's constant, $\mathfrak{F}$ = 96,500 coulombs per mole of electrons

THERMOCHEMISTRY

$$\Delta S° = \Sigma S° \text{ products} - \Sigma S° \text{ reactants}$$

$$\Delta H° = \Sigma H°_f \text{ products} - \Sigma H°_f \text{ reactants}$$

$$\Delta G° = \Sigma G°_f \text{ products} - \Sigma G°_f \text{ reactants}$$

$$\Delta G° = \Delta H° - T\Delta S°$$

$$= -RT \ln K = -2.303 \, RT \log K = -n\mathfrak{F} E°$$

$$\Delta G° = \Delta G° + RT \ln Q = \Delta G° + 2.303 \, RT \log Q$$

$$q = mc\Delta T$$

$$C_p = \frac{\Delta H}{\Delta T}$$

m = mass

E = energy

ν = frequency

λ = wavelength

p = momentum

v = velocity

n = principal quantum number

Speed of light, $c = 3.0 \times 10^8 \text{ m s}^{-1}$

Electron charge, $e = -1.602 \times 10^{-19}$ coulomb

1 electron volt per atom = 96.5 kJ mol^{-1}

Planck's constant, $h = 6.63 \times 10^{-34}$ J s

Boltzmann's constant, $k = 1.38 \times 10^{-23}$ J K^{-1}

Avogadro's number = 6.022×10^{23} mol^{-1}

Information in this table may be useful in answering the questions in Section II of the Practice Tests.

GASES, LIQUIDS, AND SOLUTIONS

$$PV = nRT$$

$$\left(P + \frac{n^2 a}{V^2}\right)(V - nb) = nRT$$

$$P_A = P_{total} \times X_A, \text{ where } X_A \frac{\text{moles A}}{\text{total moles}}$$

$$P_{total} = P_A + P_B + P_C + \ldots$$

$$n = \frac{m}{M}$$

$$K = {}^\circ C + 273$$

$$\frac{P_1 V_1}{T_1} = \frac{P_2 V_2}{T_2}$$

$$D = \frac{m}{V}$$

$$u_{rms} = \sqrt{\frac{3kT}{m}} = \sqrt{\frac{3RT}{M}}$$

$$KE \text{ per molecule} = \frac{1}{2} m v^2$$

$$KE \text{ per mole} = \frac{3}{2} RT$$

$$\frac{r_1}{r_2} = \sqrt{\frac{M_2}{M_1}}$$

molarity, M = moles solute per liter solution

molality = moles solute per kilogram solvent

$$\Delta T_f = iK_f \times \text{molality}$$

$$\Delta T_b = iK_b \times \text{molality}$$

$$\pi = iMRT$$

OXIDATION-REDUCTION; ELECTROCHEMISTRY

$$Q = \frac{[C]^c [D]^d}{[A]^a [B]^b}, \text{ where } a A + b B \to c C + d D$$

$$I = \frac{q}{t}$$

$$E_{cell} = E^\circ{}_{cell} - \frac{RT}{n\mathscr{F}} \ln Q = E^\circ{}_{cell} - \frac{0.0592}{n} \log Q @ 25^\circ C$$

$$\log K = \frac{nE^\circ}{0.0592}$$

OXIDATION-REDUCTION; ELECTROCHEMISTRY (CONT.)

P = pressure

V = volume

T = temperature

n = number of moles

D = density

v = velocity

E° = standard reduction potential

K = equilibrium constant

KE = kinetic energy

r = rate of effusion

M = molar mass

t = time (seconds)

π = osmotic pressure

i = van't Hoff factor

K_f = molal freezing-point depression constant

K_b = molal boiling-point elevation constant

Q = reaction quotient

I = current (amperes)

q = charge (coulombs)

u_{rms} = root-mean-square speed

Gas constant, R = 8.31 J mol^{-1} K^{-1}

$\qquad$ = 0.0821 L atm mol^{-1} K^{-1}

$\qquad$ = 8.31 volt coulomb mol^{-1} K^{-1}

Boltzmann's constant, k = 1.38 × 10^{-23} J K^{-1}

K_f for H_2O = 1.86 K kg mol^{-1}

K_b for H_2O = 0.512 K kg mol^{-1}

1 atm = 760 mm Hg

$\qquad$ = 760 torr

STP = 0.000°C and 1.000 atm

Faraday's constant, $\mathscr{F}$ = 96,500 coulombs per mole of electrons

HOW TO TAKE THE PRACTICE TESTS

The next section of this book consists of Practice Tests. Taking a practice AP exam gives you an idea of what it's like to answer these test questions for a longer period of time, one that approximates the real exam. You'll find out which areas you're strong in and where additional review may be required. Any mistakes you make now are ones you won't make on the actual exam, as long as you take the time to learn where you went wrong.

The two full-length Practice Tests in this book each include 75 multiple-choice questions and six free-response questions. You will have 90 minutes for the multiple-choice questions and 90 minutes to answer the six free-response questions. Before taking a Practice Test, find a quiet place where you can work uninterrupted for three hours. Time yourself according to the time limit at the beginning of each section. It's okay to take a short break between sections, but for the most accurate results, you should approximate real test conditions as much as possible.

As you take the Practice Tests, remember to pace yourself. Train yourself to be aware of the time you are spending on each question. Try to be aware of the general types of questions you encounter, and be alert to certain strategies or approaches that help you to handle the various question types more effectively.

After taking a Practice Test, be sure to read the detailed answer explanations that follow. These will help you identify areas that could use additional review. Even when you've answered a question correctly, you can learn additional information by looking at the answer explanation.

Finally, it's important to approach the exam with the right attitude. You're going to get a great score because you've reviewed the material and learned the strategies in this book.

Good luck!

HOW TO COMPUTE YOUR SCORE

SCORING THE MULTIPLE-CHOICE QUESTIONS

To compute your score on the multiple-choice portion of the two sample tests, calculate the number of questions you got right on each test, then divide by 75 to get the decimal percentage score for the multiple choice portion of that test.

SCORING THE FREE-RESPONSE QUESTIONS

The reviewers have specific points that they will be looking for in each free-response answer. Each piece of information that they are able to check off in your essay is a point toward a better score. Be careful with your spelling—don't write *bromate* if you mean *bromite*!

To figure out your approximate score for the free-response questions, look at the key points found in the sample response for each question. For each key point you included, add a point. Add together all the points you received for each question, and add together all the possible points available for each question. Divide the total number of points you received by the total number of possible points to get the decimal percentage score for the free-response portion of that test.

CALCULATING YOUR COMPOSITE SCORE

Your score on the AP Chemistry exam is a combination of your scores on the multiple-choice portion of the exam and the free-response section. The free-response section and the multiple-choice section are worth one-half of the exam score.

Add together your score on the multiple-choice portion of the exam and your approximate score on the free-response section of the exam. Divide this sum by 2 and multiply by 100 to obtain your approximate score for each full-length exam. Round up to a whole number if your score is a decimal.

Remember, however, that much of this depends on how well all of those taking the AP test do. If you do better than average, your score would be higher. The numbers here are just approximations.

The approximate score range is as follows:

5 = 70–100 (extremely well qualified)

4 = 55–69 (well qualified)

3 = 40–54 (qualified)

2 = 30–39 (possibly qualified)

1 = 0–29 (no recommendation)

If your score falls between 55 and 100, you're doing great—keep up the good work! If your score is lower than 55, there's still hope—keep studying and you will be able to obtain a much better score on the exam before you know it.

Practice Test One Answer Grid

1. Ⓐ Ⓑ Ⓒ Ⓓ Ⓔ	26. Ⓐ Ⓑ Ⓒ Ⓓ Ⓔ	51. Ⓐ Ⓑ Ⓒ Ⓓ Ⓔ
2. Ⓐ Ⓑ Ⓒ Ⓓ Ⓔ	27. Ⓐ Ⓑ Ⓒ Ⓓ Ⓔ	52. Ⓐ Ⓑ Ⓒ Ⓓ Ⓔ
3. Ⓐ Ⓑ Ⓒ Ⓓ Ⓔ	28. Ⓐ Ⓑ Ⓒ Ⓓ Ⓔ	53. Ⓐ Ⓑ Ⓒ Ⓓ Ⓔ
4. Ⓐ Ⓑ Ⓒ Ⓓ Ⓔ	29. Ⓐ Ⓑ Ⓒ Ⓓ Ⓔ	54. Ⓐ Ⓑ Ⓒ Ⓓ Ⓔ
5. Ⓐ Ⓑ Ⓒ Ⓓ Ⓔ	30. Ⓐ Ⓑ Ⓒ Ⓓ Ⓔ	55. Ⓐ Ⓑ Ⓒ Ⓓ Ⓔ
6. Ⓐ Ⓑ Ⓒ Ⓓ Ⓔ	31. Ⓐ Ⓑ Ⓒ Ⓓ Ⓔ	56. Ⓐ Ⓑ Ⓒ Ⓓ Ⓔ
7. Ⓐ Ⓑ Ⓒ Ⓓ Ⓔ	32. Ⓐ Ⓑ Ⓒ Ⓓ Ⓔ	57. Ⓐ Ⓑ Ⓒ Ⓓ Ⓔ
8. Ⓐ Ⓑ Ⓒ Ⓓ Ⓔ	33. Ⓐ Ⓑ Ⓒ Ⓓ Ⓔ	58. Ⓐ Ⓑ Ⓒ Ⓓ Ⓔ
9. Ⓐ Ⓑ Ⓒ Ⓓ Ⓔ	34. Ⓐ Ⓑ Ⓒ Ⓓ Ⓔ	59. Ⓐ Ⓑ Ⓒ Ⓓ Ⓔ
10. Ⓐ Ⓑ Ⓒ Ⓓ Ⓔ	35. Ⓐ Ⓑ Ⓒ Ⓓ Ⓔ	60. Ⓐ Ⓑ Ⓒ Ⓓ Ⓔ
11. Ⓐ Ⓑ Ⓒ Ⓓ Ⓔ	36. Ⓐ Ⓑ Ⓒ Ⓓ Ⓔ	61. Ⓐ Ⓑ Ⓒ Ⓓ Ⓔ
12. Ⓐ Ⓑ Ⓒ Ⓓ Ⓔ	37. Ⓐ Ⓑ Ⓒ Ⓓ Ⓔ	62. Ⓐ Ⓑ Ⓒ Ⓓ Ⓔ
13. Ⓐ Ⓑ Ⓒ Ⓓ Ⓔ	38. Ⓐ Ⓑ Ⓒ Ⓓ Ⓔ	63. Ⓐ Ⓑ Ⓒ Ⓓ Ⓔ
14. Ⓐ Ⓑ Ⓒ Ⓓ Ⓔ	39. Ⓐ Ⓑ Ⓒ Ⓓ Ⓔ	64. Ⓐ Ⓑ Ⓒ Ⓓ Ⓔ
15. Ⓐ Ⓑ Ⓒ Ⓓ Ⓔ	40. Ⓐ Ⓑ Ⓒ Ⓓ Ⓔ	65. Ⓐ Ⓑ Ⓒ Ⓓ Ⓔ
16. Ⓐ Ⓑ Ⓒ Ⓓ Ⓔ	41. Ⓐ Ⓑ Ⓒ Ⓓ Ⓔ	66. Ⓐ Ⓑ Ⓒ Ⓓ Ⓔ
17. Ⓐ Ⓑ Ⓒ Ⓓ Ⓔ	42. Ⓐ Ⓑ Ⓒ Ⓓ Ⓔ	67. Ⓐ Ⓑ Ⓒ Ⓓ Ⓔ
18. Ⓐ Ⓑ Ⓒ Ⓓ Ⓔ	43. Ⓐ Ⓑ Ⓒ Ⓓ Ⓔ	68. Ⓐ Ⓑ Ⓒ Ⓓ Ⓔ
19. Ⓐ Ⓑ Ⓒ Ⓓ Ⓔ	44. Ⓐ Ⓑ Ⓒ Ⓓ Ⓔ	69. Ⓐ Ⓑ Ⓒ Ⓓ Ⓔ
20. Ⓐ Ⓑ Ⓒ Ⓓ Ⓔ	45. Ⓐ Ⓑ Ⓒ Ⓓ Ⓔ	70. Ⓐ Ⓑ Ⓒ Ⓓ Ⓔ
21. Ⓐ Ⓑ Ⓒ Ⓓ Ⓔ	46. Ⓐ Ⓑ Ⓒ Ⓓ Ⓔ	71. Ⓐ Ⓑ Ⓒ Ⓓ Ⓔ
22. Ⓐ Ⓑ Ⓒ Ⓓ Ⓔ	47. Ⓐ Ⓑ Ⓒ Ⓓ Ⓔ	72. Ⓐ Ⓑ Ⓒ Ⓓ Ⓔ
23. Ⓐ Ⓑ Ⓒ Ⓓ Ⓔ	48. Ⓐ Ⓑ Ⓒ Ⓓ Ⓔ	73. Ⓐ Ⓑ Ⓒ Ⓓ Ⓔ
24. Ⓐ Ⓑ Ⓒ Ⓓ Ⓔ	49. Ⓐ Ⓑ Ⓒ Ⓓ Ⓔ	74. Ⓐ Ⓑ Ⓒ Ⓓ Ⓔ
25. Ⓐ Ⓑ Ⓒ Ⓓ Ⓔ	50. Ⓐ Ⓑ Ⓒ Ⓓ Ⓔ	75. Ⓐ Ⓑ Ⓒ Ⓓ Ⓔ

PRACTICE TEST ONE

SECTION I
Part A
Time: 90 minutes

Directions: Each set of letter choices below refers to the numbered statements immediately following it. Select the one lettered choice that best fits each statement. A choice may be used once, more than once, or not at all in each set.

Questions 1 and 2 refer to an atom of the following elements:

(A) F
(B) Cl
(C) C
(D) Mg
(E) Al

1. The atom that is most likely to give up its valence electrons

2. The atom that is most likely to form a molecule with hydrogen according to the following formula: XH_4

Questions 3–6 refer to the following gases:

(A) CO_2
(B) He
(C) NO
(D) HCl
(E) N_2

3. Has the slowest average atomic or molecular speed at 1 atm and 273 K

4. Has the weakest attractive forces between particles

5. Has a rate of effusion from a pinhole closest to that of F_2

6. Pollutant that is a major contributor to acid rain

GO ON TO THE NEXT PAGE ⟹

Questions 7–10 refer to the following elements or compounds:

(A) HO_2

(B) CO

(C) I_2

(D) CaI_2

(E) Ag_2Hg_3

7. Is formed from incomplete combustion of hydrocarbons

8. Is soluble in CCl_4

9. Exhibits the bond with the most ionic character

10. Is an amalgam

Questions 11–14 refer to the following reactions:

(A) $heat + NH_4Cl\ (s) \rightarrow NH_3\ (g) + HCl\ (g)$

(B) $2SO_2\ (g) + O_2\ (g) \rightarrow 2SO_3\ (g) + heat$

(C) $heat + H_2\ (g) + CO_2\ (g) \rightarrow H_2O\ (g) + CO\ (g)$

(D) $2NO\ (g) \rightarrow N_2\ (g) + O_2\ (g) + heat$

(E) $2HNO_2(aq) \rightarrow NO\ (g) + NO_2\ (g) + H_2O\ (l) + heat$

11. Is most likely to have both a positive ΔH and ΔS

12. Has a negative change in entropy

13. Does *not* involve reduction or oxidation

14. Is a disproportionation reaction

GO ON TO THE NEXT PAGE

Part B

Directions: Each of the questions or incomplete statements below is followed by five suggested answers or completions. Select the one that is best in each case.

15. Which of the following *must* be true for a sample of supercritical uranium (^{235}U)?

    I. The density exceeds that found at ambient temperatures and pressures.

    II. The rate of production of neutrons exceeds their rate of consumption.

    III. The sample is pure.

    (A) I only

    (B) II only

    (C) III only

    (D) I and II only

    (E) II and III only

16. In water with a pH of 7, the amino acid glycine ($CH_2(NH_2)CO_2H$) primarily exists in which form?

    (A) ($CH_2(NH_3^+)CO_2H$)

    (B) ($CH_2(NH_2)CO_2^-$)

    (C) ($CH_2(NH_2)CO_2H$)

    (D) ($CH_2(NH_3^+)CO_2^-$)

    (E) Glycine is not appreciably soluble in neutral water.

17. Which of the following statements best explains why white phosphorus is potentially dangerous?

    I. It is pyrophoric.

    II. It is a caustic base.

    III. It is a strong acid.

    (A) I only

    (B) II and III only

    (C) I and II only

    (D) I and III only

    (E) I, II, and III

18. Diamond and graphite are

    (A) isomers.

    (B) allotropes.

    (C) organic molecules.

    (D) hydrocarbons.

    (E) carbohydrates.

19. A 0.1 M solution of an unknown substance is prepared and shown to have a pH of 4. The unknown substance is most likely a

    (A) strong base.

    (B) strong acid.

    (C) 1:1 salt of a strong acid and a weak base.

    (D) 1:1 salt of a strong base and a weak acid.

    (E) 1:1 salt of a strong acid and a strong base.

20. Mercury (Hg) forms molecules with other metals that are called amalgams. Which of the following does NOT form an amalgam with Hg?

    (A) Silver

    (B) Gold

    (C) Copper

    (D) Iron

    (E) Lead

GO ON TO THE NEXT PAGE

21. The noble gases are

    (A) highly reactive with many elements.

    (B) He, Li, Ar, Na, Kr, Xe, and Rn.

    (C) nonreactive with many elements.

    (D) He, Ne, Ar, Br, Kr, and Rn.

    (E) highly acidic in water.

Questions 22 and 23 refer to the following unbalanced high-temperature reaction:

$$SiO_2 \text{ (l)} + Al \text{ (l)} \rightarrow Al_2O_3 \text{ (s)} + Si \text{ (l)}$$

22. When the coefficient for each of the substances is reduced to its lowest whole number, the coefficient in front of Al is

    (A) 2.

    (B) 3.

    (C) 4.

    (D) 6.

    (E) 8.

23. In the reaction, SiO_2 functions as a(n)

    (A) reductant.

    (B) electrolyte.

    (C) moderator.

    (D) catalyst.

    (E) oxidant.

24. In which of the following coordination compounds does the metal have an oxidation number of III?

    (A) $[Cu(NH_3)_4]^{2+}$

    (B) $[Co(NH_3)_5Cl]^{2+}$

    (C) $[Fe(CN)_6]^{4-}$

    (D) $[Cr(H_2O)_5NH_3]Cl_2$

    (E) $K_2[FeCl_4]$

Questions 25 and 26 refer to the following unbalanced equation:

$$KO_2 \text{ (l)} + CO_2 \text{ (s)} \rightarrow K_2CO_3 \text{ (s)} + O_2 \text{ (g)}$$

25. When the equation is balanced with the smallest whole number coefficients, what is the coefficient for O_2?

    (A) 1

    (B) 2

    (C) 3

    (D) 4

    (E) 6

26. When 142 g of KO_2 (MW = 71 g/mol) is allowed to react with 88 g of CO_2 (MW = 44 g/mol) according to the reaction above, what is the maximum yield of O_2 (MW = 32)?

    (A) 22 g

    (B) 48 g

    (C) 70 g

    (D) 96 g

    (E) 192 g

27. Which of the following molecules has a bond angle closest to 109.5°?

    (A) NH_3

    (B) NO_2

    (C) CH_4

    (D) CO_2

    (E) H_2S

GO ON TO THE NEXT PAGE

28. The CN^- ion is a weak base, $K_b = 2 \times 10^{-5}$. Which of the following expressions will give the K_a of the conjugate acid, HCN?

(A) $\dfrac{(2 \times 10^{-5})}{10^{-14}}$

(B) $(2 \times 10^{-5}) \times 10^{-14}$

(C) $\dfrac{1}{2 \times 10^{-5}}$

(D) $\dfrac{(2 \times 10^{-5})^2}{10^{-14}}$

(E) $\dfrac{10^{-14}}{(2 \times 10^{-5})}$

29. Which of the following 0.1 M solutions would form a precipitate when treated with 0.1 M NaOH?

(A) $Al(OH)_3$

(B) $FeCl_2$

(C) NH_4ClO_4

(D) $CsNO_3$

(E) LiCl

30. As one goes down the periodic table, the covalent (or atomic) radius of an element does what?

(A) Changes by orders of magnitude

(B) Decreases

(C) Remains constant

(D) Changes only one time

(E) Increases

31. Alkali metals form ionic bonds with halogens to form what kind of compounds?

(A) Amalgams

(B) These compounds do not exist.

(C) These compounds only exist theoretically.

(D) 1:1

(E) 1:2

32. Assuming ideal behavior, if a sample of a gas at 300 K is compressed from 3 L to 2 L and the temperature is lowered to 200 K, the pressure of the gas will change by a factor of

(A) $\dfrac{1}{9}$.

(B) $\dfrac{1}{3}$.

(C) 1 (remain constant).

(D) 3.

(E) 9.

33. Which of the following molecules has nitrogen in a more positive oxidation state than in N_2O_4?

(A) HNO_3

(B) N_2H_4

(C) NO_2

(D) N_2O_3

(E) N_2O

34. $4NH_3\,(g) + 5O_2\,(g) \rightarrow 4NO\,(g) + 6H_2O\,(g)$

If 0.30 mol of O_2 is consumed according to the reaction above, what volume of NO is produced at 1 atm and 273 K?

(A) 22.4 L

(B) 0.30×22.4 L

(C) $\left(0.30\left(\dfrac{5}{4}\right)\right) \times 22.4$ L

(D) $\left(0.30\left(\dfrac{4}{5}\right)\right) \times 22.4$ L

(E) $5(0.30) \times 22.4$ L

GO ON TO THE NEXT PAGE

35. If 20 mL of a 0.10 M solution of $AgNO_3$ is added to 80 mL of a 0.10 M solution of $CaCl_2$, what is the final concentration of Cl^- in solution?

    (A) 0.08 M

    (B) 0.10 M

    (C) 0.14 M

    (D) 0.16 M

    (E) 0.18 M

36. $H_2\,(g) + Cl_2\,(g) \rightarrow 2HCl\,(g)$

    If 2 L of H_2 and 3 L of Cl_2 are allowed to react according to the equation above, what is the volume of HCl produced? Assume conditions of 1 atm and 273 K.

    (A) 1.5 L

    (B) 4 L

    (C) 5 L

    (D) $\frac{2}{3} \times 22.4$ L

    (E) 22.4 L

Questions 37–39 refer to the following cooling curve for gaseous benzene starting at 110°C. Identify the letter on the curve that represents each of the following situations:

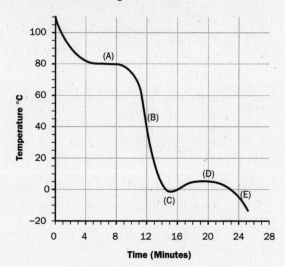

37. A supercooled liquid

38. The boiling point of benzene

39. The cooling of solid benzene

Questions 40–42 refer to a titration of 100 mL of 0.01M H_3PO_4, a triprotic acid, with a solution of 0.05M NaOH. Identify the letter on the curve that represents each of the following situations:

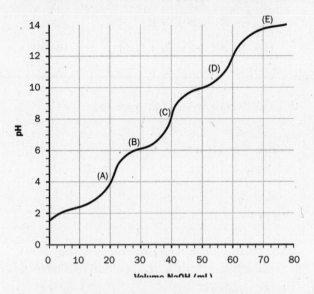

40. pK_a of HPO_4^{2-} is equal to the pH of the solution

41. The lowest concentration of H_3PO_4

42. Optimal buffering of $H_2PO_4^-$ and HPO_4^{2-}

43. To allow electrons to move toward the cathode, a battery uses which of the following as a catalyst?

    (A) Electrolyte
    (B) Electrode
    (C) Insulator
    (D) Cathode
    (E) Anode

44. $N_2 (g) + 3H_2 (g) \rightleftarrows 2NH_3 (g) + heat$

    For the reaction above, which of the following statements are true regarding an equilibrated mixture?

    I. Increasing the pressure with helium (He) shifts the equilibrium toward products.
    II. Cooling the mixture shifts the reaction toward products.
    III. Adding a catalyst shifts the equilibrium toward products.

    (A) I only
    (B) II only
    (C) III only
    (D) I and III only
    (E) I, II, and III

45. Rank the following with respect to increasing bond angle.

    I. SO_2
    II. CCl_4
    III. H_2O
    IV. SF_6

    (A) IV < III < II < I
    (B) III < IV < II < I
    (C) IV < III < I < II
    (D) III < I < II < IV
    (E) I < III < II < IV

GO ON TO THE NEXT PAGE

46. $2H_2SO_4 + P_4O_{10} \rightarrow 2SO_3 + HPO_3$

    For the reaction above, P_4O_{10} functions as a(n)

    (A) dehydrating agent.

    (B) reductant.

    (C) reducing agent.

    (D) acid.

    (E) catalyst.

47. Which combination of ΔH and ΔS describes a reaction that can never be spontaneous?

    (A) $\Delta H+$ $\Delta S+$

    (B) $\Delta H-$ $\Delta S+$

    (C) $\Delta H+$ $\Delta S-$

    (D) $\Delta H-$ $\Delta S-$

    (E) No combination of ΔH and ΔS can describe a reaction that is never spontaneous.

48. Which best describes the function of a catalyst?

    (A) It makes ΔG more negative, thus increasing the rate of reaction.

    (B) It lowers the energy of the transition state.

    (C) It stabilizes the products of the reaction, making the reaction more downhill.

    (D) It destabilizes the reactants, making them react more quickly.

    (E) It lowers the total number of steps of the reaction, making the reaction faster.

Questions 49 and 50 refer to the following reaction and mechanistic steps:

$$2I^- (aq) + H_2O_2 (aq) + 2 H_3O^+ (aq) \rightarrow I_2 (s) + 4H_2O (l)$$

$$H_2O_2 + I^- \rightarrow H_2O + IO^- \quad \text{(slow)}$$

$$H_3O^+ + IO^- \rightarrow HOI + H_2O \quad \text{(fast)}$$

$$HOI + H_3O^+ + I^- \rightarrow I_2 + 2H_2O \quad \text{(fast)}$$

49. The overall order of the reaction is

    (A) -2.

    (B) -1.

    (C) 0.

    (D) 2.

    (E) 5.

50. Decreasing the pH of the reaction from 5 to 3 will affect the rate of the reaction by a factor of

    (A) 0.0001.

    (B) 1 (no change).

    (C) 2.

    (D) 100.

    (E) 10,000.

51. The mechanism for the chlorination of methane is shown below. Identify the types of reactions in steps II and III.

    I. $Cl_2 \rightarrow Cl\cdot + Cl\cdot$
    II. $CH_4 + Cl\cdot \rightarrow \cdot CH_3 + HCl$
    III. $\cdot CH_3 + Cl_2 \rightarrow CH_3Cl + Cl\cdot$
    IV. $\cdot CH_3 + \cdot CH_3 \rightarrow CH_3CH_3$

	II	III
(A)	Propagation	Propagation
(B)	Initiation	Termination
(C)	Propagation	Initiation
(D)	Termination	Propagation
(E)	Termination	Termination

52. Which of the following type of orbital is, on average, closest to the nucleus?

   (A) s
   (B) p
   (C) d
   (D) f
   (E) g

53. Metals are characterized by which of the following?

   I. Loosely held valence electrons
   II. Highly ionized atoms
   III. High ionization energies

   (A) I only
   (B) II only
   (C) III only
   (D) II and III only
   (E) I, II, and III only

54. An unknown compound of chlorine and oxygen is decomposed at constant temperature and pressure to give 7 mL of chlorine and 21 mL of oxygen. The compound is most likely

   (A) Cl_2O.
   (B) Cl_2O_3.
   (C) ClO_2.
   (D) Cl_2O_6.
   (E) Cl_2O_7.

55. Put the following in order from the strongest to the weakest acid.

   I. $HClO_2$
   II. NH_3
   III. $HClO_3$
   IV. H_2SO_4

   (A) IV > III > I > II
   (B) III > IV > II > I
   (C) I > II > III > IV
   (D) IV > II > III > I
   (E) III > II > IV > I

56. A unknown liquid reacts violently with solid NaOH and shows a white precipitate when treated with a solution of $Ba(ClO_4)_2$. The liquid is most likely

   (A) HNO_3.
   (B) H_2SO_4.
   (C) H_2O.
   (D) $(CH_3CH_2)_3N$.
   (E) $NaSO_4$ (aq).

57. Which of the following are examples of colligative properties?

   I. Osmotic pressure
   II. Entropy
   III. Boiling point elevation

   (A) I only
   (B) II and III only
   (C) I and III only
   (D) I, II, and III
   (E) None

GO ON TO THE NEXT PAGE

58. The dissolution of NH_4NO_3 causes the temperature of the solution to drop; which parameters best describe this process?

	ΔG	ΔH	ΔS
(A)	+	−	−
(B)	+	+	−
(C)	0	+	−
(D)	−	−	+
(E)	−	+	+

59. $2H_2 \text{ (g)} + O_2 \text{ (g)} \rightarrow 2H_2O \text{ (l)} + \text{heat}$

The reaction above describes the formation of water from its elements. Which conditions would favor the *reverse* reaction?

	Temperature	Pressure
(A)	High	High
(B)	High	Low
(C)	Low	High
(D)	Low	Low
(E)	The reaction can only proceed toward products.	

60. Based on the following enthalpies of formation, what is the overall $\Delta H°$ for the complete metabolism of glucose at 298 K, 1 atm?

$6O_2 \text{ (g)} + C_6H_{12}O_6 \text{ (l)} \rightarrow 6CO_2 \text{ (g)} + 6H_2O \text{ (l)}$

Standard $\Delta H°_f$	(kJ/mol)
$C_6H_{12}O_6$	−919.2
H_2O	−237.2
CO_2	−394.4

(A) $(−237.2) + (−394.4) − (−919.2)$

(B) $6(−237.2) + 6(−394.4) − (−919.2)$

(C) $−919.2 − 6(−237.2) − 6(−394.4)$

(D) $−919.2 − (−237.2) − (−394.4)$

(E) The problem cannot be solved without the $\Delta H°_f$ for O_2.

61. Given the following information:

$2Cu \text{ (s)} + S \text{ (s)} \rightarrow Cu_2S \text{ (s)}$ $\Delta G° = −86.2 \text{ kJ}$

$\Delta H° = −79.5 \text{ kJ}$

$S \text{ (s)} + O_2 \text{ (g)} \rightarrow SO_2 \text{ (g)}$ $\Delta G° = −300.1 \text{ kJ}$

$\Delta H° = −296.8 \text{ kJ}$

what are the values for $\Delta G°$ and $\Delta H°$ for the reaction below?

$Cu_2S \text{ (s)} + O_2 \text{ (g)} \rightarrow 2Cu \text{ (s)} + SO_2 \text{ (g)}$

	$\Delta G°$ (kJ)	$\Delta H°$ (kJ)
(A)	−386.3	−376.3
(B)	−386.3	376.3
(C)	−213.9	217.3
(D)	386.3	376.3
(E)	−213.9	−217.3

62. Which of the following is *NOT* considered good laboratory practice?

   I. Storing a stock solution of 5M NaOH in a clean Pyrex flask

  II. Keeping a graduated cylinder dry in a 110°C oven

 III. Always keeping contact lenses on in the lab

(A) None

(B) I only

(C) II only

(D) II and III only

(E) I, II, and III

GO ON TO THE NEXT PAGE

Questions 63 and 64 refer to the following graph:

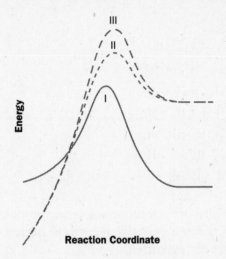

63. Which curve(s) represents a reaction that appears to have been enabled via a catalyst?

 (A) I only

 (B) III only

 (C) II only

 (D) I and II only

 (E) I, II, and III

64. Which reaction(s) is/are exothermic?

 (A) I only

 (B) III only

 (C) II only

 (D) I and II only

 (E) II and III only

65. The decomposition of ozone occurs through the following steps.

$$O_3\ (g) \rightleftarrows O_2\ (g) + O\ (g) \quad \text{(fast-equilibrium)}$$

$$O\ (g) + O_3\ (g) \rightarrow 2O_2\ (g) \quad \text{(slow)}$$

If the partial pressure of O_2 increases from 0.06 atm to 0.18 atm, the overall reaction rate (for both decomposition equations) will change by a factor of

 (A) $\frac{1}{9}$ (slower).

 (B) $\frac{1}{3}$ (slower).

 (C) 1 (remain the same).

 (D) 3 (faster).

 (E) 9 (faster).

66. $BrO_3^- + 5Br^- + 6H^+ \rightarrow 3Br_2 + 3H_2O$

Bromine (Br_2) can be formed from an acidic solution of bromate (BrO_3^-) and bromide (Br^-) ions according to the above equation.

The rate of the reaction was measured at various concentrations of reactants as shown below.

	$[Br^-]$ M	$[H^+]$ M	$[BrO_3^-]$ M	Relative Rate
I.	0.75	0.6	0.15	1
II.	1.5	0.6	0.15	2
III.	0.75	1.2	0.30	8
IV.	0.75	1.2	0.15	4

What is the rate equation for this reaction?

 (A) Rate $= k[BrO_3^-][Br^-][H^+]$

 (B) Rate $= k[BrO_3^-][Br^-]^5[H^+]^6$

 (C) Rate $= k[BrO_3^-][H^+]$

 (D) Rate $= k[BrO_3^-][H^+]^2$

 (E) Rate $= k[BrO_3^-][Br^-][H^+]^2$

GO ON TO THE NEXT PAGE

67. $2N_2O_5 (g) \rightarrow 2N_2 (g) + 5O_2 (g)$

Dinitrogen pentoxide decomposes in a first-order reaction with a half-life of 7 minutes at 230°C according to the equation above.

If N_2O_5 is added to an evacuated flask to a pressure of 800 torr at 230°C, what is the total pressure in the flask after 14 minutes?

(A) 200 torr

(B) 800 torr

(C) 1,400 torr

(D) 2,300 torr

(E) 2,800 torr

68. $H_2SO_4 (l) + NaCl (s) \rightarrow HCl (g) + NaHSO_4 (s)$

Concentrated sulfuric acid can be used to generate HCl through the reaction above. Why does an analogous reaction with NaBr fail to produce HBr as a final product?

(A) HBr is a liquid and thus there is no driving force for the forward reaction.

(B) HBr is formed initially but further oxidized by H_2SO_4.

(C) HBr is a strong acid and the equilibrium lies to the left.

(D) NaBr is more stable than NaCl and doesn't react.

(E) The activation energy is too high and the reaction requires a catalyst.

69. Which of the following has chlorine in the same oxidation state as in Cl_2O_3?

(A) NH_4ClO_4

(B) HClO

(C) Cl_2O_7

(D) $HClO_2$

(E) $AlCl_3$

70. Rank 0.5 M aqueous solutions of the following salts in order of pH from high to low.

I. NaF

II. RbCl

III. NH_4ClO_4

(A) I > II > III

(B) I > III > II

(C) II > III > I

(D) III > I > II

(E) III > II > I

71. Which of the following aqueous solutions would be expected to possess the highest osmotic pressure?

(A) 1.1 M glucose ($C_6H_{12}O_6$)

(B) 0.1 M H_3PO_4

(C) 0.25 M $Al_2(SO_4)_3$

(D) 0.05 M HCl

(E) 1 M ethanol (CH_3CH_2OH)

72. If 30.0 mL of a 0.30 M solution of $Al_2(SO_4)_3$ is added to 60.0 mL of distilled water, what is the final concentration of aluminum ion?

(A) 0.15 M

(B) 0.20 M

(C) 0.30 M

(D) 0.90 M

(E) 1.80 M

GO ON TO THE NEXT PAGE

73. Using the ideal gas law, set up the following problem: A 3.0 L flask at 25°C contains 35 g of chlorine gas. What is the pressure in the flask?

   (A) $P = \dfrac{1.0 \text{ mol} \times R \times 25°C}{3.0 \text{ L}}$

   (B) $P = \dfrac{0.50 \text{ mol} \times R \times 298K}{3.0 \text{ L}}$

   (C) $P = \dfrac{35 \text{ g} \times R \times 25°C}{32.4 \text{ L}}$

   (D) $P = 3.0 \text{ L} \times 1.0 \text{ mol} \times 22.4$

   (E) $P = 1.0$ atm

74. If 100 mL of a 2×10^{-5} M HCl solution is mixed with 100 mL of a 2×10^{-4} M NaOH solution, the final pH of the resulting solution will be closest to

   (A) 5.

   (B) 7.

   (C) 8.

   (D) 9.

   (E) 10.

75. Given the following standard reduction potentials, which reaction(s) would be expected to occur?

   $Cu^+ + e^- \rightarrow Cu \text{ (s)}$ \quad +0.52 \quad (V)

   $2H^+ + 2e^- \rightarrow H_2 \text{ (g)}$ \quad 0 \quad (V)

   $Pb^{2+} + 2e^- \rightarrow Pb \text{ (s)}$ \quad −0.13 \quad (V)

   $Cr^{3+} + e^- \rightarrow Cr^{2+}$ \quad −0.41 \quad (V)

   I. $2Cu \text{ (s)} + Pb^{2+} \rightarrow 2Cu^+ + Pb \text{ (s)}$

   II. $Pb \text{ (s)} + 2H^+ \rightarrow Pb^{2+} + H_2 \text{ (g)}$

   III. $2Cr^{2+} + Pb^{2+} \rightarrow Pb \text{ (s)} + 2Cr^{3+}$

   (A) I only

   (B) I and II only

   (C) I and III only

   (D) II and III only

   (E) I, II, and III

IF YOU FINISH BEFORE TIME IS CALLED, YOU MAY CHECK YOUR WORK ON THIS SECTION ONLY. DO NOT TURN TO ANY OTHER SECTION IN THE TEST.

STOP

SECTION II
Part A
Time: 55 minutes

Directions: Answer all three questions below. Calculators are permitted, except for those with typewriter (QWERTY) keyboards. Clearly show the method used and steps involved in arriving at your answers. Partial credit can only be given if your work is clear and demonstrates an understanding of the problem. The Section II score weighting for each question is 20 percent.

YOU MAY USE YOUR CALCULATOR FOR PART A

1. Many underground iron pipes are protected from corrosion by connecting them via a conducting wire to a block of magnesium metal buried close to the pipe.

   (a) In this situation, what is the chemical function of the magnesium metal?

   (b) Describe two chemical characteristics that a metal must have to function in place of magnesium in this situation.

   (c) $2Fe\ (s) + O_2\ (g) + 4H^+\ (g) \rightarrow 2Fe^{2+}\ (s) + 2H_2O\ (l)$

   and also: $4Fe\ (s) + 3O_2\ (g) \rightarrow 2Fe_2O_3\ (s)$

   Under a given set of conditions without magnesium, it is found that 1 g of iron would normally be corroded per meter per year according to the equation above.

   i. Assume that the rate of corrosion is metal-independent. Calculate the time it would take for a 1 kg block of magnesium attached to a 200 m length of pipe to be consumed.

   ii. How long would it take for one mole of electrons to pass through the wire connecting the pipe to the magnesium, per a single 200 m section of pipe?

   (d) Calculate the equilibrium constant for the corrosion reactions with and without magnesium present. Assume standard conditions.

   (e) What is $\Delta G°$ for the reaction if magnesium is replaced by zinc? Is the reaction spontaneous? Assume standard conditions.

2. (a) Write out a balanced equation for the complete combustion of a hydrocarbon with the formula C_3H_6.

   (b) A 1 L sample of a mixture of this hydrocarbon and argon at STP is burned with an excess of oxygen and produces 1 mL of liquid water. What is the percentage, by volume, of hydrocarbon in the original mixture?

   (c) If the reaction in (b) produces 6.51 kJ of heat, what is the ΔH/mol for combustion?

   (d) Two isomers of C_3H_6 are possible.

   i. Draw both of them, being sure to show all carbons and hydrogens.

   ii. One of these isomers has carbon atoms in two different hybridization states. Indicate which isomer this is and which carbon corresponds to which hybridization state.

GO ON TO THE NEXT PAGE

3. $HCN\ (g) \rightarrow H^+\ (g) + CN^-\ (g)$

The equilibrium constant for the above reaction has been shown to be 4.9×10^{-10}.

(a) A solution is made by dissolving 13 g of HCN gas in 1 L of water.

i. What is the pH of the solution?

ii. What is the percent dissociation of HCN?

iii. How much NaCN would be required to bring the pH of this solution to 10?

(b) A new solution is made by dissolving 10 g of HCN in 200 mL of water. What volume of 0.1 M NaOH solution is required to give $[OH^-] = 1 \times 10^{-3}$?

(c) Two different forms of CN^- are possible (cyanide and isocyanide).

i. Draw the Lewis structures of both forms.

ii. Identify the structure that is predicted to be more stable and explain the reasoning behind your answer.

GO ON TO THE NEXT PAGE

Part B
Time: 40 minutes

NO CALCULATORS MAY BE USED FOR PART B

Answer Question 4 below. The Section II score weighting for this question is 10 percent.

4. For each of the following three reactions, write a BALANCED equation in part (i) and answer the question about the reaction in part (ii). Coefficients should be in terms of lowest whole numbers in part (i). Assume that solutions are aqueous unless stated otherwise. Represent substances in solution as ions if the substances are extensively ionized. Omit formulas for any ions or molecules that are unchanged by the reaction.

Example: *A strip of magnesium is added to a solution of titanium.*

i. $Ti^{2+} + Mg \rightarrow Ti + Mg^{2+}$

ii. *Which substance is reduced in the reaction?*

Answer: *Titanium (Ti) is reduced.*

(a) Zinc metal is added to hydrochloric acid.

   i. Write a balanced equation for this reaction.

   ii. What is the oxidation number of hydrogen **before** and **after** the reaction?

(b) Carbon is reacted with hydrogen gas.

   i. Write a balanced equation for this reaction.

   ii. Is $\Delta S°$ positive or negative for this reaction? Explain your reasoning.

(c) Aqueous solutions of lead (II) nitrate and potassium sulfate are mixed.

   i. Write a balanced equation for this reaction.

   ii. Identify the precipitate.

Answer Question 5 and Question 6. The Section II score weighting for these questions is 15 percent each.

Response to these questions will be graded on the basis of accuracy and the relevance of information cited. Make your explanations clear and well organized. You may include examples and equations in your responses where appropriate. Specific responses are preferable to broad, diffuse responses.

GO ON TO THE NEXT PAGE

5. Five bottles containing 0.1 M solutions of the following chemicals are prepared.

   (A) Copper (II) nitrate

   (B) Sodium chloride

   (C) Aluminum acetate

   (D) Sucrose ($C_{12}H_{22}O_{11}$)

   (E) Ammonia

   (a) Propose five single laboratory tests that will uniquely identify each one of the five solutions from the others.

   (b) Which solution will have the lowest freezing point and why?

   (c) Adding solution A to E results in an absorption change.

      i. Describe what is happening in this reaction.

      ii. If ethylenediamine ($NH_2CH_2CH_2NH_2$) is used in place of ammonia, the equilibrium constant for this reaction is much larger. What principle does this illustrate?

   (d) Two isomeric compounds both show an empirical formula of $CrN_4H_{12}Cl_3$, and both produce one equivalent of AgCl when treated with an excess of $AgNO_3$ solution.

      i. Draw the correct molecular formula for either isomer.

      ii. Draw structures for each of the two isomers.

6. Uranium 235 can be enriched from the more abundant uranium 238 through use of a gas centrifuge. In this apparatus, both isotopes are converted into the volatile UF_6 and then separated according to their weight.

   (a) Describe two characteristics of UF_6 that would account for its high volatility.

   (b) Without computing the numbers, show the calculation that relates the rate of diffusion of both isotopes of UF_6.

   (c) UF_6 is corrosive and reacts with water to form uranyl fluoride (UO_2F_2) and hydrogen fluoride.

      i. Show the balanced equation for this reaction.

      ii. Give two reasons why this reaction would be undesirable.

   (d) A series of reactions is required to generate UF_6 from natural uranium ore. For the unbalanced reactions shown below, identify if the uranium is oxidized, reduced, or neither, and the oxidation state for uranium in the products and reactants.

      I. $[UO_2(CO_3)_3]^{4-} \rightarrow UO_3 + CO_2$

      II. $UO_3 + H_2 \rightarrow UO_2 + H_2O$

      III. $UO_2 + HF \rightarrow UF_4 + H_2O$

      IV. $UF_4 + F_2 \rightarrow UF_6$

IF YOU FINISH BEFORE TIME IS CALLED, YOU MAY CHECK YOUR WORK ON THIS SECTION ONLY. DO NOT TURN TO ANY OTHER SECTION IN THE TEST.

STOP

Practice Test One: **Answer Key**

1. D	26. B	51. A
2. C	27. C	52. A
3. A	28. E	53. A
4. B	29. B	54. D
5. D	30. E	55. A
6. C	31. D	56. B
7. B	32. C	57. C
8. C	33. A	58. E
9. D	34. D	59. B
10. E	35. C	60. B
11. A	36. B	61. E
12. B	37. C	62. E
13. A	38. A	63. C
14. E	39. E	64. A
15. B	40. D	65. B
16. D	41. E	66. E
17. A	42. B	67. D
18. B	43. A	68. B
19. C	44. B	69. D
20. D	45. A	70. A
21. C	46. A	71. C
22. C	47. C	72. B
23. E	48. B	73. B
24. B	49. D	74. E
25. C	50. B	75. D

ANSWERS AND EXPLANATIONS

SECTION I

1. D

Alkali metals and alkaline earth metals, such as calcium and magnesium, are the most likely to give up their valence electrons and become cations.

2. C

Hydrogen has an oxidation state of +1 or, rarely, –1 in compounds. In order to form a compound with the given formula, the other atom must be at a –4 oxidation state, which carbon is most likely to adopt.

3. A

The average atomic or molecular speed of gases is inversely proportional to the square root of their molecular mass. Therefore, the heaviest gas (CO_2) has the slowest speed.

4. B

The gas with the weakest attractive forces between particles will also have the lowest boiling point (He).

5. D

The rate of effusion is also inversely proportional to the square root of each gas's molecular mass, thus the correct answer is the gas (HCl, MW = 36) that has a molecular mass closest to that of fluorine (MW = 38).

6. C

Nitric oxide is formed in the internal combustion engine and is eventually converted to HNO_3, a major contributor to acid rain.

7. B

Incomplete combustion of many hydrocarbons generates carbon monoxide (CO).

8. C

CCl_4 is a nonpolar solvent and doesn't dissolve ionic or polar compounds. It would slowly react with choice (E) as well.

9. D

The biggest difference in electronegativity between constituent atoms is found in choice (D), thus this would have the most ionic bond.

10. E

Amalgams are alloys containing mercury (Hg).

11. A

Endothermic reactions have the heat on the left side and a positive ΔH. Reactions with an increase in entropy also have a positive ΔS. Because the reaction in choice (A) produces two moles of gas for each mole of solid, it would have a positive ΔS.

12. B

Only the reaction in choice (B) has a net decrease in the number of molecules of gas and thus should have a decrease in the amount of disorder and a negative ΔS.

13. A

The other reactions are all reduction-oxidation reactions.

14. E

A disproportionation reaction is one in which an element is both oxidized and reduced. In choice (E), nitrogen goes from +3 in HNO_2 to +2 in NO and +4 in NO_2.

15. B

In nuclear chemistry, a sample of a radioactive substance is said to be critical, or have critical mass, when each fission process induces one other fission process to occur. If a radioactive substance is supercritical, each fission process starts more than one fission process. The radioactive sample need not be pure. Increasing the temperature has no effect on nuclear reaction rates. Although increasing the density of the fissile material (^{235}U) could increase the efficiency of neutron capture, it is not required for the sample to be supercritical.

16. D

In neutral water, both the amine ($-NH_2$) and carboxylic acid ($-COOH$) groups of an amino acid are ionized to $-NH_3^+$ and $-CO_2^-$ respectively.

17. A

White phosphorus is pyrophoric. White phosphorus will spontaneously smoke and burn when exposed to room temperature air. A substance that self-ignites under ambient conditions is called pyrophoric. Exposure causes deterioration of bones, especially those of the jaw.

18. B

Different forms of the same element are called allotropes. Other common elements that possess different allotropes are phosphorus (white and red), oxygen (normal O_2 and ozone O_3), and sulfur.

19. C

A pH of 4 is mildly acidic, so neither a strong acid nor strong base could give this reading. Choice (E) would simply neutralize both the acid and base, giving a reading close to 7. Choices (C) and (D) represent salts of weak bases or weak acids respectively. Only choice (C) would be acidic (e.g., NH_4Cl), whereas choice (D) would be basic (e.g., NaCN).

20. D

Mercury (Hg) forms molecules with all metals except iron.

21. C

The noble gases are almost exclusively nonreactive and include all the inert gases on the far right of the periodic table (with full valence electron shells): He, Ne, Ar, Kr, Xe, and Rn. Many of the incorrect answers have either incomplete lists or include elements that are not noble gases on the list.

22. C

This reaction needs to be balanced. One way to solve this problem is to assume that the most complex molecule's coefficient is 1. This would make the coefficients $1.5SiO_2 + 2Al \rightarrow 1Al_2O_3 + 1.5Si$. These are then turned into whole numbers by multiplying all coefficients by a factor of 2, giving:

$$3SiO_2 + 4Al \rightarrow 2Al_2O_3 + 3Si$$

23. E

The simplest way to understand that SiO_2 is an oxidant is to realize that it transfers oxygen to another reactant (Al). Also, it is reduced in the reaction Si ($+4$) $\rightarrow$ Si (0), indicating that it is the oxidant. Aluminum is the reductant in the reaction. There is no catalyst indicated in this reaction, and a moderator is seen in nuclear chemistry.

24. B

The oxidation number of a metal in a coordination compound is found by balancing the charge of the entire complex with the ligands and the metal.

Neutral ligands do not affect the overall charge (H_2O, NH_3), whereas anionic ligands (Cl^-, CN^-) contribute a charge equal to that of the free anion (−1 for chloride and cyanide). Thus, the other oxidation numbers are: choice (A) Cu (II), choice (C) Fe (II), choice (D) Cr (II), and choice (E) K (I), Fe (II).

25. C

As in question 22, this problem can be answered quickly by setting the coefficient for K_2CO_3 to 1 and then solving for the other reactants. Balancing the carbon, potassium, and oxygen leaves $2KO_2 + 1CO_2 \rightarrow 1K_2CO_3 + 1.5O_2$. This equation then needs to be doubled to give the balanced equation $4KO_2 + 2CO_2 \rightarrow 2K_2CO_3 + 3O_2$. The answer can be checked by simply adding up the number of each element on both sides (e.g., 12 oxygen atoms).

26. B

First, calculate the number of moles of each reactant. There are originally two moles of each reactant (142 g/71 g/mol = 2 mol KO_2 and 88 g/44 g/mol = 2 mol CO_2). From the reaction stoichiometry, the reaction consumes twice as much KO_2 as CO_2, so KO_2 is the limiting reagent. From stoichiometry, for every four moles of KO_2 used, three moles of oxygen are produced. If the reaction is started with two moles, the reaction will end with $\frac{2}{4} \times 3 = 1.5$ moles, or 48 g of oxygen.

27. C

Only atoms with sp^3 hybridization can give bond angles of 109.5° (tetrahedral), and choices (A), (C), and (E) have central atoms with that hybridization. However, both choices (A) and (E) have lone pairs on the central atom, while choice (C) does not. According to VSEPR theory, lone pairs will cause more repulsion than bonds and push the bonds closer together (less than 109.5°). Choice (C), therefore, is the best answer.

28. E

K_b for a weak base and K_a for its conjugate acid are related by the equilibrium constant for the dissociation of water, K_w, according to the following equation: $K_w = 10^{-14} = K_a \times K_b$.

29. B

Ammonium hydroxide and all alkali metal hydroxides are soluble. Aluminum hydroxide, $Al(OH)_3$, is insoluble in neutral solution but forms the soluble $Al(OH)_4^-$ ion in basic solutions. Many transition metal hydroxides are not soluble, including iron (II).

30. E

As one goes down the periodic table, the covalent (or atomic) radius of an element increases. Electrons are added to larger orbitals that are further away from the nucleus, which is why the size of atoms increases as one goes down the periodic table. The radius changes incrementally as orbits are added but does not increase exponentially over the scale of the periodic table.

31. D

Alkali metals (sodium, lithium, etc.) always adopt an oxidation state of +1 and are strongly electropositive. Halogens normally form −1 ions and are strongly electronegative. These factors result in a 1:1 compound between alkali and halogen elements. Amalgams are compounds formed with mercury, which is neither an alkali metal nor a halogen.

32. C

The combined gas law is needed for this question: $\frac{(P_1V_1)}{T_1} = \frac{(P_2V_2)}{T_2}$. Plugging in the values given in the question produces $\frac{(3P_1)}{300} = \frac{(2P_2)}{200}$, so $P_1 = P_2$.

33. A

Nitrogen in N_2O_4 is in the +4 oxidation state, and in the answers it is in the following oxidation states: choice (A) +5, choice (B) −2, choice (C) +4, choice (D) +3, and choice (E) +1.

34. D

One mole of any ideal gas occupies 22.4 L at STP. From the reaction shown, for every 5 moles of oxygen, 4 moles of NO are produced. Thus, $\frac{4}{5}$ of 0.3 moles of NO are produced and occupy $0.3\left(\frac{4}{5}\right) \times 22.4$ L.

35. C

When dealing with solutions, remember that the molarity is equal to the number of moles divided by the volume (in liters): $M = \frac{n}{L}$. Since AgCl is insoluble, each mole of silver removes one mole of chloride from solution. Calculating the moles of each leads to Ag: $0.02 \text{ L} \times 0.1\text{M} = 0.002$ moles Ag; Cl: $0.08 \text{ L} \times 0.1\text{M} \times 2 = 0.016$ moles Cl. Notice that the chlorine was doubled because there are two atoms of Cl in each molecule of $CaCl_2$. Subtraction gives the final number of moles of chloride in solution: 0.014 moles, and dividing by the final volume (0.1 L) gives the final molarity of chloride: 0.14 M.

36. B

First, realize that there is a limiting reagent for the reaction: H_2. The equation shows that for each mole of hydrogen consumed, two moles of HCl are produced. Since the volume of gases is directly proportional to their number of moles, the volume of HCl produced would be $2 \times 2 = 4$ L.

37. C

Each phase transition is characterized by a horizontal portion of the curve, thus (A) is the condensation of gaseous benzene to liquid benzene and (D) is the solidification of liquid benzene. At point (C), liquid

benzene is temporarily below its freezing point, and thus represents a supercooled liquid.

38. A

See the answer explanation for question 37.

39. E

Since (D) represents the solidification of benzene, (E) shows the cooling of the solid.

40. D

The pK_a, or negative log of the dissociation constant, is a measure of the strength of an acid or a base. The pK_a allows you to determine the charge on a molecule at any given pH, or the pK_a can be calculated from the pH at the inflection points. With titration graphs, it is useful to calculate the equivalence points for the added acid or base. From the volumes and concentrations for each 20 mL of base added, one equivalent of acid is neutralized. Thus, at points (A), (C), and (D), 1, 2, and 3 protons in H_3PO_4 are respectively neutralized. Since we are starting with H_3PO_4 at the left side of the graph, this third inflection point (D) indicates the pK_{a3} for ionization of HPO_4^{2-}.

41. E

At the right side of the graph (point E), where many equivalents of base have been added, most of the protons have been removed from H_3PO_4 and its concentration is lowest.

42. B

At point B, 1.5 equivalents of NaOH have been added and there is an equal concentration of $H_2PO_4^-$ and HPO_4^{2-} in solution. At this point, the buffering capacity is at its highest and the pH is changed very little by the addition of base.

43. A

To facilitate electrons moving from the anode toward the cathode, a battery uses an electrolyte as a catalyst.

44. B

Addition of an inert gas does not affect the partial pressures of the reactants and thus doesn't affect the equilibrium. Adding a catalyst only changes the rate at which the mixture reaches equilibrium and doesn't affect the position of equilibrium. Heat can be thought of as a reactant and removing it (by cooling the reaction) shifts the equilibrium to the right (LeChatelier's principle).

45. A

Drawing the structure for all the molecules displays the following geometric forms: I, rigonal or equilateral triangular; II and III, tetrahedral; and IV, octahedral. Thus, the expected bond angles would be I: 120°, II and III: 109.5°, and IV: 90°. Due to the repulsion of lone pairs, III can be expected to have slightly more compressed bonds to oxygen, making choice (A) the correct answer.

46. A

The elements of water are removed from H_2SO_4, and the oxidation state of P is not changed, thus P_4O_{10} acts as a dehydrating agent.

47. C

This problem can be most easily solved by memorizing the equation $\Delta G = \Delta H - T\Delta S$. For a reaction to be spontaneous, ΔG must be negative. From the equation $\Delta G = \Delta H - T\Delta S$, we can see that only when ΔH is positive and ΔS is negative can a reaction never be spontaneous.

48. B

From the definition of a catalyst, only choice (B) is applicable.

49. D

The rate law is found from the slow step. Since the reactants are also the reagents found in the overall reaction, the rate $= k[H_2O_2][I^-]$ and is thus a second-order reaction.

50. B

The rate law does not contain any term for $[H^+]$, and thus the overall rate is unaffected by the pH, even though it is found in the overall reaction.

51. A

For both steps II and III, the number of radicals (molecules with unpaired electrons such as Cl and CH_3) are unchanged. These steps are termed propagation steps. When the number of radicals increases, the step is termed initiation (first step), and when the number of radicals decreases, it is termed termination (last step).

52. A

As the energy of the orbitals increases, so does the average distance from the nucleus, so the orbital filled first (s) is also the one closest to the nucleus.

53. A

For metals, only I applies.

54. D

The volumes of gases are directly proportional to their molar ratio. The molar ratio of chlorine to oxygen should therefore be 1:3, which is best satisfied by choice (D).

55. A

The weakest acid would be NH_3, which normally functions as a weak base. An increasing number of oxygens increases the acidity of acids so I < III, but both are considered to be weak acids. Only IV is considered to be a strong acid.

56. B

A precipitate forms when barium ions are exposed to the sulfate ion ($SO_4{}^{2-}$). Out of choices (B) and (E), only choice (B) will react violently when exposed to NaOH.

57. C

Colligative properties are properties of solutions that change with respect only to the number of ions or solute molecules in solution, not their identity. Some examples are boiling point elevation, freezing point depression, osmotic pressure, and vapor pressure.

58. E

In the spontaneous ($\Delta G = -$), endothermic ($\Delta H = +$) dissolution of ammonium nitrate, the entropy increases ($\Delta S = +$).

59. B

Le Chatelier's principle applies; a high temperature and low pressure would shift the equilibrium to the left.

60. B

This problem is solved by subtracting the sum of the $\Delta H°_f$ of the reactants (multiplied by their coefficients) from that of the products. For elements in their natural state, $\Delta H°_f = 0$.

61. E

Reversing the top equation and the signs for ΔG and ΔH and adding the reversed top equation to the bottom equation produces the last oxidation of Cu_2S. Adding up the ΔG and ΔH (after reversing their signs for the top equation) gives the answer.

62. E

Concentrated NaOH will dissolve glass (Pyrex), so storing a strong solution of this base in glass would not be good laboratory practice. The glass in graduated cylinders can slowly flow and warp in warm oven temperatures and give inaccurate readings. Finally, contacts should not be worn in the lab because chemicals can get trapped between the lens and the eye.

63. C

Catalysts don't change the energy of either the starting materials or the products; they only change the transition state (the highest energy intermediate) by lowering its energy.

64. A

Only in reaction I does the starting material (left on the reaction coordinate) have a higher energy than the product.

65. B

If the partial pressure of O_2 increases threefold, the overall rate of reaction will be $\frac{1}{3}$ as fast. This is because decomposition of ozone creates $2O_2$ for every ozone molecule, and if there is an increase in the abundance of O_2, then the reaction will be slowed by the inverse of the change in partial pressure.

Solving the rate equation for this reaction: rate $= k[O_3]^2[O_2]^{-1}$; and an increase in the concentration of oxygen by a factor of 3 would reduce the rate by a factor of 3, effectively slowing the reaction by $\frac{1}{3}$.

66. E

This problem can be solved by looking at a pair of runs in which only one of the reactants' concentrations changes. Doubling the concentration of $[Br^-]$ in runs I to II doubles the rate. So $[Br^-]$ should appear in the rate equation with an exponent of 1. Similarly, runs III and IV show that $[BrO_3{}^-]$ should appear with an exponent of 1. Finally, comparing runs I and IV shows that a doubling of $[H^+]$ increases the reaction by 4 times, thereby including $[H^+]$ in the rate with an exponent of 2.

67. D

Two half-lives occur in 14 minutes, thus $\left(\frac{1}{2}\right)^2$ or $\frac{1}{4}$ of the starting material remains, or 200 torr. Also, 600 torr of N_2O_5 have reacted. For every two moles of N_2O_5 reacting, seven moles of N_2 and O_2 are produced or $\left(\frac{600}{2}\right) \times 7 = 2,100$ torr. Because the partial pressures are additive, the total pressure would be $2,100 + 200 = 2,300$ torr.

68. B

HBr is easily oxidized to Br_2 with concentrated H_2SO_4, which is reduced to SO_2. The reaction also fails when NaI is used to produce HI.

69. D

In Cl_2O_3, chlorine has an oxidation state of +3. The answers have chlorine in the following oxidation states: choice (A) +7, choice (B) +1, choice (C) +7, choice (D) +3, and choice (E) −1.

70. A

NaF is a salt of a strong base (NaOH) and a weak acid (HF) and would dissolve to form a basic solution. RbCl is the salt of a strong base (RbOH) and a strong acid (HCl) and would be expected to form a neutral solution. NH_4ClO_4 is the salt of a weak base (NH_4OH) and a strong acid ($HClO_4$) and thus would form an acidic solution.

71. C

The colligative property of osmotic pressure is influenced by the number of ions or solute molecules in solution. Because each mole of $Al_2(SO_4)_3$ dissociates into five moles of ions, it has the equivalent of 1.25 M ions and thus would have the highest osmotic pressure.

72. B

Using the equation $M = \frac{n}{L}$, you can calculate the number of aluminum ions in the starting and final

solutions. Alternatively, you can do the calculation in your head by noticing that the solution is diluted by a factor of 3, so an initial concentration of $0.3\,M \times 2 = 0.6\,M$ would be $\frac{0.6\,M}{3} = 0.2\,M$ in the final solution.

73. B

There are several common traps that could be set by the question writers for a problem like this. First, you must solve the $PV = nRT$ equation correctly:

$$P = \frac{nRT}{V}$$

Then, you must be able to determine the correct value for n without your calculator. For this you must remember that chlorine gas is Cl_2, not Cl, so 35 grams of Cl_2 is only $\frac{1}{2}$ mole, or 0.5 mole. The third common pitfall is failing to convert temperature into the Kelvin scale; you must always use temperatures in Kelvin; but they are usually given in °C.

74. E

Not taking dilution into consideration, after reaction, there would be 1.8×10^{-4} M of NaOH. Since equal volumes of acid and base solution are added, the actual concentration is halved, or 0.9×10^{-4} M. This is very close to 1×10^{-4}, which is a pOH of 4 or pH of 10.

75. D

To determine what reactions can be expected to occur, we need to look at both sides of each reaction and determine which reagents become products and in what direction the reaction goes. Since all of these reactions appear to be going in the opposite direction of our initial table reactions, we need to reverse the sign of the net energetics of the reactions. Hence, a value of +0.52 becomes −0.52 and a reaction that will not spontaneously occur with no other added energy. The only two reactions with negative signs will become positive and thus be highly likely to be spontaneous and likely to occur with no other added energy or reagents.

SECTION II: FREE-RESPONSE QUESTIONS

PART A

1. (a) The general situation described in this question is cathodic protection of iron.

   (b) A more active metal (magnesium) is electrically connected to the iron pipe and will corrode (oxidize) in place of the iron. The magnesium serves as a sacrificial anode.

   For a metal to function as a sacrificial anode, it must be more active (a stronger reducer) than the metal that is protected and also be stable compared to the conditions present underground (sodium metal would make for a poor sacrificial anode).

   (c) i. The number of moles of iron corroded per meter per year is found by dividing

   $$\frac{1g}{55.85 \text{ g/mol}} = 0.0179 \text{ mol}.$$

   Assume that the rates of corrosion are equal, so 0.0179 mol per meter per year would be consumed, or $0.0179 \text{ mol} \times 24.30 \text{ g/mol Mg} = 0.435$ g. Because one block is attached to a 200 m pipe, multiply by 200 = 87 g per 200 m per year. Finally, take $\frac{1,000 \text{ g block}}{87 \text{ g yr}^{-1}} = 11.5$ yr.

   ii. For each 0.5 mole of magnesium oxidized, one mole of electrons must be released into the wire. We know that the rate of magnesium loss is 0.0179 mol m^{-1} yr^{-1}, and each block is 200 m apart, so we need to multiply: $0.0179 \text{ mol m}^{-1}$ $yr^{-1} \times 200 \text{ m} = 3.58 \text{ mol (Mg) yr}^{-1}$.
   Finally, calculate $\frac{0.5 \text{ mol (Mg)}}{3.58 \text{ mol (Mg) yr}^{-1}} = 0.14$ yr, or 51 days.

   (d) To calculate the equilibrium constants for the reactions, we need to calculate the voltage potential for both reactions. This is found in the standard reduction potentials table. For iron, this produces $E°_{red} = 1.23$ V $+ E°_{ox}$, and since $- E°_{red} = (-) -0.44$ V, we can substitute to make a simple equation:

   $$1.23 \text{ V} + 0.44 \text{ V} = 1.67 \text{ V}$$

   For magnesium, this produces $E°_{red} = 1.23$ V $+ E°_{ox}$, and since $- E°_{red} = (-) -2.37$ V, we again can substitute: $1.23 \text{ V} + 2.37 \text{ V} = 3.6$ V.

   Finally, we plug these values in to the Nernst equation where $\log K = \frac{nE°}{0.0592}$. The value for n is equal to the number of electrons transferred in the reaction (2). Calculating the values produces $K_{Fe} = 10^{56.4}$ for iron and $K_{Mg} = 10^{121}$ for magnesium.

   (e) We need to recalculate the voltage potential again for zinc: $E°_{red} = 1.23 \text{ V} + E°_{ox} = 1.23$ V$- E°_{red} = 1.23 \text{ V} + 0.76 \text{ V}. = 1.99$ V. Then, $\Delta G° = - nFE° = - 2 \text{ mol} \times 96,500 \text{ C/mol} \times 1.99 \text{ V} = 3.84 \times 10^6 \text{ J} = 3.84 \times 10^3 \text{ kJ}$.

2. (a) $2C_3H_6 + 9O_2 \rightarrow 6CO_2 + 6H_2O$

   (b) First, calculate the number of moles of water produced and work backwards to find the moles of hydrocarbon and then the volume of hydrocarbon. One mL of water = 1 gram of water $= \frac{1}{18}$ mole water (formula weight of $H_2O = 18$). For each three moles of water formed from combustion, one mole of hydrocarbon is present. The number of moles of hydrocarbon is therefore $\frac{1}{54}$ moles. Finally, since each mole of an ideal gas at STP occupies 22.4 L, the hydrocarbon takes up $22.4 \times \dfrac{1}{54}$ L $= 0.414$ L or 41.1% of the gas mixture.

(Also see Dalton's law of partial pressures.)

(c) If $\dfrac{1}{54}$ mole of hydrocarbon produces 6.51 kJ of heat during combustion, then one mole of hydrocarbon will produce 6.51 kJ $\times 54 = $ 351.54 kJ/mol.

(d) i. Carbon has a valence of 4 while hydrogen has a valence of 1.

Isomer A:

Isomer B:

ii. Isomer A has one carbon sp^3-hybridized (the left one), and the other two carbons are sp^2-hybridized, with the double bond between them formed from the leftover p orbital.

3. (a) i. The molarity of HCN must first be calculated. The molecular weight of HCN $= 27$, so we have $\dfrac{13}{27}$ moles, or 0.481 moles of gas, which in 1 L is 0.481 M HCN. Next, set up the ionization equation and solve for $[H^+]$. For HCN, $K_a = 4.9 \times 10^{-10}$ $= \dfrac{[H^+][CN^-]}{[HCN]}$. Setting $[H^+]$ and $[CN^-]$ both equal to x produces $4.9 \times 10^{-10} = \dfrac{x^2}{(0.481 - x)} \approx \dfrac{x^2}{(0.481)}$ since x is very small compared to 0.418. Rearranging the equation gives $2.35 \times 10^{-10} - x^2 = 0$, so $x = 1.5 \times 10^{-5} = [H^+]$. To get the pH, we need $-\log[H^+] = 4.81$.

ii. Using the value calculated for x above, calculate the percent dissociation of HCN as: $\dfrac{1.5 \times 10^{-5}}{0.481} = 3.4 \times 10^{-3}$ %.

iii. To solve this problem, use the Henderson-Hasselbalch equation:

$pH = pK_a + \log \dfrac{[\text{base}]}{[\text{acid}]}$. Plugging in the values given yields:

$10 = 9.31 + \log \dfrac{[x]}{[0.481]}$. Solving for x produces 2.36 M.

(b) Recalculating the initial concentration of HCN gives $\dfrac{10}{27} = 0.37$ mol HCN, $\dfrac{0.37 \text{ mole}}{0.2 \text{ L}} = 1.85$ M HCN. To calculate the volume of base required, the Henderson-Hasselbalch equation once again needs to be used. An $[OH^-]$ of 1×10^{-3} is the equivalent of a pH of 11. What is different from (a) iii is that for each equivalent of NaOH added, one equivalent of HCN is consumed (making one NaCN).

Therefore, the equation becomes: $11 = 9.31 + \log \frac{[x]}{[0.37 - x]}$. Rearranging the equation yields $48.97 = \frac{x}{(0.37 - x)}$, or $x = 0.362$ mol NaOH. Plugging numbers into $M = \frac{n}{L}$, the volume of 0.1 M NaOH solution required can be calculated as: $0.1 = \frac{0.362}{x}$, $x = 3.62$ L.

(c) The normal cyanide anion is:

$$-\overset{..}{C} \equiv N:$$

Each atom has an octet, and the negative charge rests on the carbon. The isocyanide anion has the form:

$$:C = \overset{..}{N}:^-$$

where the negative charge rests on the nitrogen. The nitrogen has an octet, but the carbon does not, therefore this form would be expected to be less stable than the cyanide form.

PART B

4. (a) i. $Zn + HCl \rightarrow ZnCl_2 + H_2$

    ii. The oxidation number of hydrogen in HCl is +1. In H_2, it is 0.

    (b) i. $C + 2H_2 \rightarrow CH_4$

    ii. $\Delta S°$ is negative for this reaction. Entropy is decreased when there are more moles of gas on the reactants side than on the products side.

    (c) i. $PbNO_3 + KSO_4 \rightarrow PbSO_4 + KNO_3$

    ii. $PbSO_4$ will precipitate out. KNO_3 is soluble in water.

5. (a) Color—Only choice (A) will give a colored solution (blue).

    Flame test—Only solution (B) will give a strong yellow flame test from the presence of the Na^+ ion.

    Hydroxide precipitation—Solution (C) will form a precipitate, $Al(OH)_3$, upon treatment with a solution of sodium hydroxide. Solution (A) will also give a precipitate, but only (C) will have that precipitate redissolve with excess hydroxide, forming $Al(OH)_4^-$.

    (D) Conductivity—Only the sucrose solution will not produce ions, and the solution will not conduct electricity. Taste is obviously not a valid test.

    (E) pH—Ammonia is a medium strong base and the pH of its solution would be the most basic.

    (b) This question is really asking which solution produces the most solute ions or molecules. Because each solution has the same molarity, the one with the most component ions will decrease the freezing point of the solution the most. This solution is choice (C), which, upon dissolution, produces four ions: one Al^{3+} ion and three CH_3CO_2- ions.

    (c) i. A coordination complex is being formed between the transition metal copper and the ammonia ligand. The full reaction is $Cu^{2+} + 4NH_3 \rightarrow Cu(NH_3)_4^{2+}$. The ammonia ligand changes the electronic state of the copper and also changes the absorption characteristics of the solution.

ii. The en (ethylenediamine) ligand is comprised of two ammonia ligands and forms stronger complexes because of the chelate effect. The chelate effect stabilizes complexes formed from multidentate ligands relative to equivalent monodentate ligands and is mostly an entropic effect.

(d) i. Since only one equivalent of the chloride ions is used to form AgCl, the other two are involved in coordination with the chromium. Recognizing four ammonia molecules in the remaining formula gives $[Cr(NH_3)_4Cl_2]Cl$.

ii. The isomers differ in the relative positions of the chlorine ligands:

and

6. (a) The UF_6 molecule has no net dipole and therefore experiences few intermolecular attractive forces. Additionally, fluorine is small and not polarizable; it reduces the van der Waals attractive forces.

(b) The rate of diffusion of gases is related to the square root of the masses of the gases.

The ratio of the rate of diffusion of
$$\frac{^{235}UF_6}{^{238}UF_6} = \sqrt{\frac{349}{352}}.$$

(c) i. $UF_6 + 2H_2O \rightarrow UO_2F_2 + 4HF$

ii. Generation of toxic and corrosive HF is bad for both the operators and the equipment. Also, the volatile UF_6 is converted into a more polar molecule and probably will not be volatile enough for isotropic separation.

(d) Reaction I – U(VI) → U(VI) no change

Reaction II – U(VI) → U(IV) reduction

Reaction III – U(IV) → U(IV) no change

Reaction IV – U(IV) → U(VI) oxidation

Practice Test Two Answer Grid

1. Ⓐ Ⓑ Ⓒ Ⓓ Ⓔ
2. Ⓐ Ⓑ Ⓒ Ⓓ Ⓔ
3. Ⓐ Ⓑ Ⓒ Ⓓ Ⓔ
4. Ⓐ Ⓑ Ⓒ Ⓓ Ⓔ
5. Ⓐ Ⓑ Ⓒ Ⓓ Ⓔ
6. Ⓐ Ⓑ Ⓒ Ⓓ Ⓔ
7. Ⓐ Ⓑ Ⓒ Ⓓ Ⓔ
8. Ⓐ Ⓑ Ⓒ Ⓓ Ⓔ
9. Ⓐ Ⓑ Ⓒ Ⓓ Ⓔ
10. Ⓐ Ⓑ Ⓒ Ⓓ Ⓔ
11. Ⓐ Ⓑ Ⓒ Ⓓ Ⓔ
12. Ⓐ Ⓑ Ⓒ Ⓓ Ⓔ
13. Ⓐ Ⓑ Ⓒ Ⓓ Ⓔ
14. Ⓐ Ⓑ Ⓒ Ⓓ Ⓔ
15. Ⓐ Ⓑ Ⓒ Ⓓ Ⓔ
16. Ⓐ Ⓑ Ⓒ Ⓓ Ⓔ
17. Ⓐ Ⓑ Ⓒ Ⓓ Ⓔ
18. Ⓐ Ⓑ Ⓒ Ⓓ Ⓔ
19. Ⓐ Ⓑ Ⓒ Ⓓ Ⓔ
20. Ⓐ Ⓑ Ⓒ Ⓓ Ⓔ
21. Ⓐ Ⓑ Ⓒ Ⓓ Ⓔ
22. Ⓐ Ⓑ Ⓒ Ⓓ Ⓔ
23. Ⓐ Ⓑ Ⓒ Ⓓ Ⓔ
24. Ⓐ Ⓑ Ⓒ Ⓓ Ⓔ
25. Ⓐ Ⓑ Ⓒ Ⓓ Ⓔ

26. Ⓐ Ⓑ Ⓒ Ⓓ Ⓔ
27. Ⓐ Ⓑ Ⓒ Ⓓ Ⓔ
28. Ⓐ Ⓑ Ⓒ Ⓓ Ⓔ
29. Ⓐ Ⓑ Ⓒ Ⓓ Ⓔ
30. Ⓐ Ⓑ Ⓒ Ⓓ Ⓔ
31. Ⓐ Ⓑ Ⓒ Ⓓ Ⓔ
32. Ⓐ Ⓑ Ⓒ Ⓓ Ⓔ
33. Ⓐ Ⓑ Ⓒ Ⓓ Ⓔ
34. Ⓐ Ⓑ Ⓒ Ⓓ Ⓔ
35. Ⓐ Ⓑ Ⓒ Ⓓ Ⓔ
36. Ⓐ Ⓑ Ⓒ Ⓓ Ⓔ
37. Ⓐ Ⓑ Ⓒ Ⓓ Ⓔ
38. Ⓐ Ⓑ Ⓒ Ⓓ Ⓔ
39. Ⓐ Ⓑ Ⓒ Ⓓ Ⓔ
40. Ⓐ Ⓑ Ⓒ Ⓓ Ⓔ
41. Ⓐ Ⓑ Ⓒ Ⓓ Ⓔ
42. Ⓐ Ⓑ Ⓒ Ⓓ Ⓔ
43. Ⓐ Ⓑ Ⓒ Ⓓ Ⓔ
44. Ⓐ Ⓑ Ⓒ Ⓓ Ⓔ
45. Ⓐ Ⓑ Ⓒ Ⓓ Ⓔ
46. Ⓐ Ⓑ Ⓒ Ⓓ Ⓔ
47. Ⓐ Ⓑ Ⓒ Ⓓ Ⓔ
48. Ⓐ Ⓑ Ⓒ Ⓓ Ⓔ
49. Ⓐ Ⓑ Ⓒ Ⓓ Ⓔ
50. Ⓐ Ⓑ Ⓒ Ⓓ Ⓔ

51. Ⓐ Ⓑ Ⓒ Ⓓ Ⓔ
52. Ⓐ Ⓑ Ⓒ Ⓓ Ⓔ
53. Ⓐ Ⓑ Ⓒ Ⓓ Ⓔ
54. Ⓐ Ⓑ Ⓒ Ⓓ Ⓔ
55. Ⓐ Ⓑ Ⓒ Ⓓ Ⓔ
56. Ⓐ Ⓑ Ⓒ Ⓓ Ⓔ
57. Ⓐ Ⓑ Ⓒ Ⓓ Ⓔ
58. Ⓐ Ⓑ Ⓒ Ⓓ Ⓔ
59. Ⓐ Ⓑ Ⓒ Ⓓ Ⓔ
60. Ⓐ Ⓑ Ⓒ Ⓓ Ⓔ
61. Ⓐ Ⓑ Ⓒ Ⓓ Ⓔ
62. Ⓐ Ⓑ Ⓒ Ⓓ Ⓔ
63. Ⓐ Ⓑ Ⓒ Ⓓ Ⓔ
64. Ⓐ Ⓑ Ⓒ Ⓓ Ⓔ
65. Ⓐ Ⓑ Ⓒ Ⓓ Ⓔ
66. Ⓐ Ⓑ Ⓒ Ⓓ Ⓔ
67. Ⓐ Ⓑ Ⓒ Ⓓ Ⓔ
68. Ⓐ Ⓑ Ⓒ Ⓓ Ⓔ
69. Ⓐ Ⓑ Ⓒ Ⓓ Ⓔ
70. Ⓐ Ⓑ Ⓒ Ⓓ Ⓔ
71. Ⓐ Ⓑ Ⓒ Ⓓ Ⓔ
72. Ⓐ Ⓑ Ⓒ Ⓓ Ⓔ
73. Ⓐ Ⓑ Ⓒ Ⓓ Ⓔ
74. Ⓐ Ⓑ Ⓒ Ⓓ Ⓔ
75. Ⓐ Ⓑ Ⓒ Ⓓ Ⓔ

PRACTICE TEST TWO

SECTION I
Part A
Time: 90 minutes

Directions: Each set of letter choices below refers to the numbered statements immediately following it. Select the one lettered choice that best fits each statement. A choice may be used once, more than once, or not at all in each set.

Questions 1–2 refer to the following elements:

Consider atoms of the following elements in the ground state.

(A) Fe (II)

(B) Ar

(C) C

(D) Na

(E) Cl

1. The atom that contains zero unpaired electrons

2. The atom that contains four valence shell electrons

Questions 3–5 refer to the following molecules:

(A) CO_2

(B) BF_3

(C) NH_3

(D) N_2

(E) CH_4

3. This molecule contains double bonds and has linear geometry.

4. This molecule has triangular planar geometry.

5. This molecule has a triple bond.

GO ON TO THE NEXT PAGE

Questions 6–7 refer to the following gases at standard conditions (0°C and 1 atm):

(A) O_2

(B) CO_2

(C) H_2

(D) Ar

(E) CO

6. This gas has the greatest density.

7. This gas has the fastest effusion rate.

8. Determine the density (with the correct number of significant figures) of 19.0 mL of a liquid based on a measurement of 18.004 g for its mass.

(A) 0.94757894 g/mL

(B) 9.48×10^{-1} g/mL

(C) 9.486×10^{-2} g/mL

(D) 1.05533 g/mL

(E) 1.06 g/mL

9. Which of the following compounds is the strongest base?

(A) H_2S

(B) HBr

(C) KOH

(D) NH_3

(E) KCl

10. A pure, white, and crystalline powder dissolves instantaneously in water. The resulting solution is basic and liberates a gas readily when a strong acid is added to it. This substance is

(A) KNO_3.

(B) KCl.

(C) KOH.

(D) NH_3.

(E) K_2CO_3.

11. In lab, you combine a non-volatile solute in a pure solvent at room temperature and standard pressure (1 atm). Relative to the original solvent, the solution you just made has

(A) a higher freezing point.

(B) a higher vapor pressure.

(C) a lower boiling point.

(D) a higher boiling point.

(E) the same boiling point.

GO ON TO THE NEXT PAGE

Questions 12–15 refer to the reactions represented below:

(A) $H_2SO_4(aq) + 2\,NaOH(aq) \rightarrow$
$Na_2SO_4(aq) + 2\,H_2O(l)$

(B) $2\,Fe(s) + 3\,Cl_2(g) \rightarrow 2\,FeCl_3(s)$

(C) $Zn(s) + 2\,AgNO_3(aq) \rightarrow$
$2\,Ag(s) + Zn(NO_3)_2(aq)$

(D) $2\,BaO_2(s) \rightarrow 2\,BaO(s) + O_2(g)$

(E) $C_2H_5OH(l) + 3\,O_2(g) \rightarrow$
$2\,CO_2(g) + 3\,H_2O(g)$

12. An example of a combustion reaction

13. An example of a synthesis reaction

14. An example of a single-displacement reaction

15. An example of a neutralization reaction

Part B

Directions: Select the best answer choice for each incomplete statement or question below.

16. Which of the following compounds is least soluble in water (at pH 7 and 20°C)?

(A) NaCl

(B) $LiNO_3$

(C) $CaCO_3$

(D) KBr

(E) NH_4Cl

17. STP, or Standard Temperature and Pressure, refers to what set of conditions?

(A) 25°C, 1 atm, where 1 mol of a gas = 22.4 L

(B) 25°C, 100 atm, where 1 mol of a gas = 2.4 L

(C) 0°C, 100 atm, where 1 mol of a gas = 22.4 L

(D) 0°C, 1 atm, where 1 mol of a gas = 22.4 L

(E) 25°C, 1 atm, where 1 mol of a gas = 2.4 L

18. Entropy of a system, S, is a measure of the disorder or chaos of a system, and

(A) solids, gases, and liquids at the same temperature have the same entropy.

(B) particles in solution have a higher entropy than gases.

(C) two moles of a substance have a higher entropy than three moles.

(D) liquids have a lower entropy than solids.

(E) liquids have a higher entropy than solids.

GO ON TO THE NEXT PAGE

19. $CH_3COOH(aq) \rightarrow H^+(aq) + CH_3COO^-(aq)$

    Which of the following best describes the outcome of adding HCl to the above ionization reaction?

    (A) The addition of HCl will push the equilibrium toward products.

    (B) HCl will have no effect on the ionization of CH_3COOH.

    (C) HCl will ionize completely and suppress the ionization of CH_3COOH.

    (D) HCl will increase the pH of the solution.

    (E) HCl addition will cause precipitation of CH_3COOH.

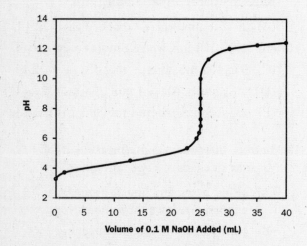

20. The plot above shows a titration curve for a 0.1 M solution of a weak acid titrated with 0.1 M NaOH. Which of the following pH values lies within the buffering region of the solution?

    (A) 2.2

    (B) 4.7

    (C) 6.5

    (D) 8.7

    (E) 9.8

21. The partial pressure of CO_2 gas above the solution in a bottle of soda at 15°C is 3 atm. How would each of the following factors change if the bottle were allowed to warm to 30°C, still sealed? Assume constant volume.

	Solubility of CO_2 in solution	Vapor pressure of water	Pressure of CO_2 gas above solution
(A)	Increase	Increase	Increase
(B)	Decrease	Increase	Increase
(C)	Decrease	Decrease	Increase
(D)	Increase	Decrease	Decrease
(E)	Decrease	Decrease	Decrease

22. Which of the following elements is the most electronegative?

    (A) Al

    (B) C

    (C) I

    (D) Na

    (E) O

23. At a pressure of 1 atm, water boils at 100°C. How would an increase in pressure affect the temperature at which water boils?

    (A) The water will boil at a higher temperature.

    (B) The water will boil at a lower temperature.

    (C) The boiling point of water is independent of pressure.

    (D) The boiling point of water is only dependent on pressure at the triple point.

    (E) The boiling point of water is only dependent on pressure at the critical point.

GO ON TO THE NEXT PAGE

24. On a phase diagram, the triple point represents

(A) the only temperature and pressure combination at which the liquid phase is in equilibrium with the gas phase.

(B) the only temperature at which the gas phase can be in equilibrium with the solid phase.

(C) the only temperature and pressure combination at which gas, liquid, and solid phases are at equilibrium.

(D) the only pressure at which the gas phase can be in equilibrium with the solid phase.

(E) the only temperature and pressure combination where the solid phase is in equilibrium with the liquid phase.

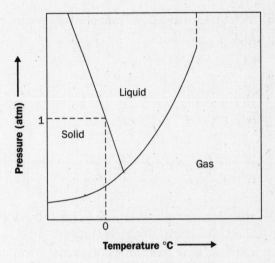

25. Given the above phase diagram for a hypothetical substance, if the temperature is held constant at 0°C and the pressure is raised above 1 atm, what physical state will eventually be reached?

(A) As the pressure is raised, the particles of the substance will continue to compact until it becomes a solid.

(B) Even as the pressure is raised to high levels, the substance will remain in its liquid phase.

(C) As the pressure is raised to very high levels, the substance will eventually reach its gas phase.

(D) As the pressure is raised to very high levels, the substance will approach equilibrium with all three phases.

(E) As the pressure is raised to very high levels, the substance will eventually reach beyond the critical point and become a supercritical fluid.

GO ON TO THE NEXT PAGE

26. What is the oxidation number of oxygen in H_2O_2?

(A) −2

(B) −1

(C) $-\frac{1}{2}$

(D) −0

(E) $+\frac{1}{2}$

27. If 90 grams of glucose are dissolved in water to a total volume of 1 liter, forming a 0.50 molar solution, how many moles of glucose are in 500 mL of that solution?

(A) 0.25 moles

(B) 2.5 moles

(C) 2.5×10^{23} moles

(D) 25 moles

(E) 2,500 moles

28. A diver wanted to calculate the density of an unusually shaped artifact found in the ocean to help classify the composition of the artifact. The diver used a graduated cylinder of water and calculated the displacement of the water as 12.3 mL when the artifact was added. What other measurement is needed to be able to calculate density?

(A) Weight

(B) Length

(C) Mass

(D) Concentration

(E) Volume

29. Octane (C_8H_{18}) is a saturated hydrocarbon that is used as a fuel. Which of the following substances has the highest solubility in octane at 21°C to form a homogenous solution?

(A) $CH_3CH_2CH_2OH$

(B) NH_4^+

(C) NaCl

(D) $NaOCH_2CH_3$

(E) KBr

30. $CaCO_3(s) \rightarrow CaO(s) + CO_2(g)$
$\Delta H_{298} = +178.3 \text{ kJ}$

The thermochemical equation for the decomposition of calcium carbonate is shown above. Which of the following statements is correct?

(A) Heat is released during the decomposition of calcium carbonate.

(B) The enthalpy of formation of calcium carbonate is zero.

(C) This reaction shows a decrease in entropy.

(D) A thermometer placed in the reaction vessel would show an increase in temperature.

(E) The reaction is endothermic.

31. Which of the following statements best describes the effect of temperature on reaction rates?

(A) In general, reaction rates increase with increasing temperature.

(B) In general, reaction rates decrease with increasing temperature.

(C) Only rates of reactions involving molecules with low-energy bonds are dependent on temperature.

(D) Rates of reactions involving catalysts are not dependent on temperature.

(E) The effect of temperature on a reaction rate is always unpredictable.

GO ON TO THE NEXT PAGE

32. $2 H_2O_2(aq) \rightarrow 2 H_2O(l) + O_2(g)$

The iodide-catalyzed decomposition of hydrogen peroxide, shown above, follows the rate law rate = $k[H_2O_2][I^-]$. What is the overall order of this reaction?

(A) Zero

(B) First

(C) Second

(D) Third

(E) Fourth

33. The reaction rate for a reagent in a reaction is the

(A) time it takes per unit volume of substance.

(B) amount of time per mole that is formed or removed.

(C) amount of time per unit volume that is formed or removed.

(D) amount (in moles or mass units) per unit volume.

(E) amount (in moles or mass units) per unit time per unit volume that is formed or removed.

34. Which of the following describes something you should *NOT* do when working with strong acids?

(A) Work under a fume hood

(B) Flush with water any acid that has splashed on the skin

(C) Add concentrated acid to water when making a dilute acid solution

(D) Store strong acids in a glass or ceramic containment tray

(E) Neutralize a strong acid by adding a strong base

35. $Al(s) + HNO_3(aq) \rightarrow Al(NO_3)_3(aq) + H_2(g)$

Aluminum metal reacts with nitric acid in the above unbalanced equation. If 1 mole of solid aluminum is mixed with an excess of nitric acid, how many moles of H_2 gas are produced?

(A) 1 mole

(B) 1.5 moles

(C) 2 moles

(D) 2.5 moles

(E) 3 moles

36. A mixture of nitrogen (N_2), helium (He), and argon (Ar) gases exerts a total pressure of 288 atm. If the partial pressures of He and Ar are each 37 atm, what is the partial pressure of N_2 gas?

(A) 74 atm

(B) 107 atm

(C) 114 atm

(D) 214 atm

(E) 288 atm

37. Which of the following combinations would have the greatest buffering capacity at pH 7.2? (pK_a values are shown in parentheses.)

  I. 1 M mixture of H_2PO_4 (7.206) and HPO_4^- (12.35)
  II. 1 mM mixture of HSO_3^- (7.22) and SO_3^{2-}
  III. 2 M mixture of HCl (strong) and NaOH (strong)

(A) I has the greatest buffering capacity of the three at pH 7.2.

(B) I and II have equal buffering capacities at pH 7.2.

(C) II is the strongest buffer at pH 7.2.

(D) II and III are equally strong at pH 7.2.

(E) III is the strongest buffer at pH 7.2.

GO ON TO THE NEXT PAGE

38. Which of the following statements best describes the reactivities of copper and zinc shown in the equations below?

$$Cu(s) + ZnSO_4(aq) \rightarrow no\ reaction$$

$$Zn(s) + CuSO_4(aq) \rightarrow ZnSO_4(aq) + Cu(s)$$

(A) Zinc metal is more likely to give up two electrons than copper metal.

(B) Aqueous zinc is more reactive than aqueous copper.

(C) Copper metal is more likely to lose two electrons than zinc metal.

(D) Copper metal is easily oxidized by aqueous zinc.

(E) Copper metal is more reactive than zinc metal.

39. $^{14}_{6}C \rightarrow ^{14}_{7}N + ^{0}_{-1}e$

What statement best describes what is occurring in the above equation?

(A) Gamma radiation of carbon-14

(B) Ionization of carbon-14

(C) Alpha particle emission from carbon-14

(D) Positron emission from carbon-14

(E) Beta particle emission from carbon-14

40. Which of the following molecules contains two sp^2-hybridized carbon atoms?

(A) C_2H_5F

(B) C_2H_2

(C) C_2H_4

(D) C_2H_6

(E) C_2H_5OH

41. Which of the following is the correct outer shell electron configuration for zinc?

(A) $5s^2 4d^{10}$

(B) $2s^2 1d^{10}$

(C) $4s^2 1d^{10}$

(D) $4s^2 3d^{10}$

(E) $3s^2 3d^{10}$

42. $K(s) + Cr^{3+}(aq) \rightarrow Cr(s) + K^+(aq)$

Which of the following equations is correctly balanced for the oxidation-reduction reaction shown above?

(A) $K(s) + Cr^{3+}(aq) \rightarrow Cr(s) + K^+(aq)$

(B) $2\ K(s) + Cr^{3+}(aq) \rightarrow Cr(s) + 2\ K^+(aq)$

(C) $K(s) + 3\ Cr^{3+}(aq) \rightarrow 3\ Cr(s) + K^+(aq)$

(D) $3\ K(s) + 2\ Cr^{3+}(aq) \rightarrow 2\ Cr(s) + 3\ K^+(aq)$

(E) $3\ K(s) + Cr^{3+}(aq) \rightarrow Cr(s) + 3\ K^+(aq)$

43. To identify an unknown metal sample, the density of the metal was determined in a laboratory. The density was determined to be 19.32 g/mL. If the volume of the sample was determined to be 0.01 L, what was the mass of the sample?

(A) 19.30 g

(B) 193.0 g

(C) 1,930 g

(D) 193.0 mg

(E) 1,930 mg

44. What is the approximate ratio of CH_3COO^- to CH_3OOH for acetic acid ($pK_a = 4.8$) at pH 4.8?

(A) Antilog (−1)

(B) Log 1

(C) 1:1

(D) 2:1

(E) 1:2

GO ON TO THE NEXT PAGE

45. The K_a of hypochlorous acid (HClO) is 3.3×10^{-2}. A solution of pH 5 contains some hypochlorous acid in equilibrium with its conjugate base. What additional information do you need to calculate the concentration of hypochlorite anion ($CHCl_2COO^-$) in the solution?

(A) No more information is needed.

(B) Only the concentration of hypochlorous acid (HClO) is needed.

(C) Only the hydrogen ion (H^+) concentration is needed.

(D) Both the concentrations of hypochlorous acid and the hydrogen ion are needed.

(E) The pK_a of hypochlorous acid is needed.

46. $Cl_2(aq) + 2\,NaBr(aq) \rightarrow 2\,NaCl(aq) + Br_2(l)$

The reaction shown above is spontaneous at 25°C. Which of the following statements referring to the reaction is correct?

(A) Cl_2 is oxidized by Br^-.

(B) Br_2 is more reactive than Cl_2.

(C) Cl_2 gains electrons more easily than Br_2.

(D) Br_2 is easier to reduce than Cl_2.

(E) NaBr is less reactive than NaCl.

47. $SO_2 + \dfrac{1}{2}O_2 \rightarrow SO_3(g)$

Calculate the standard enthalpy change for the above reaction, given the standard enthalpies of formation: $SO_2 = -297$ kJ/mol, $SO_3 = -396$ kJ/mol, O_2 (standard state).

(A) +692.5 kJ/mol

(B) +99 kJ/mol

(C) 0 kJ/mol

(D) –99 kJ/mol

(E) –692.5 kJ/mol

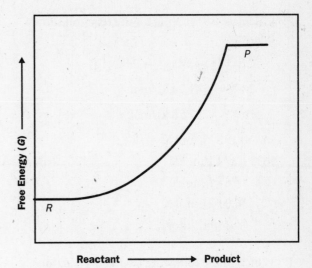

48. Which of the following statements is true based on the reaction shown in the figure above, where R is reactant and P is product?

(A) The forward reaction is spontaneous.

(B) The system is increasing in entropy.

(C) The system is proceeding spontaneously to equilibrium.

(D) The reverse reaction is spontaneous.

(E) The system is decreasing in entropy.

49. $HgO + H_2 \rightarrow Hg(s) + H_2O\ (l)$

$\Delta G^\circ_f = -58.54$ kJ, 0, 0, –237.1 kJ respectively (at 298 K)

Calculate ΔG° at 298 K for the reaction shown above.

(A) 295.6 kJ/mol

(B) 178.6 kJ/mol

(C) 4.05 kJ/mol

(D) –178.6 kJ/mol

(E) – 295.6 kJ/mol

GO ON TO THE NEXT PAGE

50. Calculate $\Delta S°$ at 30°C for the reaction shown below:

$$2\,SO_2(g) + O_2(g) \rightarrow 2\,SO_3(g)$$

$$\Delta G° = -141.74\ \text{kJ/mol}$$

$$\Delta H° = -197.78\ \text{kJ/mol}$$

(A) 0.185 kJ/mol·K

(B) 1.87 kJ/mol·K

(C) 4.76 kJ/mol·K

(D) 56.04 kJ/mol·K

(E) −0.185 kJ/mol·K

51. Which of the following is a fusion reaction?

(A) $^{10}_{5}B + ^{1}_{0}n \rightarrow ^{7}_{3}Li + ^{4}_{2}He$

(B) $^{7}_{3}Li + ^{1}_{0}n \rightarrow ^{3}_{1}H + ^{4}_{2}He + ^{1}_{0}n$

(C) $^{235}_{9}U + ^{1}_{0}n \rightarrow ^{141}_{6}Ba + ^{92}_{36}K + 3^{1}_{0}n$

(D) $^{241}_{95}Am \rightarrow ^{4}_{2}He + ^{237}_{93}Np$

(E) $^{2}_{1}H + ^{3}_{1}H \rightarrow ^{4}_{2}He + ^{1}_{0}n + \text{energy}$

52. $Zn(s) + 2\,H^+(aq) \rightarrow Zn^{2+}(aq) + H_2(g)$

A 30 g sample of Zn is dissolved in excess acid, as shown in the reaction above. How many moles of H_2 are released?

(A) 0.046 moles

(B) 0.23 moles

(C) 0.46 moles

(D) 0.92 moles

(E) 1.8 moles

53. Which of the following molecules contains an ester linkage?

(A) 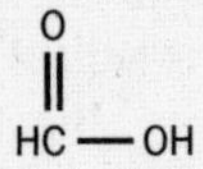
$$\begin{array}{c} O \\ \| \\ HC-OH \end{array}$$

(B) 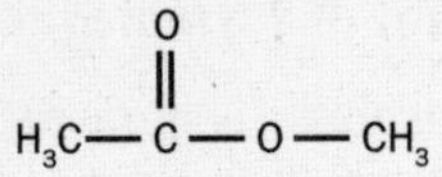
$$\begin{array}{c} O \\ \| \\ H_3C-C-O-CH_3 \end{array}$$

(C) $H_3C-O-CH_3$

(D) $H_3C-O-CH_2OH$

(E) 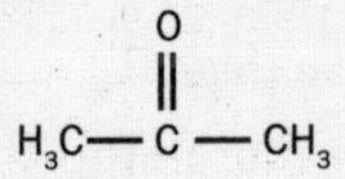
$$\begin{array}{c} O \\ \| \\ H_3C-C-CH_3 \end{array}$$

GO ON TO THE NEXT PAGE ⇨

54. $2 Al(s) + 3 Cl_2(g) \rightarrow 2 AlCl_3(s)$

If 5 moles of Al and 6 moles of Cl_2 react according to the above equation, what is the maximum number of moles of $AlCl_3$ that can be formed?

(A) 4 moles

(B) 5 moles

(C) 6 moles

(D) 11 moles

(E) 12 moles

55. What is the oxidation number of the carbon in CO_2?

(A) +2

(B) +4

(C) 0

(D) –2

(E) –4

56. On the periodic table, which group typically has an oxidation number of +3?

(A) Group 5A

(B) Group 6A

(C) Group 7A

(D) Group 3A

(E) Group 4A

57. Which of the following is the correct number of moles of O_2 present in a volume of 24 L at 37°C and 1 atm pressure?

(A) 0.01 moles

(B) 0.79 moles

(C) 0.94 moles

(D) 7.9 moles

(E) 9.4 moles

58. A sugary drink contains 20.5% sucrose. What volume of this drink (in mL) contains 10 g of sucrose? (Assume the density is 1.0 g/mL.)

(A) 2.05 mL

(B) 20.5 mL

(C) 24.4 mL

(D) 48.8 mL

(E) 97.6 mL

59. Which of the following molecules shows a peptide bond with a dark, bold line?

(A)

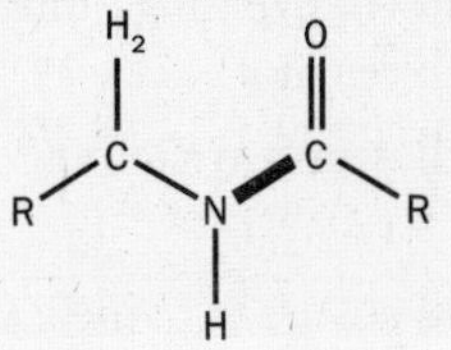

(B)

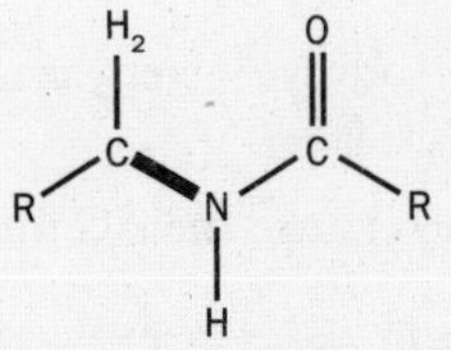

(C)

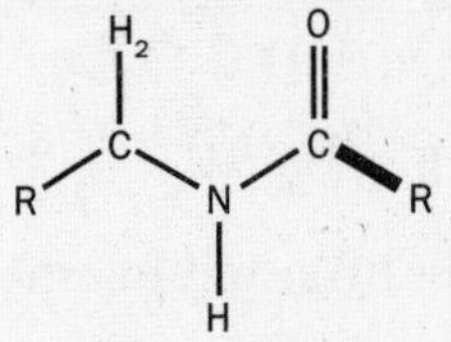

(D)

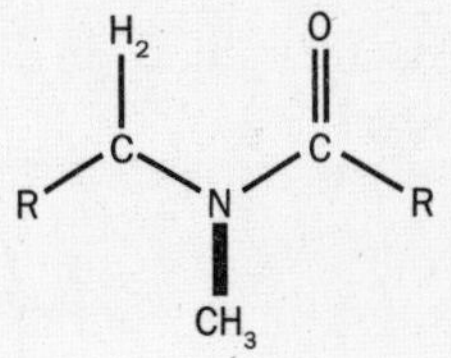

(E)

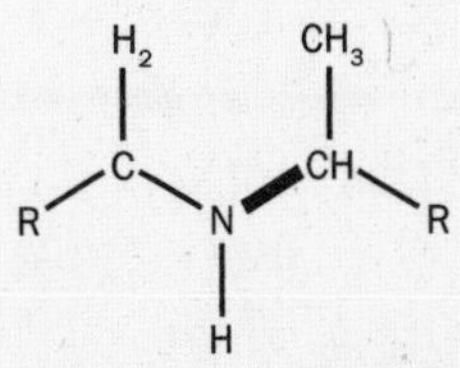

GO ON TO THE NEXT PAGE

60. How long does it take for a 2.4 mole sample of Rn-220 to decay to 0.3 moles if the half-life of Rn-220 is one minute?

    (A) One minute

    (B) Two minutes

    (C) Three minutes

    (D) Four minutes

    (E) Five minutes

61. Which of the following is a true statement about catalysts?

    (A) A catalyst will change the equilibrium constant for a reaction.

    (B) A catalyst will decrease the rate of a reaction.

    (C) A catalyst will always provide a simpler mechanism for the reaction.

    (D) A catalyst will always lower the activation energy for a reaction.

    (E) A catalyst will change the ΔG for the reaction.

$$MnO_4^-(aq) + H^+(aq) + H_2C_2O_4(aq) \rightarrow$$
$$Mn^{2+}(aq) + CO_2(g) + H_2O(l)$$

Use the equation above to answer questions 62 and 63.

62. Which element(s) is/are reduced in the reaction shown above?

    (A) C

    (B) O

    (C) Mn

    (D) H

    (E) Both C and Mn

63. Balance the reaction shown above question 62 using the lowest whole-number coefficients. The correct coefficient for oxalic acid $(H_2C_2O_4)$ is which of the following?

    (A) 2

    (B) 3

    (C) 5

    (D) 8

    (E) 16

64. Which of the following is the weakest type of intermolecular interaction?

    (A) Dipole-dipole interaction

    (B) Induced dipole–induced dipole interaction

    (C) Ion-dipole interaction

    (D) Oppositely charged ion–ion interaction

    (E) Dipole–induced dipole interaction

65. What must be true for a reaction in which the $\Delta G = 0$?

    (A) The reaction will only proceed in the forward direction.

    (B) The reaction is at equilibrium.

    (C) The reaction is spontaneous.

    (D) The reaction will only proceed in the reverse direction.

    (E) The reaction will proceed at a slow rate.

GO ON TO THE NEXT PAGE

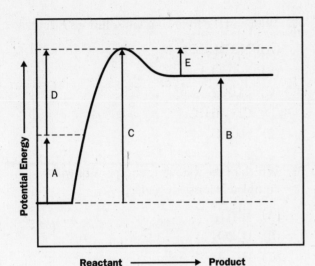

Reactant ⟶ Product

66. Given the potential energy profile for a reaction shown above, which letter corresponds to E_a for the forward reaction (Reactant → Product)?

   (A) A
   (B) B
   (C) C
   (D) D
   (E) E

67. When 1 mole of octane gas (C_8H_{18}) is burned in excess oxygen, how many moles of H_2O are produced?

   (A) 1 mole
   (B) 2 moles
   (C) 9 moles
   (D) 16 moles
   (E) 18 moles

68. A 12 M solution of HCl is diluted to make 0.5 L of 2 M HCl. How many mL of the 12 M solution are used?

   (A) 0.083 mL
   (B) 0.12 mL
   (C) 1.2 mL
   (D) 83 mL
   (E) 120 mL

69. The freezing point depression constant for water is 1.86°C kg/mol. Estimate the freezing point for a 0.2 molal solution of NaCl in water.

   (A) 0.37°C
   (B) 0.74°C
   (C) −0.37°C
   (D) −0.74°C
   (E) −1.86°C

70. The heat of solution for LiCl(s) is −37.03 kJ/mol. Which of the following statements is true?

   (A) Mixing solid LiCl in water will result in a release of heat to the surroundings.
   (B) Dissolving LiCl in water will result in a shell of water molecules surrounding each ionically bonded LiCl pair.
   (C) LiCl will dissolve more readily in benzene solvent than in water.
   (D) Dissolving LiCl in water will result in a cluster of LiCl molecules surrounded by a single shell of water.
   (E) LiCl will remain in a pure crystal form and will not dissolve in water.

GO ON TO THE NEXT PAGE

71. Which of the following molecules is a hexose?

(A)

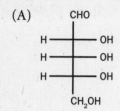

(B)

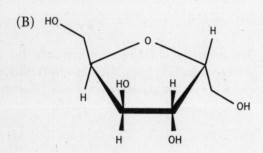

(C)

(D)

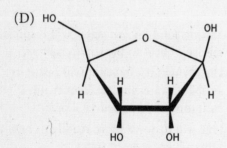

(E)

72. Which of the following salt solutions is acidic?

(A) $NaCH_3CO_2$
(B) $K_2(CO_3)$
(C) KCN
(D) CH_3NH_3Cl
(E) Na_3PO_4

73. Which of the following compounds will completely ionize in water?

(A) HNO_3
(B) H_3PO_4
(C) $HOCl$
(D) HF
(E) CH_3COOH

74. Which of the following atoms has the electron configuration of $1s^2 2s^2 2p^6 3s^2 3p^6 3d^{10} 4s^2 4p^4$?

(A) Si
(B) Se
(C) As
(D) Sb
(E) Te

75. Which of the following particles will have the largest radius?

(A) F^-
(B) Li
(C) Li^+
(D) Cl
(E) Cl^-

SECTION II

Part A
Time: 55 minutes

YOU MAY USE YOUR CALCULATOR FOR PART A

Directions: Answer all three questions below. Calculators are permitted, except for those with typewriter (QWERTY) keyboards. Clearly show the method used and steps involved in arriving at your answers. Partial credit can only be given if your work is clear and demonstrates an understanding of the problem. The Section II score weighting for each question is 20 percent.

1. The general structure of an amino acid is $NH_3^+ - CHR - COO^-$, where R is the amino acid side chain. The side chain (R) for leucine (L) is $(CH_3)_2CHCH_2$. A salt form of leucine in hydrochloric acid contains the protonated species of H_2L^+, which can dissociate twice.

    $$H_2L^+ \rightleftharpoons HL \, ; pK_{a1} = 2.329$$

    $$HL \rightleftharpoons L^- \, ; pK_{a2} = 9.747$$

    For 0.0500M of H_2L^+, calculate:

    (a) [HL]

    (b) pH

    (c) $[H_2L^+]$

    (d) $[L^-]$

    (e) For a 0.0500M solution of L^-, calculate [HL].

2. An electrochemical cell is made with Mg and Ni, using the metals and their chloride salts.

    (a) Write the two half reactions for the electrochemical cell.

    (b) Calculate the $E°$ for the Mg/Ni cell.

    (c) Draw the Mg/Ni electrochemical cell, using KCl as the salt. Clearly label the anode, cathode, and salt bridge. Indicate the direction of electron flow.

    (d) Calculate the voltage of the Mg/Ni cell at 25°C when the concentration of magnesium chloride is 1 M and the concentration of nickel chloride is 1 mM.

    (e) Which of the following combinations would produce a battery with a higher standard cell potential than that of the Mg/Ni battery described above? Justify your answer.

    Mg/Mn(s), Mg/Cu(s), or Mg/Fe(s)

GO ON TO THE NEXT PAGE

3. A student is asked to determine the caloric content of a triglyceride (fat) sample. The student first places the fat sample below an insulated container of water. The student begins to burn the fat and measure the temperature change of the water until all the fat has burned. Answer the following questions related to this experiment. (Note: Assume 4.186 joules of energy will raise 1.0 g of water 1.0°C and that the density of water is 1.0 g/mL; in other words, one calorie has the same energy value as 4.186 joules (J).)

(a) What products would be formed by the complete combustion of this triglyceride? (Note: You do not need to indicate the coefficients of the products.)

(b) Draw the mixed triglyceride that is formed by glycerol ($C_3H_8O_3$) and the three fatty acids below. Would this triglyceride be solid or liquid at room temperature? Justify your answer.

$$C_{18}H_{37}COOH$$

$$C_{15}H_{31}COOH$$

$$C_{17}H_{35}COOH$$

(c) The student begins with 200.0 mL of water and a 2.0 g sample of the triglyceride. If the initial water temperature measured was 21.3°C and the final temperature was 40.0°C, how many kilojoules were released by the combustion of the triglyceride sample?

(d) Complete combustion of the 2.0 g sample of triglyceride should ideally produce roughly 72,000 kJ. Describe the most likely sources of error in this experiment. Assume that the sample is measured correctly and that instruments such as the thermometer and the scale are functioning properly and read correctly.

GO ON TO THE NEXT PAGE >

Part B

Time: 40 minutes

NO CALCULATORS MAY BE USED FOR PART B

Answer Question 4 below. The Section II score weighting for this question is 10 percent.

4. For each of the following three reactions, write a BALANCED equation in part (i), and answer the question about the reaction in part (ii). Coefficients should be in terms of lowest whole numbers in part (i). Assume that solutions are aqueous unless stated otherwise. Represent substances in solution as ions if the substances are extensively ionized. Omit formulas for any ions or molecules that are unchanged by the reaction.

    Example: *A strip of magnesium is added to a solution of titanium.*

    i. $Ti^{2+} + Mg \rightarrow Ti + Mg^{2+}$

    ii. *Which substance is reduced in the reaction?*

    Answer: *Titanium (Ti) is reduced.*

    (a) Propane is combusted in excess air.

    　i. Write a balanced equation for this reaction.

    　ii. How many moles of CO_2 are produced for each mole of C_3H_8 burned?

    (b) Sulfur dioxide gas is bubbled through water.

    　i. Write a balanced equation for this reaction.

    　ii. Explain how pollution with sulfur dioxide is detrimental to lakes and rivers.

    (c) Sodium chloride and copper (II) sulfate are mixed.

    　i. Write a balanced equation for this reaction.

    　ii. Identify the spectator ions in the mixture.

Answer Question 5 and Question 6. The Section II score weighting for these questions is 15 percent each.

Response to these questions will be graded on the basis of accuracy and the relevance of information cited. Make your explanations clear and well organized. You may include examples and equations in your responses where appropriate. Specific responses are preferable to broad, diffuse responses.

$$H_2(g) + Br(g) \rightarrow H(g) + HBr(g) \ E_a = 78.5 \text{ kJ}$$

5. Answer the following questions about the reaction shown above, which occurs in a single step.

    (a) Draw an energy diagram for this reaction given that the transition state occurs closer to the products on the reaction pathway and that the $\Delta H°$ for the forward reaction is 69.7 kJ.

    (b) Calculate E_a for the reverse reaction.

    (c) Will the forward or reverse reaction have a higher rate constant? Justify your answer.

    (d) Write a rate equation for the reaction shown above.

    (e) How would an increase in pressure of the H_2 and Br gases affect the rate of the reaction shown above? Justify your answer.

    (f) How would the H_2 and Br pressure increase affect the equilibrium constant (K)? Justify your answer.

GO ON TO THE NEXT PAGE ⟹

6. Answer the following questions using chemical principles of reactions, atomic structure, and/or chemical bonding.

$Fe^{3+}(aq) + SCN^{-}(aq) \rightarrow FeSCN^{2+}(aq)$

(colorless) (colorless) (red)

(a) Predict the change in color, if any, that would occur if $FeCl_3$ (colorless) is added to the above reaction at equilibrium. Justify your answer.

(b) Explain why Na and Ca always have positive oxidation numbers in compounds, but the oxidation number of carbon can vary from –4 to +4.

(c) Which substance would you expect to have a higher boiling point, Ar or Xe? Justify your answer.

(d) Which substance would you expect to have a greater vapor pressure under standard conditions, H_2O or ethyl ether ($C_2H_5OC_2H_5$)? Justify your answer.

(e) Explain why the ionization energy for Group 1A elements decreases as the elements increase in atomic number in the periodic table.

IF YOU FINISH BEFORE TIME IS CALLED, YOU MAY CHECK YOUR WORK ON THIS SECTION ONLY. DO NOT TURN TO ANY OTHER SECTION IN THE TEST.

STOP

Practice Test Two: **Answer Key**

1. B	26. B	51. E
2. C	27. A	52. C
3. A	28. C	53. B
4. B	29. A	54. A
5. D	30. E	55. B
6. B	31. A	56. D
7. C	32. C	57. C
8. B	33. E	58. D
9. C	34. E	59. A
10. E	35. B	60. C
11. D	36. D	61. D
12. E	37. A	62. C
13. B	38. A	63. C
14. C	39. E	64. B
15. A	40. C	65. B
16. C	41. D	66. C
17. D	42. E	67. C
18. E	43. B	68. D
19. C	44. C	69. D
20. B	45. B	70. A
21. B	46. C	71. B
22. E	47. D	72. D
23. A	48. D	73. A
24. C	49. D	74. B
25. B	50. E	75. E

ANSWERS AND EXPLANATIONS

SECTION I

1. B

Argon (Ar) is a member of the noble gas family of elements. All the members of this family have filled valence orbitals, are unreactive, and have all electrons paired.

2. C

Carbon, choice (C), has four valence shell electrons.

3. A

$$:\overset{..}{O}=C=\overset{..}{O}:$$

In the Lewis structure of carbon dioxide, each O atom has a double bond to the central C atom. Because the molecule has a center line of symmetry through the carbon atom, the geometry is linear.

4. B

There are just three pairs of electrons around the central boron in BF_3. To achieve maximum separation of these three mutually repelling electron regions, a fluorine-to-boron-to-fluorine bond angle of 120° is predicted. This is why BF_3 has a triangular planar geometry. In NH_3, however, there are four pairs of electrons: three bonded pairs and one lone pair. This determines the triangular *pyramidal* geometry. Although the predicted hydrogen-to-nitrogen-to-hydrogen bond angle is 109.5°, the repulsion of the lone pair with the bonded pairs closes the angle a bit so that the experimentally measured angle is 103°.

5. D

Each nitrogen atom has five valence electrons. Each nitrogen contributes three valence electrons to form a triple bond. A lone pair of electrons is left on each nitrogen. A triple bond formed between the two nitrogen atoms allows each atom to have an octet, which makes this a stable molecule.

6. B

Carbon dioxide has the greatest density of the gases listed because it has the greatest molar mass. To calculate density, you divide the molar mass by the volume. Since gases have the same molar volume at standard conditions (roughly 22.4 L), the gas with the greatest molar mass will also be the most dense. The molar mass of carbon dioxide is 44.010 g/mol and its density is roughly 1.96 g/L.

7. C

Graham's Law of Effusion states that the rates of effusion of two gases at standard conditions are inversely proportional to the square roots of their densities. To put it simply, the gas with the lowest molar mass will have the fastest effusion rate.

8. B

Let's say you measured out 19.0 mL using a pipette. This volume has three significant figures. Now you weigh the liquid and get a reading of 18.004 grams. Hence, the mass of the liquid has five significant figures. When you do your density calculation, you get $\frac{18.004}{19} = 0.94757894$ g/mL.

Remember, however, that your final answer is limited to the same number of significant digits as the smallest of your incorporated figures. Since the volume measurement has only three significant figures, your final answer can have only three significant figures as well. The correct answer is 9.48×10^{-1} g/mL.

9. C

Potassium hydroxide is a strong base because it completely dissociates into K^+ and OH^- in an aqueous solution. The only other base listed, NH_3, is a weak base with a pK_a of 4.75 at 25°C.

10. E

The only substance with all of the properties listed is K_2CO_3, a pure, white crystalline powder that dissolves quickly in water to form a basic solution that liberates a gas when a strong acid is added to it.

11. D

Colligative properties of solutions depend on the number of solute particles. Colligative properties of solutions include decreasing freezing points, increasing vapor pressures, and increasing boiling points.

12. E

In a combustion reaction, a substance reacts rapidly with oxygen to produce one or more oxides. A common example of a combustion reaction is the burning of gasoline in a car engine. If the substance being combusted is a hydrocarbon or carbohydrate, the products are *always* CO_2 and H_2O. The combustion reaction shown here shows the burning of ethanol.

13. B

A formation, or synthesis, reaction occurs when a compound is formed from its elements. Here, iron chloride is formed from solid iron and chloride gas.

14. C

A single-displacement reaction is a reaction where one element replaces another element in a compound. In this case, zinc replaces silver to form zinc nitrate.

15. A

This reaction shows sulfuric acid being neutralized by the strong base, sodium hydroxide. The resulting solution of water and sodium sulfate (a salt) is

neutral. The general neutralization reaction is: acid + base = salt + water.

16. C

$CaCO_3$ is least likely to be soluble. Based on the solubility rules, compounds containing CO_3^{2-} and PO_4^{3-} are mostly insoluble while compounds containing Cl^-, Br^-, I^-, NO_3^-, Li^+, Na^+, K^+, and NH_4^+ are mostly soluble.

17. D

Standard temperature and pressure refers to 0°C and 1 atm, where 1 mol of a gas = 22.4 L.

18. E

Entropy of a system is a measure of the disorder or degree of freedom; therefore, where molecules have more disorder, they have higher entropy. For example, two moles of a substance has a higher entropy than one mole, gases have a higher entropy, or degree of freedom, than liquids, and liquids have a higher entropy than solids.

19. C

CH_3COOH, or acetic acid, is commonly used as an example of a weak acid, which is the key to answering this question. HCl, a strong acid, will ionize completely in aqueous solution. Using LaChatelier's principle, the H^+ produced from the ionization of HCl will suppress the ionization of CH_3COOH.

20. B

When a solution shows the least amount of change in pH over the greatest range of volume of NaOH added, it is acting as a buffer solution. On the graph of pH vs. volume of NaOH added, this range appears as the part of the curve with the smallest slope, from pH 4 to pH 5. (The section of the curve between pH 11 and 12.5 does not represent a buffering region, but the dilution of NaOH.) Choice (B), pH 4.7, is the only answer that falls into the buffering range of the solution. The answer (pH 4.7) is also the pK_a of the weak acid that is being titrated.

21. B

Remember that the solubility of a gas decreases as temperature is increased, the pressure of the gas above the solution increases as the gas is driven out of the solution, and the vapor pressure of water will increase in response.

22. E

Based on the Pauling electronegativity scale, electronegativity increases from left to right across a period. Similarly, it decreases from top to bottom down a group. Looking at the periodic table, oxygen is the element that is farthest to the right and closest to the top of the periodic table of the choices given. The most electronegative element is fluorine.

23. A

As the pressure increases, liquid becomes a more stable state than gas, so the boiling point for water increases. Boiling points are affected by pressure so are usually reported as "normal boiling points" for the boiling point of the substance at 1 atm pressure.

24. C

A phase diagram plots temperature on the x-axis and pressure on the y-axis, then indicates the different phases that exist at each combination of temperature and pressure. There are many temperature and pressure combinations where liquids and solids, liquids and gases, and gases and solids are in equilibrium with each other. On a phase diagram, these equilibriums are represented with lines. There is only one temperature and pressure combination

where all three phases are in equilibrium, and this is called the "triple point." The triple point is different for different substances. The triple point for water is 0.01°C and 0.0060 atm. The triple point for CO_2 is −57°C and 5.2 atm.

25. B

With the temperature held constant at 0°C, draw a vertical line up through the plot to see that the only phase available above 1 atm is the liquid phase. The supercritical fluid is only available at high temperatures and high pressures (the upper right corner). Equilibrium of all three phases only occurs at the triple point, which is below 1 atm pressure.

26. B

The oxidation number for oxygen is usually −2, except when it is bound to itself as in hydrogen peroxide, H_2O_2, H-O-O-H. Since H_2O_2 is an uncharged molecule, the oxidation numbers must add up to zero. Each hydrogen has an oxidation number of +1, leaving each oxygen atom with an oxidation number of −1.

27. A

For this question, you only need the 0.5 molar and 500 mL values. First, convert 500 mL to liters by dividing by 1,000 to get 0.5 L. Since 0.5 molar means 0.5 moles/liter, just multiply 0.5 moles/liter × 0.5 liters to obtain 0.25 moles.

28. C

Density is calculated by dividing mass (grams) by volume (mL). The displacement of water measures the volume of the artifact. The mass is estimated by weighing the artifact. Weight and mass are not equivalent since weight is the amount of gravitational pull exerted on an object and mass is the amount of matter in the object. For example, the mass of a person is the same on the earth and the moon, but the weight of the person is less on the moon since there is less gravitational force. Every substance has a characteristic density, so this measurement can be used to help identify the substance.

29. A

Propanol is the only choice that does not contain any charged ions. Saturated hydrocarbons are uncharged and are composed of covalent C-H bonds with fairly equal electron sharing, which means little to no dipole. Therefore, hydrocarbons won't mix well with charged compounds. Choices (C), (D), and (E) are all salts, so each is made up of one positively and one negatively charged ion. Choice (D) is the salt form of ethanol. Ethanol (CH_3CH_2OH) is uncharged and mixes with octane, but the salt form is made up of an ionic bond between Na^+ and $^-OCH_2CH_3$. Choice (B) is a positively charged ion, so it won't mix with octane.

30. E

Since the enthalpy of reaction is positive, the reaction is endothermic, meaning it absorbs energy from the surroundings as the reaction proceeds as written from left to right. Heat is not released and the temperature will not rise as the reaction proceeds; in fact the opposite will occur. The reaction shows an increase in entropy since there is one reactant going to two products.

31. A

Although the effect of temperature on a reaction rate is not precisely predictable, the general effect is that the reaction rate will increase with increasing temperature. Since reactions are dependent

on collisions of sufficient energy, and higher temperatures lead both to a greater total number of collisions and to more energetic collisions, the result of an increase in temperature is an increase of reaction rate. Reactions that are dependent on catalytic enzymes that are protein compounds are exceptions to this rule. Enzymes often have a preferred temperature range for maximum activity and may be denatured at higher temperatures, and reaction rates decrease with any variation from this temperature. Even with this exception, (A) is still the best choice.

32. C

The overall reaction order is determined by adding the exponents of the rate law, so $1 + 1 = 2$. The reaction is first order with respect to hydrogen peroxide and first order with respect to the iodide ion, but it is second order overall.

33. E

Reaction rate is a function of the moles or mass units per unit time per unit volume that is formed or removed in any given reaction. The moles or mass, time, and volume have to all be in the definition.

34. E

Choices (A) through (D) are all good safety practices when working with strong acids. Adding a strong base to a strong acid will neutralize the acid, but the reaction is highly exothermic and may result in an explosion. Therefore, (E) is the best choice.

35. B

The first step is to balance the equation, which gives:

$$2\,Al(s) + 6\,HNO_3(aq) \rightarrow 2\,Al(NO_3)_3(aq) + 3\,H_2(g)$$

With this equation, 1 mole of Al(s) would produce 1.5 moles of $H_2(g)$.

36. D

In a mixture of gases, the sum of the partial pressures of each gas must equal the total pressure for the mixture. In this example, the total pressure (288 atm) $- (2 \times 37$ atm$) = 214$ atm.

37. A

Both I and II would be good buffers at pH 7.2, since each contains a component with a pK_a in that region. However, choice I would have the greater buffering capacity at pH 7.2 because the concentration of the buffer is 1 M as compared to 1 mM, providing 1,000 times the buffering capacity of choice II. Choice III is a poor buffer because strong acids and bases are completely ionized in solution and cannot act as buffers.

38. A

Zn(s) and Cu(s) are understood to be elements in solid metal form and have an overall charge of zero. $CuSO_4$ and $ZnSO_4$ represent the ionic forms of each metal, so should be read as Cu^{2+}, SO_4^{2-}, and Zn^{2+}, SO_4^{2-}. The second equation shows Zn releasing two electrons to Cu^{2+}, forming Zn^{2+} and Cu. The first equation shows that the reverse will not happen. Another way to state this is that Zn(s) will reduce Cu^{2+}, but Cu^{2+} will not reduce Zn. When this occurs, it is said that Zn(s) is more reactive than Cu(s).

39. E

A beta particle (represented by $_{-1}^{0}e$) is emitted when an unstable nucleus emits an electron. As a result, a neutron is converted into a proton. Anytime the number of protons changes in the nucleus, a new element is created. In this case, carbon 14 is converted

to nitrogen 14. In the notation shown for carbon 14 ($_{6}^{14}C$), 14 is the mass number, and 6 is the number of protons. The word *emission* in a nuclear reaction always means that the named particle is a product, not a reactant.

40. C

Choice (A) can be written as CH_3-CH_2F; both carbons are sp^3-hybridized. Choice (B) is CH-CH, each carbon is sp-hybridized. Choice (C) is CH_2-CH_2; each carbon is sp^2-hybridized. Choice (D) is CH_3-CH_3; each carbon is sp^3-hybridized. Choice (E) is CH_3-CH_2OH; each carbon is sp^3-hybridized.

41. D

The question only asks for the electron configuration of the outer shell of zinc. Using the periodic table as a guide, zinc is in period 4, so the $4s$ is its outer s orbital, and it is in the d-block group, a transition metal, so it has d electrons in its outer orbital as well. The fourth period transition metals contain $3d$ electrons as their outer shell. Since zinc is the last element in the d-block, its d orbital is full, so the configuration is $4s^23d^{10}$.

42. E

Different chemistry texts use different methods to balance redox reactions. While the methods appear similar, you should remember that balancing a reaction requires that you pick a single method and stick with it all the way through the problem. Mixing and matching steps between different methods will not work.

To balance the following reaction correctly, use the half reaction method. First, assign an oxidation number to each element in the equation. For K(s), the oxidation number is 0 since it is the atom of a free element. For Cr^{3+}, the oxidation number is simply +3, given the rule that the oxidation state of a monoatomic ion is equal to its charge. Using these same rules, the oxidation state of $Cr(s)$ is 0 and $K^+(aq)$ is +1. Given these oxidation states, potassium has been oxidized by one electron and chromium has been reduced by three electrons. This is more easily seen in the half reactions below:

$$K(s) \rightarrow K^+ + 1\ e^-$$

$$Cr^{3+}(aq) + 3\ e^- \rightarrow Cr(s)$$

The next step is to make the number of electrons in each half reaction equal by multiplying:

$$3\ K(s) \rightarrow 3\ K^+ + 3\ e^-$$

$$Cr^{3+}(aq) + 3\ e^- \rightarrow Cr(s)$$

Then add the two half reactions together. Since the electrons are on different sides of the arrow, they will drop out, giving you the balanced reaction:

$$3\ K(s) + Cr^{3+}(aq) \rightarrow Cr(s) + 3\ K^+(aq)$$

It is almost always true that in a multiple-choice-format redox balancing problem, eliminating wrong answers will be faster than balancing the reaction from scratch. There is a shortcut you can frequently use in selecting a balanced redox reaction in a multiple-choice question: Check the total charge in each equation given. Typically, all but one or two of the choices will have an unbalanced charge. In this question, choice (E) is the only equation in which the charge on each side of the equation is balanced (+3 on each side). (In a case where two of the options have a balanced charge, first check to see that both reactions have their lowest whole-number coefficients. If that doesn't identify the right equation, check O next, then H. If that doesn't distinguish between the two, check the remaining atoms until

you find the unbalanced reaction.) Once you eliminate the unbalanced equations, the remaining one is the right answer.

43. B

The equation for determining density is density = mass (g) / volume (mL). The density was found to be 19.32 g/mL. The volume, 0.0100 L, must be converted to 10.0 mL since the volume must be in the same units as provided in the density value. Rearranging the equation to find mass gives you:

$$mass\ (g) = density \times volume\ (mL)$$

$$mass\ (g) = 19.32\ g/mL \times 10\ mL = 193.0\ g$$

44. C

This problem can be done without using an equation if you remember that the definition of pK_a is that the unprotonated and protonated species are present at equal concentrations when $pH = pK_a$ and are at a ratio of 1:1. The equation that represents this relationship is:

$$pH = pK_a + \log \frac{[A^-]}{[HA]}$$

Working through this equation:

$$4.8 = 4.8 + \log \frac{[CH_3COO^-]}{[CH_3COOH]}$$

$$0 = \log \frac{[CH_3COO^-]}{[CH_3COOH]}$$

$$1 = \frac{[CH_3COO^-]}{[CH_3COOH]}$$

which is the same as $\dfrac{[CH_3COO^-]}{[CH_3COOH]} = 1:1$.

45. B

The right equation to use here is:

$$K_a = \frac{[H+][A^-]}{[HA]}$$

You are given the K_a value and can calculate the $[H^+]$ ion concentration from the pH, but this still leaves two unknowns. You can calculate the ratio of $\frac{[A^-]}{[HA]}$ but not the separate values of each.

46. C

The equation shows Cl_2 being reduced by Br^-, forming Cl^- and Br_2. Since this is a redox reaction, the easiest way to see this is to look at the oxidation numbers of each species; in Cl_2 and Br_2, the oxidation state of each element is 0, while in Cl^- and Br^-, the oxidation state of each element is -1. For this reaction to occur as written, Cl_2 must gain electrons more easily than Br_2. If this were not the case, we would see no reaction when Cl_2 was mixed with NaBr. Since Cl is gaining electrons in this reaction, choice (A) cannot be true. Since this reaction is going forward as written, choices (B), (D), and (E) cannot be true.

47. D

For this problem, use the equation $\Delta H^\circ = \Delta H^\circ_f(\text{products}) - \Delta H^\circ_f(\text{reactants})$, where ΔH°_f is the standard enthalpy of formation. Since O_2 is at its standard state, its enthalpy of formation is zero. To calculate the answer:

$$(-396 \text{ kJ/mol}) - (-297 \text{ kJ/mol}) = -99 \text{ kJ/mol}$$

48. D

A reaction is spontaneous when ΔG is negative. To calculate, $\Delta G = \Delta G(\text{products}) - \Delta G(\text{reactants})$, where $G = $ Gibbs free energy. In the free energy diagram shown, the product is at a higher free energy than the reactant, so the ΔG of reaction would be positive, and not spontaneous. The reverse reaction, however, would be spontaneous. The diagram doesn't show anything about entropy or enthalpy of the reaction. ΔG is related to entropy (S) and enthalpy (H) by the equation $\Delta G = \Delta H - T\Delta S$, but we only have information on ΔG in this problem. The reaction is not shown to be proceeding spontaneously towards equilibrium, since that type of reaction would have a "U"-shaped line as you proceed from R to P in the free energy diagram, with the bottom of the "U" representing the equilibrium state, and $\Delta G = 0$ at that point.

49. D

To calculate ΔG for the reaction, calculate ΔG°_f (free energy of formation) for the products and subtract ΔG°_f for the reactants:

$$\Delta G \text{ reaction} = (-237.1 \text{ kJ}) - (-58.54 \text{ kJ})$$
$$= -178.6 \text{ kJ}$$

Note that if ΔG at any other temperature were requested, this method could not be used, because of the strong temperature dependence of ΔG. Unlike values of ΔH°_f and S°_f, which do not change much over a fairly wide range of temperatures, the values of ΔG°_f given in tables are only valid at the reported temperature.

50. E

The easiest way to solve this problem is to realize that the ΔS° will be negative for this reaction. Since two reactants are forming one product and disorder is decreasing, entropy will be negative. Only one of the answer choices is negative. To calculate the answer, memorize $\Delta G^\circ = \Delta H^\circ - T\Delta S^\circ$, with $T = $

temperature in Kelvin. Substituting the values given to you in the question and 303 K (273 + 30°C) for T, you should get −0.185 kJ/mol·K. A common mistake is to forget to transfer the negative sign for $(-T\Delta S°)$.

Remember that the people who write the test know you don't have a calculator. If it looks like you will have to do a time-consuming calculation by hand, look for a way out. There may not always be one, but often there is.

51. E

Fusion reactions involve the combination of two lighter nuclei to form one heavier nucleus and energy. Choices (A), (B), and (C) show a neutron (no proton) combining with a nucleus. Choice (D) shows a fission reaction, which is the opposite of a fusion reaction. Choice (E) is the only correct answer, showing the combination of two nuclei.

52. C

This is essentially a stoichiometric problem. Since you are given an amount in grams and must give an answer in moles, you will need to convert grams to moles using atomic mass. The atomic mass of zinc is found in the periodic table as 65.39 g/mole. The other information you need is the ratio of moles of zinc:moles of $H_2(g)$ in the reaction. First, check to see that the equation is balanced. In this case, the equation is already balanced, so one mole Zn(s) produces one mole $H_2(g)$. To solve the problem, set up ratios that allow units to cancel, with the only remaining units being moles H_2 as shown below. To solve this equation, divide 30 g by 65.39 g. Since calculators are not allowed on this portion of the test, it is easiest to first approximate the answer and see which choice is closest to your approximation. If only one answer is close to the approximation,

which is often the case, you can save yourself some time and some long division. For example, 30 is approximately half of 65.39, so 30 divided by 65.39 will be slightly less than 0.5. In this case, 0.46, or choice (D), is the only close answer and is correct.

$$30 \text{ g } \cancel{Zn} \times \frac{1 \text{ moles Zn}}{65.3 \text{ g } \cancel{Zn}} \times \frac{1 \text{ moles } H_2}{1 \text{ moles Zn}} =$$
0.46 moles H_2

53. B

An ester linkage is R_1-CO-OR_2, which is shown in choice (B). The placement of the R groups is very important in this formula. R groups represent a carbon-containing group. If the R_2 group was an H (as it is in choice (A)), it would no longer be called an ester linkage. In the formula shown in choice (B), the ester bond is between the oxygen and the R_2 group. Choices (C) and (D) are ethers, and choice (E) is a ketone.

54. A

This problem can be solved using stoichiometry. The key to solving the problem is being able to set up ratios based on the coefficients in the balanced equation. Here are the equations needed to solve the problem:

$$5 \text{ moles Al} \times \frac{2 \text{ moles AlCl}_3}{3 \text{ moles Cl}_2} = 5 \text{ moles AlCl}_3$$

$$6 \text{ moles Cl}_2 \times \frac{2 \text{ moles AlCl}_3}{3 \text{ moles Cl}_2} = 4 \text{ moles AlCl}_3$$

Given the equations above, starting with five moles of Al and six moles of Cl_2 makes Cl_2 the limiting reagent. The maximum number of moles of product ($AlCl_3$) that can be produced is four moles of $AlCl_3$.

The concept of a limiting reagent can be a confusing one. Another way to think about it is to imagine pairing tires to cars. If you start with 24 tires and eight car bodies, you can only produce a maximum of six cars with tires, since each car needs four tires. In this example, the car tire is the limiting reagent since you run out of tires before you run out of car bodies. The number of car tires determines the maximum number of cars with tires that can be produced. If you started with 24 tires and five car bodies, you could only produce five cars with tires, and the car body becomes the limiting reagent.

55. B

The oxidation number for oxygen is usually −2, so if you start there, the overall contribution to oxidation number from O is −4. Since the molecule is uncharged, carbon must offset the oxygen oxidation number with +4 so that −4 (2 O) + +4 (C) = 0 (CO_2).

56. D

Group 3A includes elements like boron (B). When drawing the Lewis dot structure for B, there are three electrons in the valence shell. To complete an octet, B needs to give up three electrons, making it +3. Looking at the electrons in the valence shell for each element can help you predict the typical oxidation number for an element, but it cannot be used as a rule.

57. C

For this question, use the ideal gas law, $PV = nRT$. The pressure is 1 atm, the volume is 24 L, and the temperature (in Kelvin) is 273 + 37 = 310 K. Plugging these numbers in to the equation, along with the gas constant, R, and solving for n, you obtain:

$$1 \text{ atm} \times 24 \text{ L} = n \, (0.0821 \text{ L·atm/mol·K}) \times 310 \text{ K}$$

$$n = 0.94 \text{ moles}$$

Since you cannot use a calculator, the easiest way to solve this equation is to round the numbers and approximate the answer. Multiplying 300 by 0.082 = 24.6. Next, dividing 24 by 24.6 will be approximately, but slightly less than, 1. The only answer close to 1 is (C), 0.94 moles.

58. D

This is essentially a stoichiometric problem. First, the equation for mass percent is:

$$\frac{\text{mass solute}}{\text{mass solution}} \times 100 = \text{mass percent}$$

When the concentration of a solution is given in a percent, you can immediately convert it into a mass ratio by filling in the mass solution as 100 g. See below:

$$20.5\% \text{ solution} = \frac{20.5 \text{ g solute}}{100 \text{ g solution}}$$

The stoichiometric equation should look like:

$$10 \text{ g sucrose} \times \frac{100 \text{ g solution}}{20.5 \text{ g sucrose}} \times \frac{1 \text{ mL}}{1 \text{ g}} = 48.78 \text{ mL}$$

Again, since you can't use a calculator, round 20.5 to 20. Then divide 100 by 20 to get 5, and 10 g × 5 × 1mL/1g = 50 mL. Choice (D) is the only answer close to 50 mL.

59. A

A peptide bond is the covalent bond that forms between two amino acids. The bond is formed between the amine group NH_3^+ of one amino acid and the carboxylic acid of the other amino acid in a condensation reaction that releases water (as shown below).

$$R\text{-}NH_3^+ + {}^-OOC\text{-}R \rightarrow R\text{-}NH\text{-}CO\text{-}R + H_2O$$

Choice (A) shows the peptide bond in bold between the nitrogen (N) of one amino acid and the carbon (part of carbonyl group) of the second amino acid.

Choices (B) and (C) also contain peptide bonds, but different bonds in the molecule are in bold. Choices (D) and (E) do not contain peptide bonds.

60. C

The easiest way to solve these problems is to think them through as shown in the following table:

Number of Moles	Number of Half-Lives	Time
2.4	0	0
1.2	1	1 min
0.6	2	2 min
0.3	3	3 min

There are 0.3 moles left of the sample after three half-lives have passed, which is equal to three minutes.

61. D

A catalyst will lower the activation energy for the reaction, usually by providing a different mechanism than the uncatalyzed reaction, but the mechanism is not always simpler. For example, an enzyme-catalyzed reaction usually contains several additional mechanistic steps compared to the uncatalyzed reaction, but the activation energy is lower. The catalyzed reaction will increase in rate due to the lower activation energy. By definition, a catalyst never changes the equilibrium constant for the reaction or the overall change in free energy for the reaction.

62. C

To solve this problem, first determine the oxidation numbers for each atom, then determine which oxidation numbers are changing as you proceed from the reactants to products. You do not need to balance the equation to answer this question, but you will be asked to balance the equation in the next question.

Assigning oxidation numbers:

MnO_4^-: Oxygen is usually -2, and $-2 \times 4 = -8$, and since the overall charge on the molecule is (-1), Mn must be $+7 = Mn^{7+}$. $H^+ = +1$ since the oxidation number equals the charge of the ion.

$H_2C_2O_4$: Again, assign the oxygen as -2, the hydrogen as $+1$, and the overall charge of the molecule as 0. Then solve for the carbon, since it has the most flexible oxidation number in the compound:

$0 = 2(+1) + 2(C) + 4(-2)$; solving for C gives you $+3$.

$= H^{1+}, C^{3+}, O^{2-}$

Using this same method on the other side of the equation:

$Mn = +2$

$CO_2 = C^{4+}O^{2-}$

$H_2O = H^{1+}O^{2-}$

The only elements that change in oxidation number from reactants to products are Mn and C. Mn decreases in oxidation number from $+7$ to $+2$, indicating a reduction of five electrons. C increases in oxidation number from $+3$ to $+4$, indicating an oxidation of one electron. Therefore, the answer is choice (C); only Mn is reduced.

63. C

To balance the equation, use the oxidation numbers that you determined in question 62. Mn is reduced by five electrons, and C is oxidized by one electron.

Therefore:

$$Mn^{7+} + 5\,e^- \rightarrow Mn^{2+}$$

$$5\,C^{3+} \rightarrow 5\,C^{4+} + 5\,e^-$$

$$\overline{Mn^{7+} + 5\,C \rightarrow Mn^{2+} + 5\,C^{4+}}$$

The coefficients of Mn and C are provided by balancing the oxidation-reduction reaction as shown above. Mn and C must be present in a 1:5 ratio in the balanced equation as shown below:

$$2\,MnO_4^-(aq) + 6\,H^+(aq) + 5\,H_2C_2O_4(aq) \rightarrow$$
$$2\,Mn^{2+}(aq) + 10\,CO_2(g) + 8\,H_2O(l)$$

Therefore, the correct coefficient for oxalic acid is 5, or choice (C).

64. B

The strongest intermolecular interaction listed is the ion-ion interaction, between a positively charged ion and a negatively charged ion. The weakest interaction occurs when one uncharged particle induces a temporary dipole on a closely neighboring particle that gives rise to a temporary attraction, called dipole-induced dipole interaction. This interaction is also referred to as dispersion forces or London forces. An example of each type of interaction is choice (A), two water molecules; choice (B), two molecules of O_2; choice (C), Cl^- and water; choice (D), Na^+ and Cl^-; choice (E), water and O_2.

65. B

When $\Delta G = 0$, the reaction is at equilibrium, meaning that the concentrations of products and reactants are not changing over time. When ΔG is negative, the reaction is said to be spontaneous, but that does not provide any information about the rate of the reaction. When the reaction is at equilibrium, it proceeds in the forward and reverse directions at equal rates.

66. C

The activation energy (E_a) for the forward reaction is the energy difference between the reactant and the transition state, which is shown with the C arrow. The arrow designated E is the activation energy for the reverse reaction. Arrow B is the difference in energy between reactants and products. Arrows A and D are arbitrarily drawn arrows and are not relevant except that when you add the energy of A and D, the result equals the activation energy in the forward direction.

67. C

The burning of octane is a combustion reaction. Combustion reactions for hydrocarbons are written as:

$$Hydrocarbon + O_2 \rightarrow CO_2 + H_2O$$

when the hydrocarbon is burned in excess oxygen. For octane, the reaction is:

$$2\,C_8H_{18} + 25\,O_2 \rightarrow 6\,CO_2 + 18\,H_2O$$

Therefore, one mole of octane gas, as a limiting reagent, would produce nine moles of H_2O.

68. D

To solve this dilution problem, memorize the formula $M_1V_1 = M_2V_2$. Substituting the values from the question gives us:

$$12\,M \times Volume\ B = 2\,M \times 0.5\,L$$
$$Volume\ B = \frac{1.0\ mol}{12\ mol/L} = 0.083\,L = 83\ mL$$

Again, you could round the numbers to estimate the answer. Rounding 12 down to 10 would give you 100 mL as the answer. Choice (D) is the closest to 100 mL.

69. D

First, *molal* refers to the molality of solution, or "moles solute per kg solvent." Since NaCl splits into Na^+ and Cl^- ions in solution, every one mol of NaCl gives two mol of particles in solution.

Therefore, a 0.2 molal solution of NaCl contributes a total particle concentration of 0.4 molal. Using the freezing point depression constant for water, 1.86°C kg/mol, the equation for freezing point depression can be set up as shown below:

$$\frac{0.4 \text{ mol}}{1 \text{ kg}} \times \frac{1.86°C \text{ kg}}{1 \text{ mol}} = 0.74°C$$

Since water freezes at 0°C, the estimated freezing point of the 0.2 molal solution is −0.74°C. A common error is giving an elevated, rather than depressed, freezing point; remember that freezing points are always lower for solutions than for pure solvents, while boiling points are always higher.

70. A

A negative heat of solution means that heat will be released to the surroundings when LiCl is dissolved in water, so choice (A) is true. None of the other statements have anything to do with the heat of solution; additionally, choice (B) is false because in water, Li^+ separates from Cl^- and each ion is surrounded by a shell of water. This is also the reason that choices (D) and (E) are false. Choice (C) is false because LiCl is an ionic compound and will not dissolve in a nonpolar solvent like benzene.

71. B

From the name, you can determine that *hexose* is a sugar (from *-ose*) that contains six carbons (*hex-*). All of the molecules shown are sugars, but the only molecule shown with six carbons is choice (B). Counting the carbons in the other choices, choice (A) has five carbons, choice (C) has five carbons, choice (D) has five carbons, and choice (E) has seven carbons. Choice (E) is tricky since it is a six-membered ring and this form is often used to

represent hexoses, but if you count up the carbons, there are seven, not six.

72. D

First, separate each salt into its ion components. Then, determine the activity of each ion component in water. For each of the choices, only one ion will react with water and affect the pH of the solution. If the ion will accept a proton from water, the solution will be basic. If the ion will donate a proton to water, the solution will be acidic. In choice (A), the acetate ion (CH_3COO^-) will be protonated by water, so the acetate ion acts as a base. In choice (B), the CO_3^{2-} ion will react with water to form HCO_3^-, and the solution will be basic. In choice (C), CN^- will accept a proton from water to form HCN, making the solution basic. In choice (D), the component ions are Cl^- and $CH_3NH_3^+$, which acts like an ammonium ion and will donate a proton to water, making the solution acidic. In choice (E), the PO_4^{3-} ion will accept a proton from water, again making the solution basic. The only choice that results in an acidic solution is choice (D), shown in the reaction below:

$Cl^- + H_2O \rightarrow$ no reaction, since Cl^- is the anion of the strong acid, HCl

$CH_3NH_3^+ + H_2O \rightarrow CH_3NH_2 + H_3O^+$

73. A

Choice (A), nitric acid, is a strong acid and, by definition, completely dissociates into H^+ and NO_3^- in water. Choices (B)–(E) are all weak acids and will dissociate into H^+ and their weak base anion to different degrees depending on their equilibrium constants, K_a.

74. B

You can quickly solve this problem by recognizing that the $3d$ orbitals are full. Begin with the $4s^2$ orbital and use the periodic table to guide you. The first elements that use the $4s^2$ orbitals are K and Ca; then move across the transition metals (since the $3d^{10}$ orbitals are filled). Starting in on the $4p$ orbitals, count four elements over to Se, which ends with $4p^4$.

75. E

Choice (E) has the largest atomic radius at $Cl^- = 181$ picometers (1 m $= 10^{-12}$ pm). Using trends in the periodic table, atomic radii increase across periods for neutral atoms. Losing an electron results in a decrease in radius, while gaining an electron results in an increase in radius. F^- and Cl^- both gain electrons and have larger atomic radii, but Cl^- is in period 3 and F^- is in period 2, so Cl^- has the larger radius. The values of the atomic radii for the other choices are: $F^- = 152$ pm, Li $= 152$ pm, $Li^+ = 68$ pm, Cl $= 99$ pm, and $Cl^- = 181$ pm.

SECTION II: FREE-RESPONSE QUESTIONS

PART A

1. (a) We first determine the K_{a1} from pK_{a1}:

$$pK_{a1} = 2.329 = -\log(K_{a1})$$

$$-2.329 = \log(K_{a1})$$

$$10^{-2.329} = K_{a1}$$

$$K_{a1} = 4.69 \times 10^{-3}$$

Similarly, we can calculate the K_{a2} of HL = 1.79×10^{-10}.

The dissociation constant of H_2L^+ is a weak acid, and HL is an even weaker acid. As such, we can treat H_2L^+ as a monoprotic acid. Writing the mass action equation for this acid:

$$H_2L^+ \rightleftharpoons HL + H^+$$

	$[H_2L^+]$ (M)	[HL] (M)	$[H^+]$ (M)
Initial concentration	0.050	0	0
Change	$-x$	$+x$	$+x$
Final concentration	$0.050 - x$	x	x

$$K_{a1} = \frac{(x)(x)}{0.050 - x}$$

$$4.69 \times 10^{-3} = \frac{x^2}{0.050 - x}$$

By rearranging and solving the quadratic equation:

$$x^2 = (4.69 \times 10^{-3})(0.050 - x)$$

$$x^2 + (4.69 \times 10^{-3})x - (2.345 \times 10^{-4}) = 0$$

$$x = [HL] = 1.31 \times 10^{-2} \text{ M}$$

(b) $x = [H^+] = 1.31 \times 10^{-2}$ M

The pH can now be calculated using pH = $-\log(1.31 \times 10^{-2}) = 1.88$

(c) $[H_2L^+] = 0.050 - x = 3.69 \times 10^{-2}$ M

(d) $HL \rightleftharpoons H^+ + L^-$

$$K_{a2} = \frac{[L^-][H^+]}{HL}$$

However, since $[H^+] = [HL]$, these two terms cancel out and:

$$K_{a2} = [L^-] = 1.79 \times 10^{-10}$$

(e) To solve this, you need to recognize that $[L^-]$ is a weak base. Since we know the K_{a2} value of the acid, we can calculate the K_{b1} value from the water dissociation constant.

$$K_w = (K_{b1})(K_{a2})$$

$$\frac{10^{-14}}{1.79 \times 10^{-10}} = K_{b1}$$

$$K_{b1} = 5.59 \times 10^{-5}$$

Then write the mass action equation for the base:

$$L^- + H_2O \rightleftharpoons HL + OH^-$$

	$[L^-]$ (M)	[HL] (M)	$[OH^-]$ (M)
Initial concentration	0.050	0	0
Change	$-x$	$+x$	$+x$
Final concentration	$0.050 - x$	x	x

$$K_{b1} = \frac{(x)(x)}{0.050 - x}$$

$$5.59 \times 10^{-5} = \frac{x^2}{0.050 - x}$$

By cross-multiplying and rearranging, you can solve the quadratic equation:

$$[HL] = x = 1.64 \times 10^{-3} \text{ M}$$

2. (a) Using the table of standard reduction potentials provided with the exam, look up each half reaction. The half reactions with more positive reduction potentials will accept electrons from metals with more negative reduction potentials. The standard reduction potential for Ni is $E° = -0.25$ V, and the standard reduction potential for Mg is $E° = -2.37$ V, so Ni has the higher reduction potential and will accept electrons from Mg.

$$Mg(s) \rightarrow Mg^{2+} + 2\ e^-$$

$$Ni^{2+} + 2\ e^- \rightarrow Ni(s)$$

(b) $E°_{cell} = E°_{red} + E°_{ox}$

Sub in the Mg/Ni $E°$ values, remembering to reverse the sign for Mg since it's being oxidized:

$$E°_{cell} = -0.25\ V + 2.37\ V = 2.12\ V$$

(c)

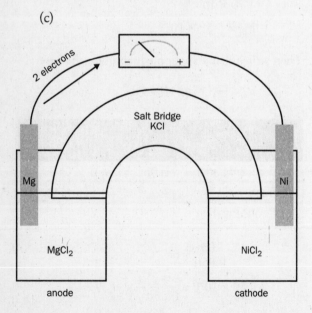

(d) Use the Nernst equation, which is provided on your equation sheet:

$$E_{cell} = E°_{cell} - \frac{0.0592}{n} \log Q$$

You have already calculated $E°_{cell}$, and n is the number of electrons that are transferred. In the Mg/Ni case, two electrons are transferred. Q is calculated by an equation that is also provided on the equation sheet:

The overall equation is:

$$Mg(s) + Ni^{2+} \rightarrow Ni(s) + Mg^{2+}$$

Since Mg(s) and Ni(s) are in solid form, their concentrations are left out of the equation shown below:

$$Q = \frac{[C]^c[D]^d}{[A]^a[B]^b} = \frac{1M}{1 \times 10^{-3}M} = 1 \times 10^3$$

$$E_{cell} = E°_{cell} - \frac{0.0592}{2} \log(1 \times 10^3)$$
$$= 2.12 - 0.089 = 2.03\ V$$

(e) Using the table of standard reduction potentials, look up each reaction (Mg/Mn, Mg/Fe, and Mg/Cu). In each case, Mg has a more negative reduction potential, so it will be oxidized in the half reaction. Since the table is organized with the highest (most positive) reduction potentials at the top and the lowest at the bottom, you can answer this problem without doing any calculations. The highest voltage for the cell will correspond with the greatest distance between the two metal pairs on the table.

Using this method, Mg is the greatest distance from Cu, so this cell will have the highest voltage. If you calculate each voltage, you should obtain the following values:

Note: $E°_{oxi}$ is the reverse sign of $E°_{red}$.

$$E°_{cell} = E°_{oxi} + E°_{red}$$

Mg/Mn

$$2.37 + (-1.18) = 1.19 \text{ V}$$

Mg/Fe

$$2.37 + (-0.44) = 1.93 \text{ V}$$

Mg/Cu

$$2.37 + 0.34 = 2.71 \text{ V}$$

3. (a) CO_2 and H_2O are always the products in the complete combustion of an organic compound.

   (b)

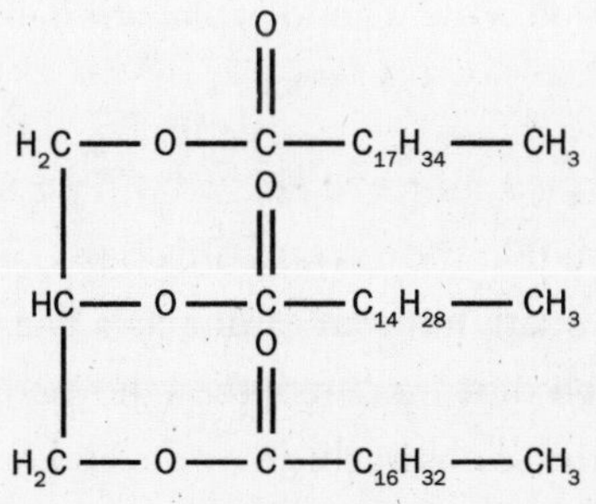

The formula for a saturated fatty acid is $C_nH_{2n+1}COOH$, so all of the fatty acids in this problem are saturated. Saturated fatty acids are solid at room temperature, so the triglyceride formed by these fatty acids is solid at room temperature. The triglyceride formed by the fatty acids is shown above. The fatty acids could be placed in any order (top, middle, bottom) in the triglyceride.

(c) Since $40.0°C - 21.3°C = 18.7°C$, we know that we have to account for the $18.7°C$ difference by using the information given about the amount of energy needed to raise 1.0 g of water $1.0°C$:

$$18.7°C \times \frac{4.186 \text{ J}}{1 \text{ g water, } 1°C} \times 200 \text{ g water} =$$
$$15{,}655.64 \text{ J} = 15.7 \text{ kJ}$$

(d) Most likely, the greatest source of error is due to the loss of heat to the ambient surroundings both from the burning fat sample and from the container of water. If heat is lost from the burning fat sample, the temperature of the water will not rise as much. Also, if the water isn't well insulated, heat will be lost from the water, and the temperature will not rise as much. Water has powerful energy storage abilities, and dissipation capacities and energy loss is often explained by this.

PART B

4. (a) i. $C_3H_8 + 5O_2 \rightarrow 3CO_2 + 4H_2O$
       ii. Every mole of C_3H_8 burned produces three moles of CO_2.

   (b) i. $SO_2 + H_2O \rightarrow H_2SO_3$
       ii. Sulfurous acid has a low pH, which is potentially deadly to fish and plants in lakes and rivers.

   (c) i. $2Cl^- + Cu^{2+} \rightarrow CuCl_2$
       ii. The sodium (Na^+) and sulfate (SO_4^{2-}) ions are spectator ions in this reaction.

5. (a)

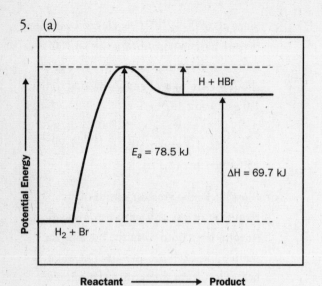

(b) Using the diagram you just made, E_a for the reverse reaction is $78.5\ kJ - 69.7\ kJ = 8.8\ kJ$.

(c) The reverse reaction will have a higher rate constant because its activation energy is lower.

(d) The primary differential rate law explains the relationship between the concentrations of reagents and the reaction rate: rate $= k[A]^x[B]^y$, where $[A]$ and $[B]$ are concentrations of the compounds A and B in moles per liter; x and y are the orders of each reagent; and k is the rate constant. We simply replace the reagents in the right spots, and the answer is: rate $= k[H_2][Br]$.

(e) The rate should increase in direct proportion to the concentrations of H_2 and Br.

(f) The equilibrium constant does *not* change due to changes in concentration. Increasing the pressure of H_2 and Br will temporarily upset the equilibrium (i.e., the system will no longer be at equilibrium), but with time, the system will again reach equilibrium, and the equilibrium constant will be the same as before.

6. General strategy hints for this part of the test: Make sure you mention each species given in the problem, and make sure you actually answer the question that was asked. A lot of students lose points when they make a fairly decent explanation but fail to link their explanation back to the question that was asked and give a final answer. For example, in (a), a student might say "the reaction will shift in the forward direction," but fail to mention that the solution would become more red, which is the answer to the question. Give a good explanation, but be sure to give an answer, too!

(a) $FeCl_3$ is a source of Fe^{3+} so will push the reaction in the forward direction (i.e., shift the equation toward products). The solution will become more red, though whether or not this is observable may depend on how red the solution already is.

(b) Na and Ca have a small number of valence electrons, which are heavily shielded from the nuclear charge by the inner-shell electrons and are therefore loosely held and easily removed. The effective nuclear charge experienced by a valence electron in a metal atom is small and too weak to attract extra electrons and form negative oxidation states. In carbon, the effective nuclear charge experienced by the valence electrons is relatively larger, so carbon can take on extra electrons from less electronegative elements or give them up to more electronegative elements. Carbon tends to form covalent, rather than ionic, compounds, so the oxidation number is not a true representation of charge.

Another option for answering this one:

In order to have a positive oxidation number in a compound, an atom must be bonded (either covalently or ionically) with a less electronegative element. Na and Ca are metals of very low electronegativity. The only elements with lower electronegativities are other metals, and metals do not form ionic or covalent compounds with each other. To form a compound, Na or Ca must combine with a nonmetal, and since all nonmetals are more electronegative, Na and Ca can only have positive oxidation states. Carbon, on the other hand, can form somewhat ionic compounds with metals (and thus take on a negative oxidation state) and commonly forms covalent compounds with other nonmetals that may have higher or lower electronegativities.

(c) Ar and Xe are both nonpolar, so the only attractive forces present are dispersion forces (also called London forces or induced-dipole forces). The strength of the interaction depends on the polarizability of the electron cloud in the substance. Since Xe has many more electrons and a much larger electron cloud, it is more polarizable and has a higher boiling point.

(d) The dominant intermolecular force in water is hydrogen bonding, which is the strongest of the intermolecular forces, so it is not easily vaporized. Ethyl ether has only (relatively weak) dispersion and dipole forces. Therefore, ethyl ether is more easily vaporized and will have a higher vapor pressure than water.

(e) Ionization energy is the amount of energy required to remove an electron from an atom. It becomes increasingly easy to remove the outer electron as it becomes more shielded from the attractive force of the protons in the nucleus. More shielding occurs as more shells are occupied and as the outer electron moves farther from the nucleus. As you move down a column in the periodic table, more shells are occupied and the outer electron is more shielded and easier to remove, so ionization energy decreases.

GLOSSARY

A

Absolute zero: The temperature at which all substances have no thermal energy; 0 K or −273.15°C.

Absorption spectrum: The series of discrete lines at characteristic frequencies representing the energy required to make an atom undergo a transition to a higher energy state.

Acid: A species that donates hydrogen ions and/or accepts electrons. See Acidic solution; Arrhenius acid; Brønsted-Lowry acid; Lewis acid.

Acid dissociation constant (K_a): The equilibrium constant that measures the degree of dissociation for an acid under specific conditions. For an acid HA:

$$K_a = \frac{[H^+][A^-]}{[HA]}$$

Acidic anhydride: An oxide that dissolves in water to form an acidic solution.

Acidic solution: An aqueous solution that contains more H^+ ions than OH^- ions. The pH of an acidic solution is less than 7 at 25°C.

Activated complex: The transition state of a reaction in which old bonds are partially broken and new bonds are partially formed. The activated complex has a higher energy than the reactants or products of the reaction.

Activation energy (E_a): The minimum amount of energy required for a reaction to occur.

Adiabatic process: A process that occurs without the transfer of heat to or from the system.

Alcohols: Organic compounds of the general formula R–OH.

Aldehydes: Organic compounds of the general formula R–CHO.

Alkali metals: Elements found in Group IA of the periodic table. They are highly reactive, readily losing their one valence electron to form ionic compounds with nonmetals.

Alkaline earth metals: Elements found in Group IIA of the periodic table. Their chemistry is similar to that of the alkali metals, except that they have two valence electrons, and thus form 2+ cations.

Alkanes: Hydrocarbons with only single bonds. The general formula for alkanes is C_nH_{2n+2}.

Alkenes: Hydrocarbons with at least one carbon-carbon double bond. Their general formula is C_nH_{2n}.

Alkynes: Hydrocarbons with at least one carbon-carbon triple bond. Their general formula is C_nH_{2n-2}.

Alpha (α) particle: A particle ejected from the nucleus in one form of radioactive decay, identical to the helium 4 nucleus.

Amines: Compounds of the general formula $R-NH_2$, RR'NH, OR RR'R"N.

Amino acids: Building blocks of proteins with the general formula $NH_2C-R-HCOOH$.

Amorphous solids: Solids that do not possess long-range order. Compare to Crystal.

Amphoteric species: A species capable of reacting either as an acid or as a base.

Anhydride: A compound obtained by the removal of water from another compound.

Anion: An ionic species with a negative charge.

Anode: The electrode at which oxidation occurs. Compare to Cathode.

Aqueous solution: A solution in which water is the solvent.

Aromatic compounds: Planar, cyclic organic compounds that are unusually stable because of the delocalization of π electrons.

Arrhenius acid: A species that donates protons (H^+) in an aqueous solution; e.g., HCl.

Arrhenius base: A species that gives off hydroxide ions (OH^-) in an aqueous solution; e.g., NaOH.

Atom: The most elementary form of an element; it cannot be further broken down by chemical means.

Atomic mass: The averaged mass of the atoms of an element, taking into account the relative abundance of the various isotopes in a naturally occurring substance. Also called the atomic weight.

Atomic mass units (amu): Units of mass defined as $\frac{1}{12}$ the mass of a carbon 12 atom; approximately equal to the mass of one proton or one neutron.

Atomic number: The number of protons in a given element.

Atomic orbital: The region of space around the nucleus of an atom in which there is a high probability of finding an electron.

Atomic radius: The radius of an atom, or the average distance between a nucleus and the outermost electron. Usually measured as one-half the distance between two nuclei of an element in its elemental form.

Aufbau principle: The principle that electrons fill energy levels in a given atom in order of increasing energy, completely filling one sublevel before beginning to fill the next.

Avogadro's number: The number corresponding to a mole. It is the number of carbon 12 atoms in exactly 12 g of carbon 12, approximately 6.022×10^{23}.

Avogadro's principle: The law stating that under the same conditions of temperature and pressure, equal volumes of different gases will have the same number of molecules.

Azimuthal quantum number (l): The second quantum number, denoting the sublevel or subshell in which an electron can be found. It reveals the shape of the orbital. This quantum number represents the orbital angular momentum of the motion of the electron about a point in space.

B

Balanced equation: An equation for a chemical reaction in which the number of atoms for each element in the reaction and the total charge are the same for the reactants and the products.

Barometer: An instrument for measuring atmospheric pressure.

Base: A species that donates hydroxide ions or electrons, or that accepts protons.

$$K_b = \frac{[B^+][OH^-]}{[BOH]}$$

See Arrhenius base; Basic solution; Brønsted-Lowry base; Lewis base.

Base dissociation constant (K_b): The equilibrium constant that measures the degree of dissociation for a base under specific conditions.

Basic anhydride: An oxide that dissolves in water to form a basic solution.

Basic solution: An aqueous solution that contains more OH^- ions than H^+ ions. The pH of a basic solution is greater than 7 at 25°C.

Beta (β) particle: An electron produced and ejected from a nucleus during radioactive beta decay.

Binding energy: The energy required to break a nucleus apart into its constituent neutrons and protons.

Bohr model: The model of the hydrogen atom postulating that atoms are composed of electrons that assume certain circular orbits about a positive nucleus.

Boiling point: The temperature at which the vapor pressure of a liquid is equal to the surrounding pressure. The normal boiling point of any liquid is defined as temperature at which its vapor pressure is one atmosphere.

Boiling-point elevation: The amount by which a given quantity of solute raises the boiling point of a liquid; a colligative property.

Bond energy: The energy (enthalpy change) required to break a particular bond under given conditions.

Boyle's law: The law stating that at constant temperature, the volume of a gaseous sample is inversely proportional to its pressure.

Brønsted-Lowry acid: Proton donor, e.g., H_3PO_4.

Brønsted-Lowry base: Proton acceptor, e.g., OH^-.

Buffer: A solution containing a weak acid and its salt (or a weak base and its salt), which tends to resist changes in pH.

Buffer region: The region of a titration curve in which the concentration of a conjugate acid is approximately equal to that of the corresponding base. The pH remains relatively constant when small amounts of H^+ or OH^- are added because of the combination of these ions with the buffer species already in solution.

C

Calorie (cal): A unit of thermal energy (1 cal = 4.184 J). The amount of heat needed to raise the temperature of 1g of water 1 degree Celsius.

Calorimeter: An apparatus used to measure the heat absorbed or released by a reaction.

Carbohydrates: Compounds with the general formula $C_n(H_2O)_m$.

Carbon dating: A technique for estimating the age of (ancient) objects by measuring the amount of radioactive carbon 14 remaining.

Carbonyl group: C=O group found in aldehydes, ketones, etc. The C=O bond is known as the carbonyl bond, and organic compounds containing this group are known as carbonyl compounds.

Carboxylic acids: Compounds of the general formula R–COOH.

Catalysis: Increasing a reaction rate by adding a substance (the catalyst) not permanently changed by the reaction.

Catalyst: A substance that increases the rates of the forward and reverse directions of a specific reaction but is itself left unchanged.

Cathode: The electrode at which reduction takes place. Compare to Anode.

Cation: An ionic species with a positive charge.

Celsius (°C): A temperature scale defined by having 0°C equal to the freezing point of water and 100°C equal to the boiling point of water; also the units of that scale. Otherwise known as the centigrade temperature scale. 0°C = 273.15 K.

Chain reaction: A reaction in which a change in a molecule makes many molecules change until a stable compound forms, often occurring in nuclear and radical reactions.

Charles's law: The law stating that the volume of a gaseous sample at constant pressure is directly proportional to its absolute (Kelvin) temperature.

Chemical bond: The interaction between two atoms resulting from the overlap of electron orbitals, holding the two atoms together at a specific average distance from each other.

Chemical properties: Those properties of a substance describing its reactivity.

Closed system: A system that can exchange energy but not matter with its surroundings.

Colligative properties: Those properties of solutions that depend only on the number of solute particles present but not on the nature of those particles. See Boiling-point elevation; Freezing-point depression; Vapor-pressure lowering.

Combustion: The oxidation reaction of a compound in the presence of oxygen. Heat is released and for organic compounds, carbon dioxide and water are typically formed as a result of the reaction.

Common ion effect: A shift in the equilibrium of a solution due to the addition of ions of a species already present in the reaction mixture.

Compound: A pure substance that can be decomposed to produce elements, other compounds, or both.

Concentration: The amount of solute per unit of solvent (denoted by square brackets), or the relative amount of one component in a mixture.

Conjugate acid-base pair: Brønsted-Lowry acid and base related by the transfer of a proton, e.g., H_2CO_3 and HCO_3^-.

Coordination complex: A compound in which a central metal atom or ion is bonded by coordinate covalent bonds to other atoms or groups.

Coordinate covalent bond: A covalent bond in which both electrons of the bonding pair are donated by only one of the bonded atoms.

Corrosion: The gradual destruction of a metal or alloy as a result of oxidation or the action of a chemical agent.

Covalent bond: A chemical bond formed by the sharing of an electron pair between two atoms. See Coordinate covalent bond; Nonpolar covalent bond; Polar covalent bond.

Critical pressure: The vapor pressure at the critical temperature of a given substance.

Critical temperature: The highest temperature at which the liquid and vapor phases of a substance can coexist; above this temperature, the substance does not liquefy at any pressure.

Crystal: A solid whose atoms, ions, or molecules are arranged in a regular three-dimensional lattice structure.

Cycloalkanes: Saturated cyclic compounds of the formula C_nH_{2n}.

D

d subshell: The subshells corresponding to the azimuthal quantum number $l = 2$, found in the third and higher principal energy levels; each contains five orbitals.

Dalton's law: The law stating that the sum of the partial pressures of the components of a gaseous mixture must equal the total pressure of the mixture.

Daniell cell: An electrochemical cell in which the anode is the site of Zn metal oxidation, and the cathode is the site of Cu^{2+} ion reduction.

Degenerate orbitals: Orbitals that possess equal energy.

Density (ρ): A physical property of a substance, defined as the mass contained in a unit of volume.

Diamagnetic: A condition that arises when a substance has no unpaired electrons and is slightly repelled by a magnetic field.

Diffusion: The random motion of gas or solute particles across a concentration gradient, leading to uniform distribution of the gas or solute throughout the container.

Dipole: In chemistry, a species containing bonds between elements of different electronegativities, resulting in an unequal distribution of charge in the species.

Dipole-dipole interaction: The attractive force between two dipoles whose magnitude is dependent on both the dipole moments and the distance between the two species.

Dipole moment: A vector quantity whose magnitude is dependent on the product of the charges and the distance between them. The direction of the moment is from the positive to the negative pole.

Dispersion force: A weak intermolecular force that arises from interactions between temporary and/or induced dipoles. Also called London force.

Disproportionation reaction: A chemical reaction in which a compound reacts with itself to form at least two new chemical species, both containing at least one of the same atom. Most of these reactions are redox reactions in which an atom is both oxidized and reduced.

Dissociation: The separation of a single species into two separate species; this term is usually used in reference to salts or weak acids or bases.

Dissolution: The process of dissolving a substance. The opposite of precipitation.

Dynamic equilibrium: A state of balance (no macroscopic change observable) that arises when opposing processes occur at equal rates.

E

Electrochemical cell: A cell within which a redox reaction takes place, containing two electrodes between which there is an electrical potential difference. See Electrolytic cell; Voltaic cell.

Electrode: An electrical conductor through which an electric current enters or leaves a medium.

Electrolysis: The process in which an electric current is passed though a solution, resulting in chemical changes that would not otherwise occur spontaneously.

Electrolyte: A compound that ionizes in water.

Electrolytic cell: An electrochemical cell that uses an external voltage source to drive a nonspontaneous redox reaction.

Electromagnetic radiation: A self-propagating wave composed of electric and magnetic fields oscillating perpendicular to each other and to the direction of propagation.

Electromagnetic spectrum: The range of all possible frequencies or wavelengths of electromagnetic radiation.

Electromotive force (EMF): The potential difference developed between the cathode and the anode of an electrochemical cell.

Electron (e–): A subatomic particle that remains outside the nucleus of an atom and carries a single negative charge. In most cases, its mass is considered to be negligible ($\frac{1}{1,837}$ that of the proton).

Electron affinity: The amount of energy that is released when an electron is added to an atom.

Electron configuration: The symbolic representation used to describe the electron occupancy of the various energy sublevels in a given atom.

Electronegativity: A measure of the ability of an atom to attract the electrons in a bond.

Electron spin: The intrinsic angular momentum of an electron, having arbitrary values of $+\frac{1}{2}$ and $-\frac{1}{2}$. See Spin quantum number.

Electroplating: The electrolytic process of coating an object with a metal to prevent corrosion.

Element: A substance that cannot be further broken down by chemical means. All atoms of a given element have the same number of protons.

Emission spectrum: The spectrum produced by a species emitting energy as it relaxes from an excited to a lower energy state.

Empirical formula: The simplest whole number ratio of the different elements in a compound.

Endothermic reaction: A reaction that absorbs heat from the surroundings as the reaction proceeds (positive ΔH).

End point: The point in a titration at which the indicator changes color, showing that enough reactant has been added to the solution to complete the reaction.

Enthalpy (H): The heat content of a system at constant pressure. The change in enthalpy (ΔH) in the course of a reaction is the difference between the enthalpies of the products and the reactants.

Entropy (S): A property related to the degree of disorder in a system. Highly ordered systems have low entropies. The change in entropy (ΔS) in the course of a reaction is the difference between the entropies of the products and the reactants.

Equilibrium: The state of balance in which the forward and reverse reaction rates are equal. In a system at equilibrium, the concentrations of all species will remain constant over time unless there is a change in the reaction conditions. See Le Châtelier's principle.

Equilibrium constant: The ratio of the concentration of the products to the concentration of the reactants for a certain reaction at equilibrium, all raised to their stoichiometric coefficients.

Equivalence point: The point in a titration at which the number of equivalents of the species being added to the solution is equal to the number of equivalents of the species being titrated.

Esters: Compounds of the general formula R–COO–R'.

Ethers: Compounds of the general formula R–O–R'.

Excess reagent: In a chemical reaction, any reagent whose amount does not limit the amount of product that can be formed. Compare to Limiting reagent.

Excited state: An electronic state having a higher energy than the ground state.

Exothermic reaction: A reaction that gives off heat (negative ΔH) to the surroundings as the reaction proceeds.

F

f subshell: The subshells corresponding to the azimuthal quantum number $l = 3$, found in the fourth and higher principal energy levels, each containing seven orbitals.

Faraday (F): The total charge on one mole of electrons (1 F = 96,487 coulombs).

Fatty acids: Carboxylic acids with long hydrocarbon chains, derived from the hydrolysis of fats.

First law of thermodynamics: The law stating that the total energy of a system and its surroundings remains constant. Also expressed as $\Delta E = Q - W$: the change in energy of a system is equal to the heat added to it minus the work done by it.

Formal charge: The conventional assignment of charges to individual atoms of a Lewis formula for a molecule, used to keep track of valence electrons. Defined as the total number of valence electrons in the free atom minus the total number of nonbonding electrons minus one-half the total number of bonding electrons.

Freezing point: At a given pressure, the temperature at which the solid and liquid phases of a substance coexist in equilibrium.

Freezing-point depression: Amount by which a given quantity of solute lowers the freezing point of a liquid. A colligative property.

G

Galvanic cell: An electrochemical cell that uses a spontaneous redox reaction to do work, i.e., produce an electrical current. Also called a voltaic cell.

Gamma (γ) radiation: High energy radiation often emitted in radioactive decay.

Gas: The physical state of matter possessing a high degree of disorder, in which molecules interact only slightly; found at relatively low pressure and high temperatures. Also called vapor. See Ideal gas.

Gas constant (R): A proportionality constant that appears in the ideal gas law, $PV = nRT$. Its value depends upon the units of pressure, temperature, and volume used in a given situation.

Geiger counter: An instrument used to measure radioactivity.

Gibbs free energy (G): The energy of a system available to do work. The change in Gibbs free energy, ΔG, is determined for a given reaction from the equation $\Delta G = \Delta H - T\Delta S$. ΔG is used to predict the spontaneity of a reaction: a negative ΔG denotes a spontaneous reaction, while a positive ΔG denotes a nonspontaneous reaction.

Graham's law: The law stating that the rate of effusion or diffusion of a gas is inversely proportional to the square root of the gas's molecular weight.

Gram-equivalent weight (GEW): The amount of a compound that contains one mole of reacting capacity when fully dissociated. One GEW equals the molecular weight divided by the reactive capacity per formula unit.

Group: A vertical column of the periodic table containing elements that are similar in their chemical properties.

H

Half-life: The time required for the amount of a reactant to decrease to one-half of its former value.

Half-reaction: Either the reduction half or oxidation half of a redox reaction. Each half-reaction occurs at one electrode of an electrochemical cell.

Halogens: The active nonmetals in Group VIIA of the periodic table; they have high electronegativities and highly negative electron affinities.

Heat: The energy representing the kinetic energy of molecules that is transferred spontaneously from a warmer sample to a cooler sample. See Temperature.

Heat of formation (ΔH_f): The heat absorbed or released during the formation of a pure substance from the elements in their standard states.

Heat of fusion (ΔH_{fus}): The ΔH for the conversion of a solid to a liquid.

Heat of sublimation (ΔH_{sub}): The ΔH for the conversion of a solid directly to a gas.

Heat of vaporization (ΔH_{vap}): The ΔH for the conversion of a liquid to a vapor.

Heisenberg uncertainty principle: The principle that states that it is impossible to simultaneously determine with perfect accuracy both the momentum and position of a particle.

Henry's law: The law stating that the mass of a gas that dissolves in a solution is directly proportional to the partial pressure of the gas above the solution.

Hess's law: The law stating that the energy change in an overall reaction is equal to the sum of the energy changes in the individual reactions which comprise it.

Heterogeneous: Nonuniform in composition.

Homogeneous: Uniform in composition.

Hund's rule: The rule that electrons will occupy all degenerate orbitals in a subshell with single electrons having parallel spins before entering half-filled orbitals.

Hybridization: The combination of two or more atomic orbitals to form new orbitals for bonding purposes.

Hydrate: A compound with associated water molecules.

Hydrocarbons: Organic compounds containing only carbon and hydrogen.

Hydrogen bonding: The strong attraction between a hydrogen atom bonded to a highly electronegative atom, such as fluorine or oxygen in one molecule and a highly electronegative atom in another molecule.

Hydrolysis: A reaction between water and a species in solution.

Hydronium ion: The H_3O^+ ion in aqueous solution.

Hydroxide ion: The OH^- ion.

I

Ideal gas: A hypothetical gas whose behavior is described by the ideal gas law under all conditions. An ideal gas would have particles of zero volume that do not exhibit interactive forces.

Ideal gas law: The law stating that $PV = nRT$, where R is the gas constant. It can be used to describe the behavior of many real gases at moderate pressures and temperatures significantly above absolute zero. See Kinetic molecular theory.

Indicator, acid-base: A substance used in low concentration during a titration that changes color over a certain pH range. The color change, which occurs as the indicator undergoes a dissociation reaction, is used to identify the end point of the titration reaction.

Inert gases: The elements located in Group 18 (or Group VIIIA) of the periodic table. They contain a full octet of valence electrons in their outermost shell; this electron configuration makes them the least reactive of the elements. Also called noble gases.

Intermolecular forces: The attractive and repulsive forces between molecules. See Van der Waals forces.

Intramolecular forces: The attractive forces between atoms within a single molecule.

Ion: A charged atom or molecule that results from the loss or gain of electrons.

Ionic bonding: A chemical bond formed through electrostatic interaction between positive and negative ions.

Ionic solid: A solid consisting of positive and negative ions arranged into crystals that are made up of regularly repeated units and held together by ionic bonds.

Ionization product: The general term for the dissociation of salts or of weak acids or bases; the ratio of the concentration of the ionic products to the concentration of the reactant for a reaction, all raised to their stoichiometric coefficients.

Ionization energy: The energy required to remove an electron from the valence shell of a gaseous atom.

Isobaric process: A process that occurs at constant pressure.

Isolated system: A system that can exchange neither matter nor energy with its surroundings.

Isomers: Compounds with the same molecular formula but different structures.

Isothermal process: Process that occurs at constant temperature.

Isotopes: Atoms containing the same number of protons but different numbers of neutrons, e.g., nitrogen 14 and nitrogen 15.

J

Joule (J): A unit of energy; $1 J = 1 kg \cdot m^2/s^2$.

K

Kelvin (K): A temperature scale with units equal in magnitude to the units of the Celsius scale and absolute zero defined as 0 K; also the units of that temperature scale. Otherwise known as the absolute temperature scale. $0 K = -273.15°C$.

Ketones: Compounds of the general formula R–CO–R'.

Kinetic energy: The energy a body has as a result of its motion, equal to $\frac{1}{2}mv^2$.

Kinetic molecular theory: The theory proposed to account for the observed behavior of gases. The theory considers gas molecules to be pointlike, volumeless particles, exhibiting no intermolecular forces and in constant random motion, undergoing only completely elastic collisions with the container or other molecules. See Ideal gas law.

L

Law of conservation of mass: The law stating that in a given reaction, the mass of the products is equal to the mass of the reactants.

Law of constant composition: The law stating that the elements in a pure compound are found in specific weight ratios.

Le Chatelier's principle: The observation that when a system at equilibrium is disturbed or stressed, the system will react in such a way as to relieve the stress and restore equilibrium. See Equilibrium.

Lewis acid: A species capable of accepting an electron pair, e.g., BF_3.

Lewis base: A species capable of donating an electron pair, e.g., NH_3.

Lewis structure: A method of representing the shared and unshared electrons of an atom, molecule, or ion.

Limiting reagent: In a chemical reaction, the reactant present in such quantity as to limit the amount of product that can be formed.

Liquid: The state of matter in which intermolecular attractions are intermediate between those in gases and in solids, distinguished from the gas phase by having a definite volume and from the solid phase in that the molecules may mix freely.

Litmus: An organic substance that is used as an acid-base indicator, most often in paper form. It turns red in acidic solution and blue in basic solution.

London force: See Dispersion force.

M

Magnetic quantum number (m_l): The third quantum number, defining the particular orbital of a subshell in which an electron resides. It conveys information about the orientation of the orbital in space (e.g., p_x versus p_y).

Manometer: An instrument used to measure the pressure of a gas.

Mass: A physical property representing the amount of matter in a given sample.

Mass defect: The difference between the sum of the masses of neutrons and protons forming a nucleus and the mass of that nucleus; the mass equivalence of binding energy, with the two related via the equation $E = mc^2$.

Mass number: The total number of protons and neutrons in a nucleus.

Maxwell-Boltzmann distribution: The distribution of the molecular speeds of gas particles at a given temperature.

Melting point: The temperature at which the solid and liquid phases of a substance coexist in equilibrium.

Metal: One of a class of elements located on the left side of the periodic table, possessing low ionization energies and electronegativities. Metals readily give up electrons to form cations; they possess relatively high electrical conductivity and are lustrous and malleable.

Metallic bonding: The type of bonding in which the valence electrons of metal atoms are delocalized throughout the metallic lattice.

Metalloid: An element possessing properties intermediate between those of a metal and those of a nonmetal. Also called a semimetal.

Miscible: Able to mix in any proportion.

Molality (m): A concentration unit equal to the number of moles of solute per kilogram of solvent.

Molar mass: The mass in grams of one mole of an element or compound.

Molarity (M): A concentration unit equal to the number of moles of solute per liter of solution.

Mole (mol): One mole of a substance contains Avogadro's number of molecules or atoms (6.022×10^{23}). The mass of one mole of substance in grams is the same as the mass of one molecule or atom in atomic mass units.

Mole fraction (χ): A unit of concentration equal to the ratio of the number of moles of a particular component to the total number of moles for all species in the system.

Molecular formula: A formula showing the actual number and identity of all atoms in each molecule of a compound.

Molecular weight: The sum of the atomic weights of all the atoms in a molecule.

Molecule: The smallest polyatomic unit of an element or compound that exists with distinct chemical and physical properties.

Monoprotic acid: An acid that can donate only one proton, e.g., HNO_3. The molarity of a monoprotic acid solution is equal to its normality.

Monosaccharides: Simple sugars that cannot be hydrolyzed to simpler compounds.

N

Net ionic equation: A reaction equation showing only the species actually participating in the reaction.

Neutral solution: An aqueous solution in which the concentration of H^+ and OH^- ions are equal (pH = 7).

Neutralization reaction: A reaction between an acid and base in which H^+ ions and OH^- ions combine to produce water and a salt solution.

Neutron: A subatomic particle contained within the nucleus of an atom. It carries no charge and has a mass very slightly larger than that of a proton.

Noble gases: See Inert gases.

Nonelectrolyte: A compound that does not ionize in water.

Nonmetal: One of a class of elements with high ionization potentials and very negative electron affinities that generally gains electrons to form anions. Nonmetals are located on the upper right side of the periodic table.

Nonpolar covalent bond: A covalent bond between elements of the same electronegativity. There is no charge separation, and the atoms do not carry any partially positive or partially negative charge. Compare to Polar covalent bond.

Nonpolar molecule: A molecule which exhibits no net separation of charge, and therefore no net dipole moment. Compare to Polar molecule.

Normality (N): A concentration unit equal to the number of gram equivalent weights of solute per liter of solution.

Nucleon: A particle found in the nucleus of an atom; can be either a neutron or a proton.

Nucleus: The small central region of an atom; a dense, positively charged area containing protons and neutrons.

O

Octet: Eight valence electrons in a subshell around a nucleus.

Octet rule: A rule stating that bonded atoms tend to undergo reactions that will produce a complete octet of valence electrons. Applies without exception only to C, N, O, and F with zero or negative formal charges.

Open system: A system that can exchange both energy and matter with its surroundings.

Orbital: A region of electron density around an atom or molecule, containing no more than two electrons of opposite spin. See Atomic orbital.

Order of reaction: In a calculation of the rate law for a reaction, the sum of the exponents to which the concentrations of reactants must be raised.

Osmosis: The movement of a solvent or solute through a semipermeable membrane across its concentration gradient, i.e., from a container in which the concentration is high to a container in which the concentration is low.

Osmotic pressure: The pressure that must be applied to a solution to prevent the passage of a pure solvent through a semipermeable membrane across its concentration gradient.

Oxidation: A reaction involving the net loss of electrons or, equivalently, an increase in oxidation number.

Oxidation number: The number assigned to an atom in an ion or molecule that denotes its real or hypothetical charge. Atoms, alone or in molecules, of standard state elements have oxidation numbers of zero. Also called the oxidation state.

Oxidizing agent: In a redox reaction, a species that gains electrons and is thereby reduced.

P

p subshell: The subshells corresponding to the azimuthal quantum number $l = 1$, found in the second and higher principal energy levels. Each subshell contains three dumbbell-shaped p orbitals oriented perpendicular to each other and referred to as the p_x, p_y, and p_z orbitals.

Paired electrons: Two electrons in the same orbital with assigned spins of $+\frac{1}{2}$ and $-\frac{1}{2}$. See Orbital; Hund's rule.

Paramagnetism: A property of a substance that contains unpaired electrons, whereby the substance is attracted by a magnetic field.

Partial pressure: The pressure that one component of a gaseous mixture would exert if it were alone in the container.

Pauli exclusion principle: The principle stating that no two electrons within an atom may have an identical set of all four quantum numbers.

Peptides: Molecules that consist of two or more amino acids linked to each other by peptide bonds.

Percent composition: The percentage of the total formula weight of a compound attributed to a given element.

Percent yield: The percentage of the theoretical product yield that is actually recovered when a chemical reaction occurs.

Period: A horizontal row of the periodic table, containing elements with the same number of electron shells.

Periodic law: The law stating that the chemical properties of an element depend on the atomic number of the element and change in a periodic fashion.

Periodic table: The table displaying all known chemical elements arranged in rows (periods) and columns (groups) according to their electronic structure.

pH: A measure of the hydrogen ion content of an aqueous solution, defined to be equal to the negative log of the H^+ concentration.

Phase: One of the three states of matter: solid, liquid, or gas. (Plasma is often considered a fourth phase of matter.)

Phase diagram: A plot, usually of pressure versus temperature, showing which phases a compound will exhibit under any set of conditions.

Phase equilibrium: For a particular substance, any temperature and pressure at which two or three phases coexist in equilibrium. See Triple point.

Photon: A quantum of energy in the form of light with a value of Planck's constant multiplied by the frequency of the light.

Physical property: A property of a substance related to its physical, not chemical, characteristics; e.g., density, smell, color.

Pi (π) bond: A covalent bond formed by parallel overlap of two unhybridized atomic p orbitals.

pOH: A measure of the hydroxide (OH^-) ion content of an aqueous solution, defined to be equal to the negative log of the OH^- concentration.

Polar covalent bond: A covalent bond between atoms with different electronegativities in which electron density is unevenly distributed, giving the bond positive and negative ends. Compare to Nonpolar covalent bond.

Polar molecule: A molecule possessing one or more polar covalent bond(s) and a geometry that allows the bond dipole moments to add up to a net dipole moment, e.g., H_2O. Compare to Nonpolar molecule.

Polyprotic acid: An acid capable of donating more than one proton, e.g., H_2CO_3.

Positron: An antielectron: it has the same mass as an electron but has an opposite charge and is emitted during a particular form of radioactive decay.

Potential energy diagram: An energy diagram that relates the potential energy of the reactants and products of a reaction to details of the reaction pathway. By convention, the x-axis shows the progression of the reaction, and the y-axis shows potential energy.

Precipitate: An insoluble solid that separates from a solution, generally the result of mixing two or more solutions or of a temperature change.

Pressure: Average force per unit area measured in atmospheres (atm), torr (mm Hg), or pascals (Pa). One atm = 760 torr = 760 mm Hg = 1.01×10^2 kPa.

Primary structure: The amino acid sequence of a protein.

Principal quantum number (n): The first quantum number, defining the energy level or shell occupied by an electron.

Proteins: Long-chain polypeptides with high molecular weights.

Proton (H^+): A subatomic particle that carries a single positive charge and has a mass defined as one or as the hydrogen ion, H^+, which is simply a hydrogen nucleus, consisting of one proton. These species are considered to be equivalent.

Q

Quantum: A discrete bundle of energy, such as a photon.

Quantum number: A number used to describe the energy levels available to electrons. The state of any electron is described by four quantum numbers. See Principal quantum number; Azimuthal quantum number; Magnetic quantum number; Spin quantum number.

R

Radioactivity: A phenomenon exhibited by certain unstable isotopes in which they undergo spontaneous nuclear transformations via emission of one or more particle(s).

Raoult's law: A law stating that the partial pressure of a component in a solution is proportional to the mole fraction of that component in the solution.

Rate constant: The proportionality constant in the rate law of a reaction; specific to a particular reaction under particular conditions.

Rate-determining step: The slowest step of a reaction mechanism. The rate of this step limits the overall rate of the reaction.

Rate law: A mathematical expression giving the rate of a reaction as a function of the concentrations of the reactants. The rate law of a given reaction must be determined experimentally.

Reaction intermediate: A species that does not appear among the final products of a reaction but is present temporarily during the course of the reaction.

Reaction mechanism: The series of steps that occur in the course of a chemical reaction, often including the formation and destruction of reaction intermediates.

Reaction rate: The speed at which a substance is produced or consumed by a reaction.

Real gas: A gas that exhibits deviations from the ideal gas law.

Redox reaction: A reaction combining reduction and oxidation processes. Also called oxidation-reduction reaction.

Reducing agent: In a redox reaction, a species that loses electrons and is thereby oxidized.

Reduction: A reaction involving the net gain of electrons or, equivalently, a decrease in oxidation number.

Resonance: Delocalization of electrons within a compound that cannot be adequately represented by Lewis structures.

S

***s* subshell:** Subshell corresponding to the azimuthal quantum number $l = 0$ and containing one spherical orbital; found in all energy levels.

Salt: An ionic substance, i.e., one consisting of anions and cations but not hydrogen or hydroxide ions. Any salt can be formed by the reaction of the appropriate acid and base, e.g., KBr from HBr and KOH.

Saturated hydrocarbon: A hydrocarbon with only single bonds.

Saturated solution: A solution containing the maximum amount of solute that can be dissolved in a particular solvent at a particular temperature.

Scintillation counter: An instrument used to measure radioactivity by the amount of fluorescence produced.

Second law of thermodynamics: The law stating that all spontaneous processes lead to an increase in the entropy of the universe.

Semimetal: See Metalloid.

Semipermeable: A quality of a membrane allowing only some components of a solution, usually including the solvent, to pass through while limiting the passage of other species.

Sigma (σ) bond: Bond formed by head-to-head overlap of orbitals from separate atoms.

Solid: The phase of matter possessing the greatest order, in which molecules are fixed in a rigid structure.

Solubility: A measure of the amount of solute that can be dissolved in a solvent at a certain temperature.

Solubility product (K_{sp}): The equilibrium constant for the ionization reaction of a slightly soluble electrolyte.

Solute: The component of a solution that is present in lesser amount than the solvent.

Solution: A homogeneous mixture of two or more substances.

Solvation: The aggregation of solvent molecules around a solute particle in the process of dissolution.

Solvent: The component of a solution present in the greatest amount; the substance in which the solute is dissolved.

Specific heat: The amount of heat required to raise the temperature of a unit mass of a substance by 1°C.

Spectator ions: Ions that are present in a solution in which a reaction is taking place but that do not participate in the reaction.

Spectrum: The characteristic wavelengths of electromagnetic radiation emitted or absorbed by an object, atom, or molecule.

Spin quantum number (m_s): The fourth quantum number, indicating the orientation of the intrinsic angular momentum of an electron in an atom. The spin quantum number can only assume values of $+\frac{1}{2}$ or $-\frac{1}{2}$.

Spontaneous process: A process that will occur on its own without energy input from the surroundings.

Standard conditions: Conditions defined as 25°C and 1 M concentration for each reactant in solution and a partial pressure of 1 atm for each gaseous reactant. Used for measuring the standard Gibbs free energy, enthalpy, entropy, and cell EMF.

Standard free energy ($G°$): The Gibbs free energy for a reaction under standard conditions. See Gibbs free energy.

Standard hydrogen electrode (SHE): The electrode defined as having a potential of zero under standard conditions. All redox potentials are measured relative to the standard hydrogen electrode. The potentials are measured relative to the standard hydrogen electrode at 25°C and with 1 M of each ion in solution.

Standard potential: The voltage associated with a half-reaction of a specific redox reaction. Generally tabulated as a reduction potential, compared to the SHE.

Standard temperature and pressure (STP): 0°C (273 K) and 1 atm. Used for measuring gas volume and density.

State: The set of defined macroscopic properties of a system that must be specified in order to reproduce the system exactly. Sometimes also used as a synonym for Phase.

State function: A function that depends on the state of a system but not on the path used to arrive at that state.

Strong acid: An acid that undergoes complete dissociation in an aqueous solution, e.g., HCl.

Strong base: A base that undergoes complete dissociation in an aqueous solution, e.g., KOH.

Sublimation: A change of phase from solid to gas without passing through the liquid phase.

Subshell: The division of electron shells or energy levels defined by a particular value of the azimuthal quantum number, e.g., s, p, d, and f subshells. Composed of orbitals. Also called Sublevels. See Orbitals.

Supersaturated solution: A solution that holds more dissolved solute than is required to reach equilibrium at a given temperature.

Surroundings: All matter and energy in the universe not included in the particular system under consideration.

System: The matter and energy under consideration.

T

Temperature: A measure of the average energy of motion of the particles in a system.

Third law of thermodynamics: The law stating that the entropy of a perfect crystal at absolute zero is zero.

Titrant: A solution of known concentration that is slowly added to a solution containing an unknown amount of a second species to determine its concentration.

Titration, acid-base: A method used to determine the concentration of an unknown solution.

Titration curve: A plot of the pH of a solution versus the volume of acid or base added in an acid-base titration.

Torr: A pressure unit equal to 1 mm Hg. 760 torr = 1 atm.

Transition metal: Any of the elements in the B groups of the periodic table, all of which have partially filled d sublevels.

Triple point: The pressure and temperature at which the solid, liquid, and vapor phases of a particular substance coexist in equilibrium. See Phase equilibrium.

U

Unit cell: A three-dimensional representation of the repeating units in a crystalline solid.

Unsaturated compound: A compound with double or triple bonds.

Unsaturated solution: A solution into which more solute may be dissolved.

V

Valence electron: An electron in the highest occupied energy level of an atom, whose tendency to be held or lost determines the chemical properties of the atom.

Van der Waals forces: The weak forces that contribute to intermolecular bonding, including hydrogen bonding, dipole-dipole interactions, and dispersion forces.

Vapor pressure: The pressure exerted by a vapor when it is in equilibrium with the liquid or solid phase of the same substance; the partial pressure of the substance in the atmosphere above the liquid or solid.

Vapor-pressure lowering: The decrease in the vapor pressure of a liquid caused by the presence of dissolved solute; a colligative property. See Raoult's law.

Voltaic cell: See Galvanic cell.

VSEPR theory: Stands for valence shell electron-pair repulsion theory. It predicts/explains the geometry of molecules in terms of the repulsion that electron pairs have that causes them to be as far apart as possible from one another.

W

Water dissociation constant (K_w): The equilibrium constant of the water dissociation reaction at a given temperature; 1.00×10^{-14} at 25.00°C.

Weak acid: An acid that undergoes partial dissociation in an aqueous solution, e.g., CH_3COOH.

Weak base: A base that undergoes partial dissociation in an aqueous solution, e.g., NH_4OH.

Y

Yield: The amount of product obtained from a reaction.

Z

Z: Nuclear charge. Equivalent to atomic number.

Z_{eff}: Effective nuclear charge; the charge perceived by an electron from its orbital. Applies most often to valence electrons and influences periodic properties such as atomic radius and ionization energy.